SOMATIC GENE THERAPY

SOMATIC GENE THERAPY

Edited by

Patricia L. Chang

CRC Press
Taylor & Francis Group
Boca Raton London New York

CRC Press is an imprint of the
Taylor & Francis Group, an **informa** business

First published 1995 by CRC Press
Taylor & Francis Group
6000 Broken Sound Parkway NW, Suite 300
Boca Raton, FL 33487-2742

Reissued 2018 by CRC Press

Library of Congress Cataloging-in-Publication Data

Somatic gene therapy / edited by Patricia L. Chang
p. cm.
Includes bibliographical references and index.
ISBN 0-8493-2440-8
1. Gene Therapy.
[DNLM: 1. Gene Therapy. QZ 50 S693 1994]
RB155.8.S56 1994
616'.042—dc20
DNLM/DLC
for Library of Congress 94-17728

A Library of Congress record exists under LC control number: 94017728

Publisher's Note
The publisher has gone to great lengths to ensure the quality of this reprint but points out that some imperfections in the original copies may be apparent.

Disclaimer
The publisher has made every effort to trace copyright holders and welcomes correspondence from those they have been unable to contact.

ISBN 13: 978-1-315-89766-0 (hbk)
ISBN 13: 978-1-351-07676-0 (ebk)

Visit the Taylor & Francis Web site at http://www.taylorandfrancis.com and the
CRC Press Web site at http://www.crcpress.com

PREFACE

After a long gestation, with assistance from a large number of prenatal attendants, gene therapy was born in September 1990. In honor of its fourth birthday, I have been asked to write a preface for the "Baby Book" (though I wonder whether the editor and publishers would think of *Somatic Gene Therapy* in quite this way).

How is the baby doing? As this volume richly documents, gene therapy is growing at a spectacular rate. Today there are over 100 approved clinical protocols worldwide; approximately two thirds of the protocols are directed at cancer, while the remainder cover primarily genetic disease and AIDS.

However, the approved clinical protocols are only the tip of the iceberg. Gene therapy requires extensive studies that range throughout the research/development spectrum. Clinical protocols are supported by a broad-based pyramid of fundamental research which includes molecular genetics, virology, and animal studies. In order to encompass the full expanse of gene therapy, this volume is organized around target organs and the leaders who helped pioneer each of these areas have authored the relevant chapters. Although a number of books have been published about human gene therapy, *Somatic Gene Therapy* is a truly comprehensive volume that documents progress on essentially all the major fronts.

So, at four years of age the baby is doing extraordinarily well. Could such rapid growth have been predicted? Maybe by somebody, but certainly not by me. Having barely survived the immense skepticism, and even hostility, that surrounded our efforts to bring about the birth, I would never have anticipated that a volume as sophisticated as *Somatic Gene Therapy* would have been possible by year four.

It is with great pleasure and personal pride that I contribute this brief preface to our baby's book. Congratulations to the editor and all the authors, not only for their efforts in producing this fine volume, but also for their research successes that have helped make gene therapy very much alive and well!

W. French Anderson, M.D.

THE EDITOR

Patricia L. Chang, Ph.D., is a Professor of Pediatrics in the Faculty of Health Sciences at McMaster University in Hamilton, Ontario, Canada.

Dr. Chang received her B.Sc. degree from the University of Hong Kong in 1967 and obtained her Ph.D. degree in 1971 from the Department of Biochemistry, University of Western Ontario, Canada. After doing post-doctoral work at University of Waterloo and Hospital for Sick Children, she was appointed an Assistant Professor at McMaster University in 1979, then Associate Professor in 1984, and Professor since 1990. She is also an associate member of the departments of Biology and Biomedical Sciences at McMaster University.

Dr. Chang is a member of the American Association for the Advancement of Science, American Society of Human Genetics, American Society of Biological Chemists, Association for Women in Science, Society of Chinese Bioscientists in American and Canadian Societies of Biochemistry and Genetics. She has been the recipient of a March of Dimes Birth Defect Foundation Basil O'Connor Award, a Velleman and Ontario Mental Health Scholarship, Medical Research Council of Canada Visiting Scientist award, and a National Institute of Health Visiting Scientist award. She has also been awarded research grants from the Medical Research Council of Canada, the March of Dimes Birth Defect Foundation, Natural Sciences and Engineering Council of Canada, Canadian Hemophilia Society, and Ontario Mental Health Foundation. Her current research interests are in human gene therapy and molecular genetics of neurological diseases.

CONTRIBUTORS

Julie K. Andersen
Division of Neurogerontology
Andrus Gerontology Center
University of Southern California
Los Angeles, CA

Brian J. Aneskievich
Division of Plastic Surgery
Robert Wood Johnson Medical School
University of Medicine and Dentistry of New Jersey
New Brunswick, NJ

John W. Belmont
Department of Molecular and Human Genetics
Baylor College of Medicine
Houston, TX

Malcolm K. Brenner
Division of Bone Marrow Transplantation
Department of Hematology-Oncology
St. Jude Children's Research Hospital
Memphis, TN

Kenneth W. Culver
Human Gene Therapy Research Institute
Iowa Methodist Medical Center
Des Moines, IA

Lisa J. Fisher
Department of Neurosciences
University of California, San Diego
La Jolla, CA

Elaine Fuchs
Department of Molecular Genetics and Cell Biology
Howard Hughes Medical Institute
University of Chicago
Chicago, IL

Fred H. Gage
Department of Neurosciences
University of California, San Diego
La Jolla, CA

Michael W. Heartlein
Department of Molecular Biology
Transkaryotic Therapies, Inc.
Cambridge, MA

Roland Jurecic
Department of Molecular and Human Genetics
Baylor College of Medicine
Houston, TX

Donald B. Kohn
Division of Research Immunology/Bone Marrow Transplantation
Children's Hospital Los Angeles
Los Angeles, CA

Kotoku Kurachi
Department of Human Genetics
University of Michigan Medical School
Ann Arbor, MI

Fred D. Ledley
GENEMEDICINE INC.
Houston, TX

Anthony G. Letai
Pritzker School of Medicine
University of Chicago
Chicago, IL

Louis M. Messina
Department of Surgery
Section of Vascular Surgery
University of Michigan
Ann Arbor, MI

Jasodhara Ray
Department of Neurosciences
University of California, San Diego
La Jolla, CA

Richard F Selden
Transkaryotic Therapies, Inc.
Cambridge, MA

James C. Stanley
Department of Surgery
Section of Vascular Surgery
University of Michigan
Ann Arbor, MI

Hideaki Tahara
Department of Surgery
University of Pittsburgh
Pittsburgh, PA

Douglas A. Treco
Transkaryotic Therapies, Inc.
Cambridge, MA

Kenneth I. Weinberg
Division of Research Immunology/
Bone Marrow Transplantation
Children's Hospital Los Angeles
Los Angeles, CA

Shou-Nan Yao
Department of Human Genetics
University of Michigan Medical School
Ann Arbor, MI

CONTENTS

1 Overview

Inder M. Verma

Gene therapy is a novel form of molecular medicine that will have a major impact on human health in the next century. Although the advent of recombinant DNA technology in modern medicine will allow fetal genetic screening and genetic counseling, the vast majority born with disease are likely to be helped by gene therapy approaches. The scope and definition of gene therapy has expanded in the past few years. In addition to the possibility of correcting inherited genetic disorders like cystic fibrosis, hemophilia, and familial hypercholesterolemia, gene therapy approaches are being used to combat acquired diseases like cancer, AIDS, Parkinson's disease, Alzheimer's disease, and infectious diseases. At present, germ line gene therapy is not being contemplated due to the complex technical and ethical issues. The scientific community is interested in pursuing somatic cell gene therapy, which is exclusively for the benefit of the individual and cannot be passed on to the succeeding generation. The minimum requirement for gene therapy is sustained production of the therapeutic gene product without any harmful side effects.[1-5]

I. THE SOMATIC TISSUE

There are a number of somatic tissues that have been used to transfer genes, but the precise choice of the tissue will be dictated by the nature of the disease. **Bone marrow** (see Chapters 2, 3, and 13) has been a favorite target because of the possibility of introducing the foreign gene in the pluripotent stem cell, but it has not been easy to obtain either therapeutic levels or sustained amounts of the foreign protein.[6] Several innovations using specific growth factors and lymphokines will improve the availability of stem cells, and therefore allow sustained production of the foreign protein. Many researchers have also taken advantage of differentiated hematopoietic cells like circulating T cells, lymphocytes, etc., for gene expression. Due to extensive knowledge in handling primary fibroblast cells, many investigators have used **fibroblasts** (Chapter 4) as a choice tissue for introducing foreign genes where the therapeutic product can be secreted in the plasma, e.g., hemophilia proteins.[7] The big drawback to date has been the lack of sustained production of the foreign protein, primarily due to "shut off" of the transgene.[8,9] Lack of appropriate promoters-enhancers specific for fibroblasts is partly responsible for the lack of progress in successfully using fibroblasts. Due to extensive advances in engineering of artificial skin *in vitro* for grafting at burn sites, **keratinocytes**

0-8493-2440-8/95/$0.00+$.50

(Chapter 6) offer a very viable target for introducing genes where the therapeutic product is secreted, but unfortunately, poor vascularization of keratinocytes has hampered their use. Nearly 40% of body weight is made up of **muscle tissue** (Chapter 8) and, therefore, it is an attractive target. Furthermore, myoblasts can be cultured *in vitro* and transplanted in the muscle to fuse with the existing myofibers. Muscle cells can also be directly infected and are very well vascularized. Not surprisingly, many investigators have relied on the use of muscle cells for producing therapeutic proteins.[10] A very large number of metabolic diseases involve proteins synthesized in the **liver** (Chapter 5). Because liver is not an actively dividing tissue, usually *ex vivo* approaches have been employed. Hepatocytes obtained by partial hepatectomy have been transduced with foreign genes and then perfused into the liver. Recently, direct *in vivo* gene delivery approaches have also been used.[11,12] Because of their close proximity to vasculature, many investigators have used **endothelial cells** (Chapter 7) to introduce genes either *ex vivo* or through a catheter *in vivo*. However, a major limitation of endothelial cells is their relatively small number, which renders them unlikely to produce therapeutic levels of the foreign gene products. The lack of postmitotic activity of neuronal cells has hindered the use of **neuron/glia** (Chapters 9, 10, and 14) for introducing foreign gene, but the advent of new vectors that may not require dividing cells should allow this problem to be overcome.

There are many other cell types, e.g., cardiac myocytes, lung air valves, macrophages, etc. but they have not been explored as extensively and, therefore, the discussion in this book has been confined to the commonly used cell types.

II. THE DELIVERY SYSTEM

Although a wide variety of somatic tissues has been identified for delivering genes, the major hurdle is still the mode of delivery of the therapeutic genes. Briefly, the currently available gene delivery approaches can be divided into two main types.

A. Physico-Chemical Methods

This approach involves DNA transfection with calcium phosphate, DEAE dextran, cationic lipids, etc., direct DNA injection with or without the aid of cationic lipids, electroporation, ballistic guns delivering gold-coated DNA particles, microinjection, liposomes, and receptor-mediated gene transfer (see Chapter 11). All these approaches have benefits, but it is fair to say they are not very successful by and large in delivering large and sustained amounts of therapeutic gene product. For instance, an average individual secretes 5 μg/ml of factor IX protein in the plasma, and the present methodology is just not quite up to par.

B. Biological Vectors

For any large-scale introduction of a biological entity to cells, the most likely approach will employ the use of viral vectors. Since high-titer viruses (up to 10^6 to 10^{13}

virus particles/ml) can be obtained (depending on the type of virus), it is easy to infect large numbers of cells. Additionally, pluripotential stem cells, which are rare (1 per 10^3 to 10^4 cells), can also be infected with high-titer viruses. Although there are many viral vectors, attention has been focused mainly on three types of viral vectors.

1. Retroviral Vectors

They have been used most extensively in gene therapy approaches[1-6,14] because of ease of manipulation, integration in the host chromosome and thus sustained expression, ability to infect a wide variety of cell types from a variety of species, ability to accommodate foreign genes ranging from 1 to 7 kb, and ability to generate replication deficient viruses because the proteins necessary for viral replication can be provided in *trans*. **Limitations:** (1) The major limitation of retroviruses is their inability to infect postmitotic cells, which limits their use for most *in vivo* gene delivery approaches. Usually the somatic tissue needs to be manipulated *in vitro*. The ability of some lentiviruses to infect nondividing cells offers some hope of overcoming this limitation. (2) Although titers of 10^5 to 10^6 can be generally achieved, higher titers will be required for attaining therapeutic levels. In this regard the recent availability of recombinant retroviruses containing vesicular stomatitis virus (VSV) glycoprotein allows concentration of the recombinant retrovirus by centrifugation to obtain titers of 10^{8-9} virus particles/ml. (3) There is some risk of integration of retroviruses in the vicinity of a proto-oncogene and may cause neoplasia.

2. Adenoviral Vectors

Adenovirus types 2 and 5 (Ad2 and Ad5) cause respiratory disease in humans but belong to a subclass of adenovirus that is not associated with human malignancies.[15] The safety of recombinant adenovirus is supported by the extensive and effective use of live adenovirus vaccines in human population, mostly military recruits in the U.S. The genome of adenovirus is composed of linear, double-stranded DNA of approximately 36 kb in length with short inverted terminal repeats (ITR) at each end of the genome required for viral DNA replication. Adenoviruses have a lytic cycle, and infection is carried out by entry into the cell by as yet unidentified receptors, followed by entry into endosome, release in the cytoplasm with loss of accompanying structural proteins, and migration of the viral DNA to the nucleus, where it is retained as a linear structure. The initial hesitation of the relatively large size of adenoviruses (36 kb) has largely been overcome by very successful use of adenoviral vectors for both *ex vivo* and *in vivo* gene delivery (for a review see Reference 16). The major advantage of adenoviral vectors are (1) high titer (10^{11} to 10^{12} virus particle/ml) or recombinant viruses, (2) ability to infect postmitotic cells, (3) ability to infect a wide variety of cell types, (4) ability to accommodate up to 10 kb of foreign DNA, and (5) they are known to be acutely pathogenic in humans because many adenovirus based vaccines have been inoculated in human without any serious side effects. **Limitations:** The major limitation of the adenoviral vectors are (1) lack of sustained expression, because the viral DNA does not integrate; (2) antigenicity against viral proteins, both humoral and cytotoxic T-lymphocytes (CTL); and (3) possible toxicity at high doses.

3. Adeno-Associated Viral Vectors

One alternative vector for stable gene expression is the adeno-associated virus (AAV) vector.[17] The AAV, a defective human parvovirus, can infect various types of mammalian cells. None of the tested primary cultures and established cell lines is resistant to AAV infection, and there are some preliminary data suggesting that AAV can integrate into non-dividing cells. AAV depends on the co-infection of a helper-virus, usually an adenovirus, for efficient replication. Without a helper-virus, AAV establishes latency *in vivo* and *in vitro* by integrating its genome, mostly into a single locus on chromosome 19. Although about 80% of adult population is sero-positive to AAV exposure, AAV has not been found to be associated with any diseases. The AAV genome contains a single-stranded DNA of 4675 bases. It has two large open reading frames for a replication (*rep*) gene and a capsid gene.[17] The Rep proteins encoded by the *rep* gene are essential for AAV DNA replication and gene regulation. They also inhibit gene expression from heterologous promoters and cellular transformation by neomycin-resistant gene or oncogenes. There are 145 bases inverted terminal repeat sequences (ITRs) at both ends of the AAV genome, which form T-shaped hairpins serving as viral origins of DNA replication. The AAV ITRs are the only *cis* elements required for the packaging of AAV genome and integration of AAV DNA into host chromosome. All of the internal AAV sequences except the ITRs can be removed to accommodate foreign DNA of up to 4.5 kb. AAV vectors with the *rep* gene have very high efficiency of transduction (50 to 80%) and predictable gene expression based on the type of promoter. Although these vectors may lose the chromosome 19 specificity of integration, they have been successfully used to transduce marker genes or to express γ-globin gene and antisense RNA against HIV LTR *in vitro*. Long-term *in vivo* expression of transduced gene has been demonstrated when the cystic fibrosis (CF) gene was introduced into the rabbit airway via an AAV vector.[49,50] More than 50% of the airway epithelium of the lobar bronchus are positive for the presence of AAV vectors. Although the life span of airway epithelial cells is not well understood, it is likely that a large fraction of them are terminally differentiated and nonreplicating. Long-term expression in these cells suggests that AAV can integrate in nondividing cells.

AAV vectors offer unique opportunities which include (1) ability to infect a very wide variety of mammalian cells; (2) although 80% of the adult population is sero-positive to AAV exposure, it has not been found to be associated with any disease; (3) in the absence of helper-adenovirus, AAV can maintain latency and integrate into a single locus on chromosome 19; (4) there are reports that AAV can infect postmitotic cells and hence is a candidate for *in vivo* gene delivery; (5) AAV particles are remarkably stable and can be concentrated to titers as high as 10^{13} particles/ml without losing infectivity; and (6) since the integrated recombinant AAV-vector generates no viral antigens, it is unlikely to be immunogenic and thus to allow superinfection with a variety of recombinant AAV vectors. The major drawbacks of AAV at present are (1) lack of suitable packaging cell lines and hence, lower titers (10^5 to 10^6 virus particles/ml) (part of the difficulty is the toxicity of AAV Rep protein); (2) if Rep protein is not provided in *cis*, the AAV does not integrate in a defined chromosomal locus; and (3) relatively small size of the inserted foreign DNA (~4 to 4.5 kb DNA).

There are other biological vectors, in particular, herpes virus vector (Chapter 9), but they have not yet been extensively employed.

III. THE DISEASES

As I mentioned earlier, the choice of vector delivery system will be dictated by the nature of the disease. Briefly, one can divide the most treatable diseases into three categories.

1. *Systemic delivery*. The product of the therapeutic gene is secreted in the plasma, e.g., hemophilia proteins factors VIII or IX, growth hormone, etc. In such cases any of the somatic tissues can be used to introduce the foreign gene.
2. *Tissue- or organ-specific delivery*. In this case the therapeutic gene has to be delivered in a specific tissue, e.g., in hypercholestrolemia, the LDL receptor has to be made in hepatic cells or CFTR protein (cystic fibrosis) has to be delivered in airway epithelial cells. One has to either use direct *in vivo* gene delivery or make the viruses tissue-specific.
3. *Localized delivery*. The transduced somatic tissue can be implanted where only local delivery of the therapeutic protein is required. For instance, fibroblasts or myoblasts producing tyrosine hydroxylase gene product can be grafted in the brain to provide local delivery of L-dopa for treatment of Parkinson's disease (see Chapter 10). Ultimately, the nature and severity of the disease, availability of the appropriate vectors, immunological concerns, and other factors will figure heavily into the type of gene therapy applied to the patient.

There is little doubt that gene therapy has come of age, and there is tremendous excitement over its potential. Over 100 clinical trials are currently under way world-wide. There are still many hurdles, e.g., sustained expression, immunogenicity, lack of appropriate therapeutic levels, etc. However, the readers of this book will find that efforts of a large number of researchers are slowly removing these obstacles. The use of gene therapy approaches for treating diseases like cancer does not require sustained expression and hence offers an ideal opportunity for intervention. I believe that in the coming years we will witness an explosion of research activity in gene therapy, making it a routine medical application. Gene therapy approaches must be affordable and economical, however. With vast social problems of society, a manageable health care system will require that the technology we developed be available not only to the affluent Western world, but also to less privileged populations.

REFERENCES

1. Mulligan, R. C., The basic science of gene therapy, *Science*, 260, 926, 1993.
2. Miller, A. D., Human gene therapy comes of age, *Nature*, 357, 455, 1992.
3. Friedmann, T., Progress towards human gene therapy, *Science*, 244, 1275, 1989.
4. Verma, I. M., Human gene therapy, *Sci. Amer.*, 262, 68, 1990.

5. Anderson, W. F., Human gene therapy, *Science*, 256, 808, 1992.
6. Miller, A. D., Progress towards human gene therapy, *Blood*, 76, 271, 1990.
7. St. Louis, D. and Verma, I. M., A novel approach to somatic cell gene therapy, *Proc. Natl. Acad. Sci. U.S.A.*, 85, 3150, 1988.
8. Scharfmann, R., Axelrod, J. H., and Verma, I. M., Long-term *in vivo* expression of retrovirus-mediated gene transfer in mouse fibroblast implants, *Proc. Natl. Acad. Sci. U.S.A.*, 88, 4626, 1991.
9. Palmer, T. C., Rosman, G. J., Osborne, W. R. A., and Miller, A. D., Genetically modified skin fibroblasts persist long after transplantation but gradually inactivate introduced genes, *Proc. Natl. Acad. Sci. U.S.A.*, 88, 1330, 1991.
10. Dai, Y., Roman, M., Naviaux, R. K., and Verma, I. M., Gene therapy via primary myoblasts: long term expression of factor IX protein following transplantation *in vivo*, *Proc. Natl. Acad. Sci. U.S.A.*, 89, 10892, 1992.
11. Wilson, J. M., Grossman, M., Wu, C.-H., Chowdhury, N. R., Wu, G.-Y., and Chowdhury, J. R., Hepatocyte-directed gene transfer *in vivo* leads to transient improvement of hypercholesteolemia in low density lipoprotein receptor deficient rabbits, *J. Biol. Chem.*, 267, 963, 1992.
12. Herz, J. and Gerard, R. D., Adenovirus-mediated transfer of low density lipoprotein receptor gene acutely accelerates cholesterol clearance in normal mice, *Proc. Natl. Acad. Sci. U.S.A.*, 90, 2812, 1993.
13. Bajocchi, G., Feldman, S. H., Crystal, R. G., and Mastrangeli, A., Direct *in vivo* gene transfer to ependymal cells in the central nervous system using recombinant adenovirus vectors, *Nature Genet.*, 3, 229, 1993.
14. Miller, A. D., Retrovirus packaging cells, *Hum. Gene Ther.*, 1, 5, 1990.
15. Strauss, S. E., Adenovirus infection in humans, in *The Adenoviruses*, Ginsberg, H. S., Ed., Plenum Press, New York, 451–496, 1984.
16. Kozarsky, K. F. and Wilson, J. M., Gene therapy: adenovirus vectors, *Cur. Opin. Genet. Dev.*, 3, 449, 1993.
17. Muzyczka, N., Use of adeno-associated virus as a general transduction vector for mammalian cells, in *Current Topics in Microbiology and Immunology*, Vol. 158, Springer-Verlag, 97–123, 1992.

2 Human Hematopoietic Cells for Gene Therapy

Roland Jurecic and John W. Belmont

I. INTRODUCTION

The bone marrow has been considered a key target tissue from the earliest days of gene therapy research.[1] Experience with bone marrow transplantation demonstrated that within the adult marrow there are cells capable of completely and permanently reconstituting the entire blood and immune systems. Progress in the understanding of human hematologic diseases at the molecular level, development of gene transfer technology, and increasing experience in isolation, *ex vivo* manipulation, and transplantation of primitive hematopoietic cells all contribute to continued interest in gene transfer into hematopoietic stem cells.[2-7] Initial efforts to introduce genes into blood cells were frustrated by the poor efficiency of gene transfer and uncertainties about safety.[8,9] Two paths of investigation have drastically improved the long-term prospects of stem cell gene therapy. The first is the development of viral vectors capable of integrating DNA stably into target cells. Stable integration appears to be necessary in this system because of the extensive cell proliferation (up to 50 cell doublings) that takes place between the stem cell and the development of terminally differentiated blood cells. The demonstration of retroviral vector transduction of murine hematopoietic progenitors provided the starting point for numerous preclinical models of stem cell gene therapy.[10] A second important contribution has come from studies addressing stem cell phenotype, their enumeration and enrichment, *ex vivo* maintenance, and developmental capacity after transplantation.

Although reproducible and efficient gene transfer into bone marrow stem cells has been accomplished in murine models, efficient transduction of most primitive hematopoietic cells and high-level expression of introduced genes have not been achieved in large animal models (canines and primates). Gene transfer into peripheral blood cells has been successfully performed in human gene marking trials, confirming that recombinant retroviral vectors can be safely used to introduce new genetic information to the hematopoietic system. Although the clinical application of stem cell gene therapy is still limited by low numbers of stem cells available from adult bone marrow and their quiescence, several stem cell gene therapy trials are under way.[4-7,11]

Further studies of phenotypic and developmental heterogeneity of stem cells that will enable their more efficient isolation, determination of growth factors that regulate stem cell self-renewal, and development of cultures in which stem cells could be propagated as

0-8493-2440-8/95/$0.00+$.50

Table 1

Candidate Genetic and Acquired Hematologic Diseases for Stem Cell Gene Therapy

Disease	Defective Gene	Chromosomal Locus
Adenosine deaminase deficiency	Adenosine deaminase	20q13.11
Purine nucleoside phosphorylase deficiency	Purine nucleoside phosphorylase (PNP)	14q13.1
Ataxia telangiectasia	—	11q22–23
Chronic granulomatous disease (CGD)	Neutrophil cytosolic factor 1 (p47-phox)	7q11.23
X-linked CGD	Cytochrome b_{558} (gp91-phox)	Xp21.1
X-linked agammaglobulinemia (XLA)	Bruton's tyrosine kinase (BTK)	Xq21.3–22
Wiskott-Aldrich syndrome	—	Xp11.2–11.3
X-linked hyper IgM syndrome	CD40 ligand	Xq24–27
X-linked severe combined immunodeficiency (SCID)	IL-2 receptor α chain	Xq13.1–21.1
Gaucher's disease	Glucocerebrosidase	1q21–23
Thalassemia	β-globin	11p14–15
Sickle cell anemia	β-globin	11p14–15
Leukocyte adhesion deficiency (LAD)	CD18	21q21–22
Paroxysmal nocturnal hemoglobinuria (PNH)	Phosphatidylinositol glycan class A (PIG-A)	Xp22.1
AIDS	—	—

undifferentiated cells will greatly advance the clinical application of stem cell gene therapy in the treatment of genetic and acquired hematologic diseases (Table 1).

In this chapter we review current progress in preclinical gene therapy models and clinical gene therapy trials, and consider some of the most important issues limiting the clinical introduction of stem cell gene therapy. We hope to outline some of the areas for further research that may ultimately make this a routine option of medical treatment.

II. ISOLATION, CHARACTERIZATION, AND EX VIVO MAINTENANCE OF HUMAN PRIMITIVE HEMATOPOIETIC CELLS

The isolation and characterization of human hematopoietic progenitor and stem cells represented a long-term challenge due to the low frequency of these cells and lack of markers for positive selection. Identification of a 115,000 Da cell surface glycoprotein (CD34) and subsequent cloning and characterization of the gene encoding CD34 have been the turning point in efforts to isolate human primitive hematopoietic cells. Human CD34 antigen is a highly glycosylated integral membrane protein of unknown function, with pattern of expression restricted to hematopoietic progenitors, a subset of bone marrow stromal cells, and the small vessel endothelium of a variety of tissues. It has no sequence similarity to any previously characterized protein but is very similar to murine CD34 gene, which has a broader expression pattern (hematopoietic progenitors, brain and embryonic fibroblasts).[12-15] Functional studies have shown that the CD34 antigen, found on the surface of 1 to 4% of human bone marrow (BM) cells, is expressed by (1) virtually all hematopoietic progenitors (CFC, colony-forming cells) analyzed by *in vitro* colony assays, (2) more primitive long-term bone marrow culture initiating cells (LTC-IC), and (3) cells with *in vivo* reconstitutive capacity but not by mature hematopoietic cells.[16-25]

Subsequent cell surface phenotyping of heterogeneous CD34$^+$ cell population has identified several CD34$^+$ cell subsets expressing T-lineage markers (CD3, CD4, CD7), B-lineage markers (CD10, CD19, CD20, CD40), myeloid lineage markers (CD33), nonlineage markers (CD35, CD38, CD45, CD71, HLA-DR, Thy-1, *c-kit*), and adhesion molecules (LFA-1, LFA-3, VLA-4, VLA-5, ICAM-1). In particular, cells that express B-lineage (CD19) and myeloid (CD33) markers alone constitute 80 to 90% of the CD34$^+$ population. Thorough phenotypic characterization of CD34$^+$ cells has enabled the isolation and characterization of a variety of more primitive subsets of CD34$^+$ cells (CD34$^+$CD33$^-$HLA-DR$^-$, CD34$^+$CD38$^-$HLA-DR$^-$, CD34$^+$Lin$^-$HLA-DR$^-$, CD34$^+$CD45^{lo}CD71^{lo}, CD34$^+$CD33$^-$LFA-1$^-$ and CD34$^+$HLA-DR$^-$c-kit$^+$ cells), that are highly enriched for multipotent and committed progenitors and LTC-IC cells.[26-35] Based upon differential expression of Thy-1 molecule in CD34$^+$ cells, a candidate human hematopoietic stem cell population (CD34$^+$Lin$^-$Thy-1$^+$) was recently isolated, comprising 0.05 to 0.1% of human BM. According to both *in vitro* (colony assays, long-term culture assay) and *in vivo* (SCID-hu mice) assays CD34$^+$Lin$^-$Thy-1$^+$ are highly enriched for pluripotent progenitors and LTC-IC cells and represent functionally and phenotypically the most primitive human hematopoietic cell population isolated so far.[36,37]

CD34$^+$ cells are present also in human umbilical cord blood (CB) and peripheral blood (PB). Clonogenic colony assays, blast cell colony assays, and limiting dilution analysis in micro-stromal cultures have shown that circulating CD34$^+$ cells from CB and PB include both committed and early hematopoietic precursors as well as LTC-IC cells. The frequency of CD34$^+$ cells appears to be equal in BM and CB, whereas PB contains no more than one-tenth of CD34$^+$ cells present in BM. Comparative expression analysis of cell surface antigens on BM-, CB-, and PB-derived CD34$^+$ cells has revealed that although CD34$^+$ cells from BM constitute a more heterogeneous population, resident and circulating CD34$^+$ cells largely display the same cell-surface antigens. Nevertheless, recent immunofluorescence studies have revealed differences in the expression of several surface antigens between these two populations. Four-color flow cytometry analysis has demonstrated that PB-CD34$^+$ cells express higher levels of CD33 and CD45 antigens and lower levels of CD71 antigen than BM-CD34$^+$ cells, whereas the expression of B- and T-lineage markers was similar in both populations. Although resident and circulating CD34$^+$ cells exhibit similar phenotypic and functional characteristics, the circulating population has the higher content of already committed progenitors. It still remains uncertain whether circulating CD34⁺ cells are functionally equivalent to their BM counterpart or whether they represent a selected population of primitive hematopoietic cells with differential growth, self-renewal capacity and colony-forming ability.[32,35,38-44]

A. Techniques for the Enrichment of Human Primitive Hematopoietic Cells

The expression of CD34 antigen on primitive human hematopoietic progenitors and stem cells has led to the development of various techniques for positive selection of $CD34^+$ cells from fetal and adult BM, PB, umbilical CB, and fetal liver. These techniques include immunoadsorption (panning), immunoabsorbance columns, immunomagnetic separation, and multi-parameter fluorescence activated cell sorting (FACS).[45]

Enrichment of human primitive hematopoietic cells from heparinized BM and CB samples starts with isolation of low density (1.07 g/ml) mononuclear cells (MNC) by density gradient centrifugation in Ficoll-Hypague (Sigma Chemical Co., St. Louis, MO). MNC cells are then used for enrichment of $CD34^+$ cells by immunoabsorbance columns, immunomagnetic beads, immunoadsorption (panning), FACS, or a combination of these techniques. An immunoabsorption column technique relies on the high affinity between the protein avidin and the vitamin biotin. MNC cells are incubated with biotinylated anti-CD34 monoclonal antibody (12.8) and then passed through a column of avidin-coated beads. Labeled cells are retained in the column, whereas nonlabeled cells pass through. Bound cells are then removed from the column by mechanical agitation of beads. Initial studies showed that CD34-enriched population was 35 to 92% pure (assessed by flow cytometry) and highly enriched for granulocyte-macrophage colony-forming units (CFU-GM).[22]

In the immunomagnetic bead separation protocol, low-density MNC cells are incubated with anti-CD34 monoclonal antibody (mAb), followed by paramagnetic beads (Dynabeads M-450, Dynal, Oslo, Norway) coupled to a secondary antibody, with a bead-to-cell ratio of 5:1. $CD34^+$ cells are isolated by magnetic separation (Dynal magnetic particle concentrator) and detached from the beads by chymopapain treatment. This technique yields more than 90% pure population of $CD34^+$ cells.[45,46]

Positive selection of $CD34^+$ cells by immunoadsorption or panning involves incubation of low density MNC cells with murine anti-CD34 mAb. Cells are then placed in tissue culture dishes coated with secondary antibody (goat anti-mouse IgG). After discarding cells left in suspension, $CD34^+$ cells are recovered from the dish surface by vigorous pipetting. This technique yields a cell population with 80 to 95% purity.[41,45,47]

Enrichment of $CD34^+$ cells by FACS includes incubation of MNC cells with anti-CD34 mAb, followed with secondary antibody conjugated to phycoerythrin (PE) or fluorescein isothiocyanate (FITC). $CD34^+$ cells are then sorted on a FACStar-Plus flow cytometer (Becton Dickinson Immunocytometry Systems, San Jose, CA) within the lymphocyte/blast cell window on the basis of intermediate forward scatter, low side scatter, and high levels of fluorescence. In order to enrich different subsets of $CD34^+$

cells ($CD34^+CD33^-HLA\text{-}DR^-$, $CD34^+CD38^-HLA\text{-}DR^-$, $CD34^+Lin^-HLA\text{-}DR^-$, $CD34^+CD45^{lo}CD71^{lo}$, etc.) a negative selection of mature cells expressing lineage and nonlineage markers is performed prior to or after CD34 selection. For this purpose, cells are labeled with a cocktail of monoclonal antibodies against lineage (CD2, CD3, CD4, CD7, CD10, CD19, CD33) and nonlineage markers (CD38, CD45, CD71, HLA-DR), followed by secondary antibody. Mature cells expressing lineage and nonlineage markers are then removed by immunomagnetic depletion, immunoadsorption (panning) or FACS.[26-28,31,37,43,45]

Enrichment of $CD34^+$ cells from PB starts with leukopheresis, a process where MNC cells are collected by semicontinuous or continuous flow centrifugation, recycling the leftover cells (erythrocytes, platelets, and granulocytes) back to the patient. Due to their low frequency in PB, enrichment of a sufficient number of $CD34^+$ cells remains cumbersome and requires multiple leukophereses. In order to increase cell yield, mobilization of stem cells from BM into blood, temporarily increasing their circulating amount ("rebound" phase), is now generally performed. Peripheral blood stem cells (PBSC) can be mobilized by chemotherapy, administration of hematopoietic growth factors (GM-CSF, G-CSF, IL-3) or by combined administration of both. In healthy adults, whole blood contains about 2.000 $CD34^+$ cells/ml, whereas mobilization with chemotherapy alone or combined with growth factors increases the number of circulating $CD34^+$ cells tenfold.[35,40,48,49] MNC cells obtained from peripheral blood by leukopheresis are further enriched for CD34 cells by one of the techniques described above.

The ability to purify primitive human hematopoietic cells ($CD34^+$ cells and their subsets) has greatly facilitated (1) analysis of the engraftment and long-term repopulating ability of BM- or PB-derived $CD34^+$ cells after autologous or allogeneic transplantation, (2) development of *in vitro* culture systems for maintenance and expansion of these cells, and (3) development of protocols for efficient transfer and expression of recombinant genes in stem cells, a technology that could play an important role in the treatment of some genetic and acquired diseases.

B. Ex Vivo Maintenance and Expansion of Human Primitive Hematopoietic Cells

The analysis of response to a variety of hematopoietic growth factors has shown that phenotypically defined human hematopoietic progenitor and stem cells ($CD34^+$ cells and their subsets) require multiple stimuli for proliferation and differentiation *in vitro*. $CD34^+$ cells and their more primitive subsets, enriched from BM, PB, or umbilical CB, require multiple growth factors (IL-1, IL-3, IL-6, G-CSF, M-CSF, GM-CSF, Epo, SCF) for (1) formation of granulocyte-macrophage (CFU-GM), erythroid (BFU-E), megakaryocyte (CFU-Meg), multilineage (CFU-GEMM), and high-proliferative potential colony-forming cell (HPP-CFC) colonies, (2) *ex vivo* expansion in suspension cultures, and (3) establishment of long-term cultures and maintenance of LTC-IC cells.[44,48,50-55] Simultaneous formation of fibroblastoid cells and hematopoietic progenitors by single $CD34^+HLA\text{-}DR^-CD38^-$ cells, treated with IL-3, IL-6, SCF, GM-CSF, Epo, IGF-1, and bFGF, emphasizes the developmental potential of primitive stem cells and the complex balance of stimuli that supports cell growth and lineage commitment.[29]

While proliferation and differentiation require multiple growth factors, it is not known whether these cells simultaneously express all appropriate receptors or whether the synergistic mode of action occurs by up-regulation or cascade transactivation of receptor expression. Recent receptor binding studies have shown that CD34$^+$ cells purified from peripheral blood express receptors for IL-3, IL-6, GM-CSF, and Epo. Studies of cross-reactivity between receptors and heterologous growth factors (including IL-6, IL-3, GM-CSF, Epo, and SCF) on these cells have revealed that (1) all growth factors except SCF induce a marked down-modulation of their own receptors and (2) each factor (except SCF) induces the transactivation of the receptors for a "distal" factor. For example, IL-6 induces transactivation of IL-3R, IL-3 induces transactivation of GM-CSF and Epo receptors, whereas GM-CSF transactivates Epo receptor.[56] This sequential chain transactivation of GF (growth factor) receptors could account for the synergistic action of HGFs (hematopoietic growth factors) on primitive hematopoietic cells in clonogenic cultures.

There is a growing consensus that an improvement of clinical practice in the areas of BM transplantation and gene therapy will depend on the ability to maintain and expand early progenitors and stem cells *ex vivo*. Although sustained production of committed progenitors in human long-term bone marrow cultures (LTBMCs) is well documented, evidence for the generation and expansion of more primitive cells (HPP-CFC, LTC-IC) in such cultures is lacking. LTMBCs have actually always exhibited exponentially decreasing numbers of total progenitor cells with time, rendering the cultures unsuitable for large-scale cell expansion. The number of LTC-IC cells in these static cultures has always been below input value, indicating their depletion through differentiation and cell death. The rate of decline has been decreased by the addition of recombinant hematopoietic GFs, but true cell and progenitor expansion was not obtained because these nonperfused cultures cannot support high cell densities. Incubation of enriched CD34$^+$ and CD34$^+$CD38$^-$HLA-DR$^-$ cells with a synergistic combination of recombinant GFs in a stroma-free suspension culture resulted in a large increase in total cell number, accompanied by expansion of progenitor cells (CFU-GM, HPP-CFC) and increase in absolute number of starting cell population (CD34$^+$ and CD34$^+$CD38$^-$HLA-DR$^-$ cells). These results indicate that *ex vivo* expansion of CD34$^+$ cells from BM, PB, and CB is feasible and that the number of cells expanded might be sufficient for repetitive use in therapeutic gene transfer and BM transplantation.[44,48,52,55,57]

Recently, a large-scale expansion of primitive human progenitors and stem cells in high cell density continuous perfusion cultures of mononuclear BM and CB cells was reported. Development of large-scale continuous perfusion cultures relied on the observation that acceleration of medium exchange rates coupled with the provision of soluble growth factors (IL-3, IL-6, GM-CSF, Epo, SCF) significantly increases total cell number, production of progenitors, and culture longevity. In conventional cultures, weekly medium exchange leads to extensive nutrient depletion, accumulation of metabolites, and wide fluctuation of physiological parameters, which in turn limit cell growth and alter growth factor production by stromal cells. Large-scale perfusion cultures mimic *in vivo* hematopoiesis sufficiently to produce clinically meaningful numbers of hematopoietic progenitors and repopulating cells over a 2-week period. In the continuously perfused cultures, total cell number increased 10- to 25-fold, CFU-GM progenitors expanded 10- to 18-fold,

BFU-E cells expanded 2.5- to 12-fold, CFU-Mix cells expanded 5.3-fold, and LTC-IC cells expanded 7.5-fold. These cultures allow large-scale expansion of human hematopoietic cells while maintaining and expanding primitive progenitors and repopulating cells, which are the primary targets for gene therapy.[58-60]

III. DEVELOPMENT OF PRECLINICAL MODELS FOR GENE THERAPY

A. Murine Models for Gene Therapy of Hematologic Diseases

There is extensive evidence in murine models that retroviral vectors can introduce foreign genes at high efficiency into hematopoietic progenitors and reconstituting stem cells. Gene transfer to murine hematopoietic cells has been achieved using retroviruses containing a variety of genes, including the bacterial neomycin-resistance gene (Neo^R),[61,62] human dihydrofolate reductase (DHFR) and multidrug resistance 1 (MDR1) genes,[63,64] human β-globin,[65-67] human adenosine deaminase (ADA),[68-77] CD18, glucocerebrosidase (GC), and argininosuccinate synthetase.[78-80] Several groups have demonstrated retroviral-mediated gene transfer of human ADA into mouse PHSC (pluripotent hematopoietic stem cells) in the absence of replication-competent retrovirus. Stable and potentially therapeutic levels of *in vivo* expression of the human ADA were achieved in all hematopoietic tissues of recipent mice.[70-76] Although some early studies suggested that expression in stem cells was limited by regulation of the retroviral long terminal repeat promoters, multiple examples have subsequently shown that such simple vectors do express introduced genes. These studies, in aggregate, demonstrate that the single most important factor limiting long-term expression is the efficiency with which stem cells are transduced.

Long-term *in vivo* expression of human DHFR, MDR1, β-globin, glucocerebrosidase, and argininosuccinate synthetase has been demonstrated also in hematopoietic cells of recipient mice after retroviral transduction of hematopoietic progenitors and stem cells.[63-67,79,80] A number of variables appeared to be important for efficient transduction of murine primitive progenitors and stem cells, such as (1) 5-fluorouracil pretreatment of the donor animals prior to BM harvest; (2) coculture of the target BM cells with the retrovirus-producing cells instead of with purified viral-containing supernatant or viral supernatant infection of BM cells on stromal cell layers; (3) use of very high titer viruses ($>10^6$ infectious particles/ml), and (4) inclusion of hematopoietic GFs (recombinant IL-1, IL-3, IL-6, SCF, LIF) in the culture media before or during transduction. The presence of growth factors (particularly IL-1, IL-6, SCF, and LIF) that increase survival and number of cycling progenitors and stem cells appears to be critical for efficient transduction and expression of the transgene in the progeny of transduced progenitors and stem cells.[63,64,66,68-77,79-82] Recent reports on efficient transduction of highly enriched murine long-term repopulating cells and spleen colony-forming cells (CFU-S) were the first to demonstrate the feasibility of introducing heterologous genes into purified HSC, an approach that could represent a major advance in the field of gene therapy.[83,84]

B. Large Animal Models for Gene Therapy of Hematologic Diseases

Gene transfer into hematopoietic cells of large animals is thought to represent a model that more closely reproduces the conditions for human gene therapy, where a patient's BM cells are subjected to genetic modification prior to autologous BM transplantation. Because of their phylogenetic proximity to humans and similar markers for primitive hematopoietic cells (CD34), monkeys constitute a particularly useful animal model. Early experiments in dogs resulted in gene transfer only into committed hematopoietic progenitors.[85] The efficiency of gene transfer and expression of selectable genes (Neo and DHFR) in canine progenitor cells (CFU-GM) has been increased fourfold (19 to 87% of transduced progenitors) by combining cocultivation with long-term BM culture as compared with results obtained with cocultivation only (6 to 16% of transduced progenitors).[86] Increased efficiency of retroviral gene transfer (supernatant infection) into neonatal progenitors (CFU-GM), relative to fetal or adult progenitors, was also demonstrated.[87] More recent studies show that 0.1 to 10% of circulating cells and progenitor cells of multiple lineages (CFU-GM, BFU-E, CFU-Mix) are transduced for up to 2 years after transplantation in the absence of helper virus.[88,89] Although the expression of neomycin resistance was documented in progenitor cells,[89] ADA expression was not detected. Retroviral vectors have also been used to transduce hematopoietic cells in cats. Provirus presence and drug-resistant progenitor cells were observed in some of the animals for over 2 years. Expression of human ADA, however, was not detected. Detection of contaminating helper-virus in these animals, accompanied by the development of diabetes mellitus in two cats, has made it difficult to conclude with certainty that stem cells were transduced.[90]

In utero transfer of the *neo* gene into fetal hematopoietic cells of sheep has been accomplished by isolating hematopoietic cells from fetal circulation during midgestation period, infecting these cells *in vitro*, and transplanting them back into the fetus by an intravascular catheter.[91] Persistence of infected cells after birth was demonstrated by detection of *neo* sequence in blood, BM, spleen, and thymus samples from several animals 1 week after birth and by the presence of G418-resistant progenitor cells in BM of one sheep for more than 2 years after birth. Unfortunately, virus-producing cells used to transduce sheep hematopoietic cells were also contaminated with helper virus. The same infection protocol was not successful with BM cells from adult sheep, indicating that actively proliferating fetal hematopoietic progenitors and stem cells are more amenable to efficient transduction with replication-defective retroviral vectors. Another group has performed *in utero* gene transfer by injecting high titer replication-defective retrovirus carrying *neo* and *E. coli* DNA repair gene *ada* into the livers of 11-, 14-, 16-, and 18-day gestation rats. The efficiency of BM transduction was highest in rats infected at day 14 to 16 of gestation. In rats killed at 1 to 26 weeks of age, gene transfer was detected by Southern analysis in 48% and by PCR in 86% of BM samples. The provirus was also detected in white blood cells, CFU-GM cells, thymus, spleen, liver, and lung. Northern and SDS-PAGE analysis have failed to detect expression of transduced *ada* gene. The presence of the transgene in hematopoietic tissues of adult rats suggested that fetal hematopoietic progenitors are susceptible targets for gene transfer and demonstrated that these cells become resident in the BM of adult animals. These experiments introduced the *in vivo* model for gene transduction of hematopoietic cells with potential application in the correction of genetic defects *in utero*.[92]

The first primate studies using short exposure to low-titer viral supernatant without addition of growth factors demonstrated low levels of expression of human ADA and bacterial neomycin phosphotransferase (Neo^R) in the PB cells for up to 4 months, but there was no evidence for transduction of long-term repopulating cells.[93] Introduction of modifications found to increase gene transfer efficiency in murine models has also improved the results in primates. All primates transplanted with 5-FU pretreated marrow transduced in co-culture with a high titer, helper-contaminated producer line in the presence of IL-3 and IL-6 were positive for proviral DNA in hematopoietic tissue for up to 6 to 12 months.[94] Three out of seven animals transplanted with CD34-enriched transduced BM had the marker gene detectable in BM, PB, purified granulocytes, and T cells at levels of 1 to 10% for longer than 100 days post-transplant. However, contamination with helper virus and chronic viremia in recipient animals suggests that these results could be explained by persistent propagation of the vector due to to the presence of helper virus. Three of these animals subsequently developed T cell lymphomas related to the helper virus contamination.[95,96] Recently, a low level expression of human ADA for up to 1 year in rhesus monkeys has been reported. In this study BM marrow cells were cocultivated with helper virus-free virus producing cells in the presence of IL-3.[97,98]

In conclusion, high efficiency gene transfer and stable long-term expression of transduced genes in hematopoietic progenitor and stem cells, observed in mice, has not been achieved in larger animals. More studies that will optimize conditions for *in vitro* culture, transduction, and transplantation of primitive hematopoietic cells are needed to improve the efficiency of gene transfer and expression of transgenes in large animal models. Only then will large animals provide a solid experimental model for extending gene therapy protocols to trials in human hematopoietic stem cells.

C. Gene Transfer into Human Hematopoietic Cells

Genes that have been transferred and expressed in human hematopoietic cells via retroviral vectors include neomycin phosphotransferase, dihydrofolate reductase (DHFR) conferring resistance to methotrexate, β-globin, ADA, glucocerebrosidase, CD18, and argininosuccinate synthetase.[7,99-118] Real progress has been made in determining optimal conditions for efficient transduction and expression of vector-derived genes in human primitive hematopoietic cells, especially in the case of ADA gene therapy model. After cocultivation of BM cells with virus-producing cells in the presence of GFs (IL-3, IL-6), an average of 83% of clonogenic progenitors (CFU-E and CFU-GM) were transduced with ADA vector. Furthermore, 24 to 44% of the clonogenic progenitors, derived from 9-week-old myeloid long-term cultures (LTCs), contained vector sequence, thus demonstrating efficient transduction of LTC-IC cells. ADA enzyme was found to be expressed in both normal and ADA-deficient erythroid colonies and in the nonadherent cells from LTC for at least 2 weeks.[7,106,111] After supernatant infection of normal and ADA-deficient BM, a 50% transduction efficiency of human clonogenic progenitors from 6-week-old myeloid LTC was achieved. ADA expression was detected in 30% of colonies, derived from transduced progenitors, and the ADA activity increased by 3.7-fold in the nonadherent cells from 9-week-old myeloid LTCs.[115]

One limitation of such studies is the uncertainty about the relationship of LTC-IC cells and the *in vivo* long-term repopulating cells. Since in some models relatively high transduction efficiency has been achieved in LTC-IC cells which has not been obtained after transplantation, there is a reasonable concern that LTC-IC cells are more similar to committed progenitors in their capacity for retroviral vector transduction.

More recent studies have taken a different approach and have evaluated primary lymphocytes as cellular targets for ADA gene therapy. Primary T-lymphocytes from an ADA-deficient patient were successfully transduced with an ADA vector and had their sensitivity to 2′-deoxyadenosine restored to normal levels. Furthermore, transduced T-cells were found to have the same phenotype, proliferative capacity, and cytotoxic potential as T-cells derived from healthy humans. These studies have provided evidence for the safety of transducing T-cells from ADA patients with a defective retroviral vectors.[102,107] Recently, lymphocytes from another ADA-deficient patient were shown to survive in immunodeficient mice only if they were transduced with an ADA vector. Expression of the vector-derived ADA also restored specific immune functions of transduced T-cells.[108] Similarly, transduction of leukocytes from LAD patients with a CD18-containing retroviral vector has demonstrated normal levels of LFA-1 expression and reconstitution of LFA-1-dependent adhesive function in infected cells.[105] However, efficient transduction of human CFC and LTC-IC cells remains a primary goal in further development of human gene therapy models. Several groups have demonstrated that up to 50% of CFC cells from 5- to 9-week-old bone marrow cultures initiated with retrovirally-transduced marrow are marked with the provirus.[111-116,118] Other studies have found that pretreatment of human BM cells for at least 2 days with hematopoietic GFs (combinations of IL-1, IL-3, IL-6, and SCF) or supernatant transduction on autologous stroma (endogenous source of GFs) is necessary for efficient transduction of clonogenic progenitors (CFU-GM, BFU-E and CFU-GEMM) and LTC-IC cells.[116,117] Recently, another study has also shown that pretreatment of $CD34^+$ BM cells with GFs (IL-1, IL-3, IL-6, SCF) greatly enhances transduction of progenitors (CFU-GM, BFU-E, CFU-GEMM) and LTC-IC cells, probably by recruiting quiescent cells into active proliferation.[118] These studies have demonstrated that high-efficiency gene transfer can be obtained in the most primitive class of human hematopoietic cells detectable *in vitro*. Further work, including *in vivo* comparisons, will be required to determine whether stroma-supported retrovirus vector infection or combined growth factor stimulation is superior in transducing long-term repopulating cells. Increased cell survival during manipulation *ex vivo* and efficient transduction of primitive hematopoietic cells would enhance successful engraftment of transduced stem cells and stable expression of exogenous gene in their progeny. The ability to efficiently transduce human primitive hematopoietic cells (derived either from MNC cells or from purified BM cells enriched for HSC) in the absence of virus-producing cells will greatly improve applicability of gene therapy in the treatment of hematologic diseases.

Recently, interest has been focused on peripheral blood and cord blood as a source of transplantable hematopoietic stem cells ($CD34^+$ cells), which could be used as targets for genetic manipulation prior to transplantation. Successful transfer of neomycin resistance gene was documented by PCR in 93% of day 14 myelomonocytic colonies, after supernatant infection of mononuclear cells from PB. Neo gene was documented also in CD34⁺/cyclophosphamide-resistant precursors to CFU-GM cells.[119] Highly efficient transduction of clonogenic progenitors and LTC-IC cells with TKNEO and PGKmADA vectors was also demonstrated after cocultivation of low-density CB cells with virus-producing cell lines. In addition, stable expression of the introduced genes was demonstrated in the progeny of infected LTC-IC cells after 5 weeks in LTCs.[120] In another study, a replication-defective retrovirus encoding $p47^{phox}$ gene was used to transduce PB-derived $CD34^+$ cells from patients with $p47^{phox}$-deficient chronic granulomatous disease, which resulted in significant correction of microbicidal superoxidase generation in mature neutrophils and monocytes.[121] These studies have provided the first evidence for efficient infection of PB- and CB-derived progenitors and stem cells, suggesting that circulating hematopoietic stem cells could represent an alternative target population in gene therapy of some hematologic diseases.

In addition to inherited hematologic disorders, somatic gene therapy will find application in the treatment of acquired disorders, such as infectious diseases. Several strategies are currently being deployed in the development of preclinical models for the gene therapy of acquired immunodeficiency syndrome (AIDS). One of the strategies for inhibition of HIV infection and propagation relies on the T-cell antigen CD4, which mediates entry of HIV into the cell. It was shown that soluble forms of the CD4 antigen can bind to and inhibit HIV infection of $CD4^+$ cells. Therefore, soluble CD4 could be produced *in vivo* in AIDS patients by retroviral transduction of the hematopoietic cells.[122] Another approach involves introduction of HIV-inducible genes that would directly or indirectly cause cell death after infection with HIV. This strategy is based on the induction of HIV gene expression by *tat* protein (transactivator of transcription) and the *tat*-responsive RNA sequence, called TAR element, which is located at the 5′ end of all HIV messenger RNAs. Thus, the expression of heterologous gene would be inducible by *tat* if the TAR region is inserted into the gene at the required position. Induction of a suicide gene, when a cell is infected by HIV, would eliminate the initial pool of infected cells and prevent virus propagation. Among several suicide genes, thymidine kinase and diphtheria toxin A genes have been extensively evaluated for use in anti-HIV gene therapy. Introduction of TK and DT-A genes under the control of HIV promoter and TAR element into human cells expressing *tat* has, upon HIV infection, resulted in death of infected cells and arrest of virus spreading. The problems posed by the use of these genes for AIDS gene therapy include required administration of antiherpetic drugs (gancyclovir, acyclovir) and low basal expression of DT-A gene.[123,124]

Alternative approaches involve introduction of HIV-inducible genes encoding products that can induce a cytocidal immune response (antibody or cytotoxic T-cell response) and HIV-transcript targeted ribozymes, small RNA molecules that allow sequence-specific cleavage of target RNA in a catalytic manner. Recently, several groups have developed defective HIV retroviral vectors for targeted delivery and expression of anti-HIV genes in $CD4^+$ cells.[125,126]

Table 2

Approved Human Clinical Trials for Gene Marking of Hematopoietic Cells

Disease	Gene Inserted	Target Cells
Malignant melanoma	Neomycin phosphotransferase	Tumor-infiltrating lymphocytes (TIL)
Pediatric AML	Neomycin phosphotransferase	Leukemic BM cells
CML	Neomycin phosphotransferase	Leukemic BM cells
Adult AML and ALL	Neomycin phosphotransferase	Leukemic BM cells
AIDS	Hygromycin phosphotransferase-thymidine kinase	HIV antigen-specific T-cells

IV. HUMAN GENE MARKING CLINICAL TRIALS

Characterization of human tumor infiltrating lymphocytes as vehicles for retroviral-mediated gene transfer[127] has provided the basis for the first approved human gene marking clinical trial, initiated in 1989. This clinical trial used the *neo*-transducing retroviral vector to mark tumor infiltrating lymphocytes (TIL), thus allowing the study of their traffic and survival *in vivo* after they had been readministered to patients with advanced melanoma (Table 2). Small numbers of transduced TILs persisted in the circulation of all 5 patients for at least 3 weeks. In two patients, transduced cells were detectable in the circulation for 2 months after infusion. Transduced cells were also recovered from tumor deposits in three of five patients 2 months after reinfusion, confirming the ability of TILs to home to tumor sites. Most importantly, no side effects due to gene transduction were noted, and all safety studies provided negative results. The results of this trial have suggested that only a subset of TILs is homing to tumor sites. Other gene-marking protocols are being developed (Table 2) to identify specific subsets ($CD4^+$ and $CD8^+$) of TILs and to further analyze applicability of TIL therapy.[4-6,11,128,129]

One of the recently approved cell-marking protocols involves marking of human HIV antigen-specific $CD8^+$ T-cells (Table 2). These cells will be used along with BM transplantation treatment of AIDS patients with non-Hodgkin's lymphoma. Patients will be given normal matched BM after high-dose chemotherapy and total body irradiation to eradicate lymphoma. In an attempt to protect transplanted normal BM cells from residual HIV infection, Zidovudine (AZT) and HIV antigen-specific killer T-cells will be administered to block virus spread and to kill residual HIV-infected cells. Patient-derived T-cells in this study will be transduced with a retroviral vector expressing hygromycin phosphotransferase-thymidine kinase fusion protein. Transduced cells can thus be positively selected for hygromycin resistance before reinfusion, and negatively selected *in vivo*, in a case in which marked cells attack other HIV-infected cells (CNS microglial cells, alveolar macrophages).[130]

Molecular analyses of retroviral transduction of cells from the BM of chronic myelogenous leukemia patients[131,132] have provided the basis for subsequent human gene-marking protocols for detection of residual leukemic cells in autologous BM transplants

Table 3

Approved Human Clinical Trials for Gene Therapy of Hematologic Diseases

Disease	Gene Inserted	Target Cells
ADA deficiency	ADA	T-lymphocytes from ADA-deficient patients
ADA deficiency	ADA	Peripheral blood T-lymphocytes + progenitor-enriched BM cells
ADA deficiency	ADA	Bone marrow cells
ADA deficiency	ADA	T-lymphocytes and peripheral blood stem cells ($CD34^+$ cells)
Malignant melanoma	Tumor necrosis factor (TNF)	TILs

used for treatment of patients with chronic (CML) and acute myelogenous leukemia (AML) (Table 2). The aim of this study is to determine whether the relapse of disease in patients is caused by residual leukemic cells in the patient or in the infused BM cells. Based on the results of this trial, improved strategies for ablation of leukemic cells in the patients or for removal of these cells from the infused marrow will be developed.[4,5,133-135]

V. HUMAN GENE THERAPY CLINICAL TRIALS

The first human gene therapy trial, initiated in September 1990, involves treatment of ADA deficiency by retroviral transfer of ADA gene into the patient's peripheral T-cells (Table 3, Figure 1). In this trial, two ADA-deficient patients with severe combined

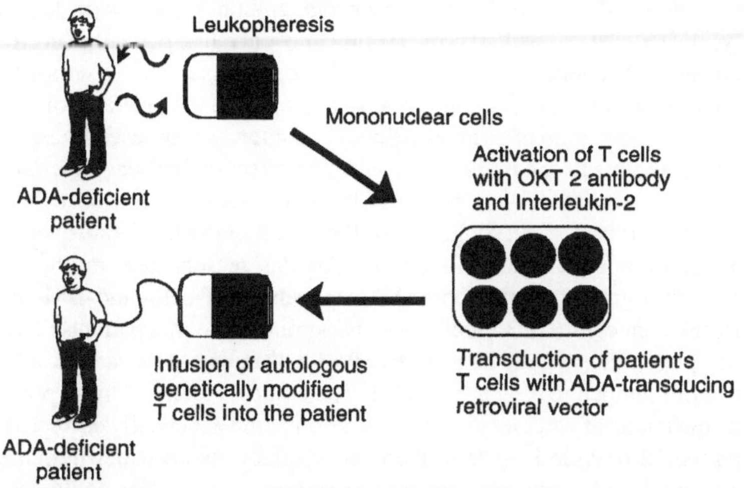

FIGURE 1. Schematic of the human ADA gene therapy protocol involving transduction of T-lymphocytes from peripheral blood of ADA-deficient patients.

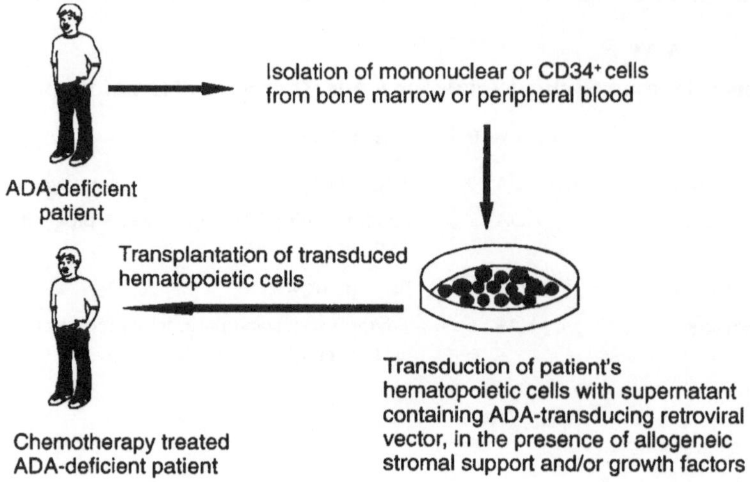

FIGURE 2. Schematic of the human ADA gene therapy protocol involving transduction of primitive progenitor cells from bone marrow or peripheral blood of ADA-deficient patients.

immunodeficiency (SCID) were treated by infusion of peripheral T-cells that had been transduced with a retroviral vector that expresses human ADA and neomycin as a selectable marker. The patient's mononuclear cells were isolated by leukopheresis and Ficoll gradient and grown in culture under conditions that stimulate T-lymphocyte activation and proliferation (OKT3 antibody and interleukin-2 stimulation). Dividing T-cells were then transduced with ADA vector and reinfused into patients. Both patients received multiple infusions of genetically modified autologous T-cells. In both patients, the ADA level in circulating cells increased from < 2 to 20% of normal, and lymphocyte count reached normal range. In addition, the majority of immune functions in both patients improved significantly as measured by isohaemagglutinin titers, skin tests for antigen sensitivity, and cytotoxic T-cell assays. Twenty to twenty-five percent of corrected T-cells persisted in the first patient for over 6 months after cell infusions were stopped. There have been no significant side effects from the cell infusions and the presence of the transferred ADA gene itself. The issue of clinical efficacy is somewhat confused by the continuation of PEG-ADA administration in these patients. The effect on immune function of repeated administration of activated T-lymphocytes by themselves is also not resolved. Clinical concern has been raised about longevity of therapeutic effects of transduced T-cells and complete restoration of immune functions. For this reason, new protocols have been proposed by two groups, where both ADA-transduced T-cells and BM- or peripheral blood-derived stem cells ($CD34^+$ cells) will be administered into patients (Table 3, Figure 2). Two cell populations would be transduced with different ADA vectors, allowing direct comparison of therapeutic effects provided. The clinical results of these protocols would answer the questions of whether the administration of the genetically corrected $CD34^+$ cell population would provide long-term therapeutic effects and broader immune protection than the mature T-cell population. Equally important, these studies could provide *in vivo* evidence for gene transfer into human hematopoietic stem cells.[11,136-138]

The second group of gene therapy trials involves the transfer of genes that would stimulate immune response or otherwise cause the destruction of tumor cells (Table 3). In these trials, a gene for tumor necrosis factor (TNF), a powerful anticancer cytokine, was introduced into tumor-infiltrating lymphocytes (TILs). Localized secretion of TNF by TILs, which home into tumor deposits, would enhance immune destruction of tumor cells. However, because a number of TILs are trapped in liver, spleen, and lungs, the production of ectopic TNF with toxic effects is possible. Twelve patients have been treated, and so far there have been no side effects and organ toxicity from secreted TNF.[11,139]

VI. CONCLUSION

In the last 5 years, gene therapy has made progress from the laboratory and preclinical stage to actual clinical trials. The first gene marking and gene therapy trials have shown that the introduction of new genetic information to the human hematopoietic system by recombinant retroviral vectors is feasible. Meanwhile, clinical trials aiming at stem cell gene transfer have been initiated, and most recent studies have started to address functional heterogeneity of human hematopoietic stem cells and mechanisms of their self-renewal and lineage commitment.[3-5,7,31,140,141] Along with development of improved and alternative gene transfer techniques, a better understanding of the biology of long-term reconstituting stem cells and factors that control expression of introduced genes will ultimately lead to gene therapy trials for various hematologic disorders. Although gene therapy now requires specialized technologies and expertise, it will surely have a major impact on the treatment of hematological diseases, not just as an optional curative therapy but possibly as a preventive one as well.

REFERENCES

1. Anderson, W. F., Prospects for human gene therapy, *Science*, 226, 401, 1984.
2. Kantoff, P. W., Freeman, S. M., and Anderson, W. F., Prospects for gene therapy for immunodeficiency diseases, *Annu. Rev. Immunol.*, 6, 581, 1988.
3. Beutler, E. and Sorge, J., Gene transfer in the treatment of hematologic disease, *Exp. Hematol.*, 18, 857, 1990.
4. Anderson, W. F., Human gene therapy, *Science*, 256, 808, 1992.
5. Miller, A. D., Human gene therapy comes of age, *Nature*, 357, 455, 1992.
6. Mulligan, R. C., The basic science of gene therapy, *Science*, 260, 926, 1993.
7. Cournoyer, D. and Caskey, C. T., Gene therapy of the immune system, *Annu. Rev. Immunol.*, 11, 297, 1993.
8. Mercola, K. E., Stang, H. D., Browne, J., Salser, W., and Cline, M. J., Insertion of a new gene of viral origin into bone marrow cells of mice, *Science*, 208, 1033, 1980.
9. Cline, M. J., Stang, H., Mercola, K., Morse, L., Ruprecht, R., Brown, J., and Salser, W., Gene transfer in intact animals, *Nature*, 284, 422, 1980.
10. Joyner, A., Keller, G., Phillips, R. A., and Bernstein, A., Retrovirus transfer of a bacterial gene into mouse haematopoietic progenitor cells, *Nature*, 305, 556, 1983.

11. Human gene marker/therapy clinical protocols, *Hum. Gene Ther.*, 4, 391, 1993.
12. Civin, C. I., Strauss, L. C., Brovall, C., Fackler, M. J., Schwartz, J. F., and Shaper, J. H., Antigenic analysis of hematopoiesis. III. A hematopoietic progenitor cell surface antigen defined by a monoclonal antibody raised against KG-1a cells, *J. Immunol.*, 133, 157, 1984.
13. Strauss, L. C., Rowley, S. D., La Russa, V. F., Sharkis, S. J., Stuart, R. K., and Civin, C. I., Antigenic analysis of hematopoiesis. V. Characterization of My-10 antigen expression by normal lymphohematopoietic progenitor cells, *Exp. Hematol.*, 14, 878, 1986.
14. Simmons, D., Satterthwaite, A. B., Tenen, D. G., and Seed, B., Molecular cloning of a cDNA encoding CD34, a sialomucin of human hematopoietic stem cells, *J. Immunol.*, 148, 267, 1992.
15. Burn, T. C., Satterthwaite, A. B., and Tenen, D. G., The human CD34 hematopoietic stem cell antigen promoter and a 3′ enhancer direct hematopoietic expression in tissue culture, *Blood*, 80, 3051, 1992.
16. Andrews, R. G., Singer, J. W., and Bernstein, I. D., Monoclonal antibody 12-8 recognizes a 115-kd molecule present on both unipotent and multipotent hematopoietic colony-forming cells and their precursors, *Blood*, 67, 842, 1986.
17. Brandt, J. E., Baird, N., Lu, L., Srour, E., and Hoffman, R., Characterization of a human hematopoietic progenitor cell capable of forming blast cell containing colonies in vitro, *J. Clin. Invest.*, 82, 1017, 1988.
18. Andrews, R. G., Singer, J. W., and Bernstein, I. D., Precursors of colony-forming cells in humans can be distinguished from colony-forming cells by expression of the CD33 and CD34 antigens and light scatter properties, *J. Exp. Med.*, 169, 1721, 1989.
19. Sutherland, H. J., Eaves, C. J., Eaves, A. C., Dragowska, A., and Lansdorp, P. M., Characterization and partial purification of human marrow cells capable of initiating long-term hematopoiesis in vitro, *Blood*, 74, 1563, 1989.
20. Sutherland, H. J., Lansdorp, P. M., Henkelman, D. H., Eaves, A. C., and Eaves, C. J., Functional characterization of individual human hematopoietic stem cells cultured at limiting dilution on supportive marrow stromal layers, *Proc. Natl. Acad. Sci. U.S.A.*, 87, 3584, 1990.
21. Berenson, R. J., Andrews, R. G., Bensinger, W. I., Kalamasz, D., Knitter, G., Buckner, C. D., and Bernstein, I. D., Antigen CD34$^+$ marrow cells engraft lethally irradiated baboons, *J. Clin. Invest.*, 81, 951, 1988.
22. Berenson, R. J., Bensinger, W. I., Hill, R. S., Andrews, R. G., Garcia-Lopez, J., Kalamasz, D., Still, B. J., Spitzer, G., Buckner, C. D., Bernstein, I. D., and Thomas, E. D., Engraftment after infusion of CD34$^+$ marrow cells in patients with breast cancer or neuroblastoma, *Blood*, 77, 1717, 1991.
23. Peault, B., Weissman, I. L., Baum, C., McCune, J. M., and Tsukamoto, A., Lymphoid reconstitution of the human fetal thymus in SCID mice with CD34$^+$ precursor cells, *J. Exp. Med.*, 174, 1283, 1991.
24. Tjoonfjord, G. E., Veiby, O. P., Steen, R., and Egeland, T., T lymphocyte differentiation in vitro from adult human prethymic CD34$^+$ bone marrow cells, *J. Exp. Med.*, 177, 1531, 1993.
25. Galy, A., Verma, S., Barcena, A., and Spits, H., Precursors of CD3$^+$CD4$^+$CD8$^+$ cells in the human thymus are defined by expression of CD34. Delineation of early events in human thymic development, *J. Exp. Med.*, 178, 391, 1993.
26. Verfaillie, C., Blakolmer, K., and McGlave, P., Purified primitive human hematopoietic progenitor cells with long-term in vitro repopulating capacity adhere selectively to irradiated bone marrow stroma, *J. Exp. Med.*, 172, 509, 1990.
27. Lansdorp, P. M., Sutherland, H. J., and Eaves, C. J., Selective expression of CD45 isoforms on functional subpopulations of CD34$^+$ hemopoietic cells from human bone marrow, *J. Exp. Med.*, 172, 363, 1990.

28. Papayannopoulou, T., Brice, M., Broudy, V. C., and Zsebo, K. M., Isolation of *c-kit* receptor-expressing cells from bone marrow, peripheral blood, and fetal liver: functional properties and composite antigenic profile, *Blood*, 78, 1403, 1991.
29. Huang, S. and Terstappen, L. W. M. M., Formation of haematopoietic microenvironment and haematopoietic stem cells from single human bone marrow stem cells, *Nature*, 360, 745, 1992.
30. Gunji, Y., Nakamura, M., Hagiwara, T., Hayakawa, K., Matsushita, H., Osawa, H., Nagayoshi, K., Nakauchi, H., Yanagisawa, M., Miura, Y., and Suda, T., Expression and function of adhesion molecules on human hematopoietic stem cells: $CD34^+LFA\text{-}1^-$ cells are more primitive than $CD34^+LFA\text{-}1^+$ cells, *Blood*, 80, 429, 1992.
31. Lansdorp, P. M. and Dragowska, W., Long-term erythropoiesis from constant numbers of $CD34^+$ cells in serum-free cultures initiated with highly purified progenitor cells from human bone marrow, *J. Exp. Med.*, 175, 1501, 1992.
32. Saeland, S., Duvert, V., Caux, C., Pandrau, D., Favre, D., Valle, A., Durand, I., Charbord, P., de Vries, J., and Banchereau, J., Distribution of surface-membrane molecules on bone marrow and cord blood $CD34^+$ cells, *Exp. Hematol.*, 20, 24, 1992.
33. Pontvert-Delucq, S., Breton-Gorius, J., Schmitt, C., Baillou, C., Guichard, J., Najman, A., and Lemoine, F. M., Characterization and functional analysis of adult human bone marrow cell subsets in relation to B-lymphoid development, *Blood*, 82, 417, 1993.
34. Denkers, I. A. M., Dragowska, W., Jaggi, B., Palcic, B., and Lansdorp, P. M., Time lapse video recordings of highly purified human hematopoietic progenitor cells in culture, *Stem Cells*, 11, 243, 1993.
35. Henon, P. R., Peripheral blood stem cell transplantations: past, present and future, *Stem Cells*, 11, 154, 1993.
36. Baum, C. M., Weissman, I. L., Tsukamoto, A. S., Buckle, A., and Peault, B., Isolation of a candidate human hematopoietic stem-cell population, *Proc. Natl. Acad. Sci. U.S.A.*, 89, 2804, 1992.
37. Craig, W., Kay, R., Cutler, R. L., and Lansdorp, P. M., Expression of Thy-1 on human hematopoietic progenitor cells, *J. Exp. Med.*, 177, 1331, 1993.
38. Broxmeyer, H. E., Douglas, G. W., Hangoc, G., Cooper, S., Bad, J., English, D., Arny, M., Thomas, L., and Boyse, F. A., Human umbilical cord blood as a potential source of transplantable hematopoietic stem/progenitor cells, *Proc. Natl. Acad. Sci. U.S.A.*, 86, 3828, 1989.
39. Bender, J. G., Unverzagt, K. L., Walker, D. E., Lee, W., Van Epps, D. E., Smith, D. H., Stewart, C. C., and To, L. B., Identification and comparison of CD34 positive cells and their subpopulations from normal peripheral blood and bone marrow using multicolor flow cytometry, *Blood*, 77, 2591, 1991.
40. Lowry, P. A. and Tabbara, I. A., Peripheral hematopoietic stem cell transplantation: current concepts, *Exp. Hematol.*, 20, 937, 1992.
41. Abboud, M., Xu, F., LaVia, M., and Laver, J., Study of early hematopoietic precursors in human cord blood, *Exp. Hematol.*, 20, 1043, 1992.
42. Dooley, D. C. and Law, P., Detection and quantitation of long-term culture-initiating cells in normal human peripheral blood, *Exp. Hematol.*, 20, 156, 1992.
43. Udomsakdi, C., Lansdorp, P. M., Hogge, D. E., Reid, D. S., Eaves, A. C., and Eaves, C. J., Characterization of primitive hematopoietic cells in normal human peripheral blood, *Blood*, 80, 2513, 1992.
44. Lu, L., Xiao, M., Shen, R., Grigsby, S., and Broxmeyer, H. E., Enrichment, characterization and responsiveness of single primitive $CD34^{+++}$ human umbilical cord blood hematopoietic progenitors with high proliferative and replating potential, *Blood*, 81, 41, 1993.
45. Lansdorp, P. M. and Thomas, T. E., Selection of human hemopoietic stem cells, in *Bone Marrow Processing and Purging: A Practical Guide*, Gee, A. P., Ed., CRC Press, Boca Raton, FL, 1991, 351.

46. Sawada, K., Dai, C. H., Koury, S. T., Horn, S. T., Glick, A. D., and Civin, C. I., Purification of human blood burst-forming-units erythroid and demonstration of the evolution of erythropoietin receptors, *J. Cell Physiol.*, 142, 219, 1990.
47. Leary, A. and Ogawa, M., Blast cell colony assay for umbilical cord blood and adult bone marrow progenitors, *Blood*, 69, 953, 1987.
48. Brugger, W., Mocklin, W., Heimfeld, S., Berenson, R. J., Mertelsmann, R., and Kanz, L., Ex vivo expansion of enriched peripheral blood CD34+ progenitor cells by stem cell factor, interleukin-1β (IL-1β), IL-6, IL-3, interferon-γ, and erythropoietin, *Blood*, 81, 2579, 1993.
49. Guillaume, T., D'Hondt, V., and Symann, M., IL-3 and peripheral blood stem cell harvesting, *Stem Cells*, 11, 173, 1993.
50. Bernstein, I. D., Andrews, R. G., and Zsebo, K. M., Recombinant human stem cell factor enhances the formation of colonies by CD34+ and CD34+Lin− cells and the generation of colony-forming cell progeny from CD34+Lin− cells cultured with interleukin-3, granulocyte colony-stimulating factor, or granulocyte-macrophage colony-stimulating factor, *Blood*, 77, 2316, 1991.
51. Leary, A. G., Zeng, H. Q., Clark, S. C., and Ogawa, M., Growth factor requirements for survival in G0 and entry into the cell cycle of primitive human hemopoietic proegnitors, *Proc. Natl. Acad. Sci. U.S.A.*, 89, 4013, 1992.
52. Haylock, D. N., To, L. B., Dowse, T. L., Juttner, C. A., and Simmons, P. J., Ex vivo expansion and maturation of peripheral blood CD34+ cells into the myeloid lineage, *Blood*, 80, 1405, 1992.
53. Xiao, M., Leemhuis, T., Broxmeyer, H. E., and Lu, L., Influence of combinations of cytokines on proliferation of isolated single cell-sorted human bone marrow hematopoietic progenitor cells in the absence and presence of serum, *Exp. Hematol.*, 20, 276, 1992.
54. Debili, N., Masse, J., Katz, A., Guichard, J., Breton-Gorius, J., and Vainchenker, W., Effects of the recombinant hematopoietic growth factors interleukin-3, interleukin-6, stem cell factor, and leukemia inhibitory factor on the megakaryocytic differentiation of CD34+ cells, *Blood*, 82, 84, 1993.
55. Mayani, H., Dragowska, W., and Lansdorp, P. M., Cytokine-induced selective expansion and maturation of erythroid versus myeloid progenitors from purified cord blood precursor cells, *Blood*, 81, 3252, 1993.
56. Testa, U., Pelosi, E., Gabbianelli, M., Fossati, C., Campisi, S., Isacchi, G., and Peschle, C., Cascade transactivation of growth factor receptors in early human hematopoiesis, *Blood*, 81, 1442, 1993.
57. Srour, E. F., Brandt J. E., Briddell, R. A., Grigsby, S., Leemhuis, T., and Hoffman, R., Long-term generation and expansion of human primitive hematopoietic progenitor cells in vitro, *Blood*, 81, 661, 1993.
58. Palsson, B. O., Paek, S., Schwartz, R. M., Palsson, M., Lee, G., Silver, S., and Emerson, S. G., Expansion of human bone marrow progenitor cells in a high cell density continuous perfusion system, *Biotechnology*, 11, 368, 1993.
59. Koller, M. R., Bender, J. G., Miller, W. M., and Papoutsakis, E. T., Expansion of primitive human hematopoietic progenitors in a perfusion bioreactor system with IL-3, IL-6, and stem cell factor, *Biotechnology*, 11, 358, 1993.
60. Koller, M. R., Emerson, S. G., and Palsson, B. O., Large-scale expansion of human stem and progenitor cells from bone marrow mononuclear cells in continuous perfusion culture, *Blood*, 82, 378, 1993.
61. Williams, D. A., Lemischka, I. R., Nathan, D. G., and Mulligan, R. C., Introduction of new genetic material into pluripotent haematopoietic stem cells of the mouse, *Nature*, 310, 476, 1984.

62. Keller, G., Paige, C., Gilboa, E., and Wagner, E. F., Expression of a foreign gene in myeloid and lymphoid cells derived from multipotent haematopoietic precursors, *Nature*, 318, 149, 1985.
63. Corey, C. A., DeSilva, A. D., Holland, C. A., and Wiliams, D. A., Serial transplantation of methotrexate-resistant bone marrow: protection of murine recipients from drug toxicity by progeny of transduced stem cells, *Blood*, 75, 337, 1990.
64. Sorrentino, B. P., Brandt, S. J., Bodine, D., Gottesman, M., Pastan, I., Cline, A., and Nienhuis, A. W., Selection of drug-resistant bone marrow cells in vivo after retroviral transfer of human MDR1, *Science*, 257, 99, 1992.
65. Dzierzak, E. A., Papayannopoulou, T., and Mulligan, R. C., Lineage-specific expression of a human β-globin gene in murine bone marrow transplant recipients reconstituted with retrovirus-transduced stem cells, *Nature*, 331, 35, 1988.
66. Karlsson, S., Bodine, D. M., Perry, L., Papayannopoulou, T., and Nienhuis, A. W., Expression of the human β-globin gene following retroviral-mediated transfer into multipotential hematopoietic progenitors of mice, *Proc. Natl. Acad. Sci. U.S.A.*, 85, 6062, 1988.
67. Plavec, I., Papayannopoulou, T., Maury, C., and Meyer, F., A human β-globin gene fused to the human β-globin locus control region is expressed at high levels in erythroid cells of mice engrafted with retrovirus-transduced hematopoietic stem cells, *Blood*, 81, 1384, 1993.
68. Williams, D. A., Orkin, S. H., and Mulligan, R. C., Retrovirus-mediated transfer of human adenosine deaminase gene sequences into cells in culture and into murine hematopoietic cells *in vivo*, *Proc. Natl. Acad. Sci. U.S.A.*, 83, 2566, 1986.
69. Belmont, J. W., Henkel-Tigges, J., Chang, S. M. W., Wager-Smith, K., Kellems, R. E., Dick, J. E., Magli, M. C., Phillips, R. A., Bernstein, A., and Caskey, C. T., Expression of human adenosine deaminase in murine hematopoietic progenitor cells following retroviral transfer, *Nature*, 322, 385, 1986.
70. Belmont, J. W., MacGregor, G. R., Wager-Smith, K., Fletcher, F. A., Moore, K. A., Hawkins, D., Villalon, D., Chang, S. M., and Caskey, C. T., Expression of human adenosine deaminase in murine hematopoietic cells, *Mol. Cell. Biol.*, 8, 5116, 1988.
71. Luskey, B. D., Rosenblatt, M., Zsebo, K., and Williams, D. A., Stem cell factor, interleukin-3, and interleukin-6 promote retroviral-mediated gene transfer into murine hematopoietic stem cells, *Blood*, 80, 396, 1992.
72. Moore, K. A., Fletcher, F. A., Villalon, D. K., Utter, A. E., and Belmont, J. W., Human adenosine deaminase expression in mice, *Blood*, 75, 2085, 1990.
73. Fletcher, F. A., Williams, D. E., Maliszewski, C., Anderson, D., Rives, M., and Belmont, J. W., Murine leukemia inhibitory factor (LIF) enhances retroviral-vector infection efficiency of hematopoietic progenitors, *Blood*, 76, 1098, 1990.
74. van Beusechem, V. W., Kuklerm A., Einerhand, M. P. W., Bakx, T. A., van der Eb, A. J., van Bekkum, D. W., and Valerio, D., Expression of human adenosine deaminase in mice transplanted with hemopoietic stem cells infected with amphotropic retroviruses, *J. Exp. Med.*, 172, 729, 1990.
75. Fletcher, F. A., Moore, K. A., Ashkenazi, M., De Vries, P., Overbeek, P. A., Williams, D. E., and Belmont, J. W., Leukemia inhibitory factor improves survival of retroviral vector-infected hematopoietic stem cells in vitro, allowing efficient long-term expression of vector-encoded human adenosine deaminase in vivo, *J. Exp. Med.*, 174, 837, 1991.
76. Einerhand, M. P. W., Bakx, T. A., Kukler, A., and Valerio, D., Factors affecting the transduction of pluripotent hematopoietic stem cells: long-term expression of a human adenosine deaminase gene in mice, *Blood*, 81, 254, 1993.
77. Chertkov, J. L., Jiang, S., Lutton, J. D., Harrison, J., Levere, R. D., Tiefenthaler, M., and Abraham, N. G., The hematopoietic stromal microenvironment promotes retrovirus-mediated gene transfer into hematopoietic stem cells, *Stem Cells*, 11, 218, 1993.

78. Krauss, J. C., Bond, L. M., Todd, R. F., and Wilson, J. M., Expression of retroviral transduced human CD18 in murine cells: an *in vitro* model of gene therapy for leukocyte adhesion deficiency, *Hum. Gene Ther.*, 2, 221, 1991.
79. Correll, P. H., Colilla, S., Dave, H. P. G., and Karlsson, S., High levels of human glucocerebrosidase activity in macrophages of long-term reconstituted mice after retroviral infection of hematopoietic stem cells, *Blood*, 80, 331, 1992.
80. Demarquoy, J., Herman, G. E., Lorenzo, I., Trentin, J., Beaudet, A. L., and O'Brien, W. E., Long-term expression of human argininosuccinate synthetase in mice following bone marrow transplantation with retrovirus-transduced hematopoietic stem cells, *Hum. Gene Ther.*, 3, 3, 1992.
81. Bodine, D. M., Karlsson, S., and Nienhuis, A. W., Combination of interleukins 3 and 6 preserves stem cell function in culture and enhances retrovirus-mediated gene transfer into hematopoietic stem cells, *Proc. Natl. Acad. Sci. U.S.A.*, 86, 8897, 1989.
82. Bodine, D. M., McDonagh, K. T., Seidel, N. E., and Nienhuis, A. W., Survival and retrovirus infection of murine hematopoietic stem cells in vitro: effects of 5-FU and method of infection, *Exp. Hematol.*, 19, 206, 1991.
83. Szilvassy, S. J., Fraser, C. C., Eaves, C. J., Lansdorp, P. M., Eaves, A. C., and Humphries, R. K., Retrovirus-mediated gene transfer to purified hemopoietic stem cells with long-term lympho-myelopoietic repopulating ability, *Proc. Natl. Acad. Sci. U.S.A.*, 86, 8798, 1989.
84. Spain, L. M. and Mulligan, R. C., Purification and characterization of retrovirally transduced hematopoietic stem cells, *Proc. Natl. Acad. Sci. U.S.A.*, 89, 3790, 1992.
85. Stead, R. B., Kwok, W. W., Storb, R., and Miller, A. D., Canine model for gene therapy: inefficient gene expression in dogs reconstituted with autologous marrow infected with retroviral vectors, *Blood*, 71, 742, 1988.
86. Schuening, F. G., Storb, R., Stead, R. B., Goehle, S., Nash, R., and Miller, A. D., Improved retroviral transfer of genes into canine hematopoietic progenitor cells kept in long-term marrow culture, *Blood*, 74, 152, 1989.
87. Al-Lebban, Z. S., Henry, J. M., Jones, J. B., Eglitis, M. A., Anderson, F. W., and Lothrop, C. D., Increased efficiency of gene transfer with retroviral vectors in neonatal hematopoietic progenitor cells, *Exp. Hematol.*, 18, 180, 1990.
88. Schuening, F. G., Kawahara, K., Miller, A. D., Goehle, S., Stewart, D., Mullally, K., Graham, T. C., Appelbaum, F. R., To, R., Hackman, R., Osborne, W. R. A., and Storb, R., Retrovirus-mediated gene transduction into long-term repopulating marrow cells of dogs, *Blood*, 78, 2568, 1991.
89. Carter, R. F., Abrams-Ogg, A. C. G., Dick, J. E., Kruth, S. A., Valli, V. E., Kamel-Reid, S., and Dube, I. D., Autologous transplantation of canine long-term marrow culture cells genetically marked by retroviral vectors, *Blood*, 79, 356, 1992.
90. Lothrop, C. D., Al-Lebban, Z. S., Niemeyer, G. P., Jones, J. B., Peterson, M. G., Smith, J. R., Baker, H. J., Morgan, R. A., Eglitis, M. A., and Anderson, W. F., Expression of foreign gene in cats reconstituted with retroviral vector infected autologous bone marrow, *Blood*, 78, 237, 1991.
91. Kantoff, P. W., Flake, A. W., Eglitis, M. A., Scharf, S., Bond, S., Gilboa, E., Erlich, H., Harrison, M. R., Zanjani, E. D., and Anderson, W. F., In utero gene transfer and expression: a sheep transplantation model, *Blood*, 73, 1066, 1989.
92. Clapp, D. W., Dumenco, L. L., Hatzoglou, M., and Gerson, S. L., Fetal liver hematopoietic stem cells as a target for in utero retroviral gene transfer, *Blood*, 78, 1132, 1991.
93. Kantoff, P. W., Gillio, A. P., McLachlin, J. R., Bordignnon, C., Eglitis, M. A., Kernan, N. A., Moen, R. C., Kohn, D. B., Yu, S. F., Karson, E., Karlson, S., Zweibel, J., Gilboa, E., Blaese, R. M., Nienhuis, A., O'Reilly, R. J., and Anderson, W. F., Expression of human adenosine deaminase in nonhuman primates after retrovirus-mediated gene transfer, *J. Exp. Med.*, 166, 219, 1987.

94. Wieder, R., Cornetta, K., Kessler, S. W., and Anderson, W. F., Increased efficiency of retroviral-mediated gene transfer and expression in primate bone marrow after 5-fluorouracil-induced hematopoietic suppression and recovery, *Blood*, 77, 448, 1991.
95. Bodine, D. M., McDonagh, K. T., Brandt, S. J., Ney, P. A., Agricola, B., Byrne, E., and Nienhuis, A. W., Development of a high-titer retrovirus producer cell line capable of gene transfer into rhesus monkey hematopoietic stem cells, *Proc. Natl. Acad. Sci. U.S.A.*, 87, 3738, 1990.
96. Donahue, R. E., Kessler, S. W., Bodine, D. B., Agricola, B. A., Byrne, E. R., Metzger, M. E., McDonagh, K. T., Bacher, J. D., Zsebo, K. M., and Nienhuis, A. W., Gene transfer into primate immunoselected $CD34^+$ cells: in vivo evidence that CD34 cells are responsible for both short- and long-term reconstitution of lymphoid and myeloid populations, *Blood*, 78 (Abstr.), 79a, 1991.
97. van Beusechem, V. W., Kukler, A., Heidt, P. J., and Valerio, D., Long-term expression of human adenosine deaminase in rhesus monkeys transplanted with retrovirus-infected bone marrow cells, *Proc. Natl. Acad. Sci. U.S.A.*, 89, 7640, 1992.
98. van Beusechem, V. W., Bakx, T. A., Kaptein, L. C. M., Bart-Baumeister, J. A., Kukler, A., Braakman, E., and Valerio, D., Retrovirus-mediated gene transfer into rhesus monkey hematopoietic stem cells: the effect of viral titers on transduction efficiency, *Hum. Gene Ther.*, 4, 239, 1993.
99. Dick, J. E., Retrovirus-mediated gene transfer into hematopoietic stem cells, *Ann. N.Y. Acad. Sci.*, 507, 242, 1987.
100. Bender, M. A., Miller, D. A., and Gelinas, R. E., Expression of the human beta-globin gene after retroviral transfer into murine erythroleukemia cells and human BFU-E cells, *Mol. Cell Biol.*, 8, 1725, 1988.
101. Holland, C. A., Rothstein, L., Sakakeeny, M. A., Anklesaria, P., Griffin, J. D., Harigaya, K., Newburger, P. E., and Greenberger, J. S., Infection of hematopoietic and stromal cells in human continuous bone marrow cultures by a retroviral vector containing the neomycin resistance gene, *Acta Hematol.*, 82, 136, 1989.
102. Kohn, D. B., Kantoff, P., Zweibel, J., Gilboa, E., Anderson W. F., and Blaese, R. M., Transfer and expression of the human adenosine deaminase (ADA) gene in ADA-deficient human T lymphocytes with retroviral vectors, in *Gene Transfer and Gene Therapy*, UCLA Symposia on Molecular and Cellular Biology, New Series, Vol. 87, Verma, I., Mulligan, R. C., and Beaudet, A. L., Eds., Alan R. Liss, New York, 1989, 365.
103. Cournoyer, D., Scarpa, M., and Caskey, C. T., Gene therapy, *Curr. Opin. Biotechnol.*, 1, 196, 1990.
104. Fink, J. K., Correll, P. H., Perry, L. K., Brady, R. O., and Karlsson, S., Correction of glucocerebrosidase deficiency after retroviral-mediated gene transfer into hematopoietic progenitor cells from patients with Gaucher disease, *Proc. Natl. Acad. Sci. U.S.A.*, 87, 2334, 1990.
105. Wilson, J. M., Ping, A. Y., Krauss, J. C., Mayo-Bond, L., Rogers, C. E., Anderson, D. C., and Todd, R. F., Correction of CD18-deficient lymphocytes by retrovirus-mediated gene transfer, *Science*, 248, 1413, 1990.
106. Cournoyer, D., Scarpa, M., Moore, K. A., Fletcher, F. A., MacGregor, G. R., Belmont, J. W., and Caskey, C. T., Gene replacement therapy: human and animal model progress, in *Etiology of Human Disease at the DNA Level*, Lindsten, J. and Pettersson, U., Eds., Raven Press, New York, 1991, 199.
107. Braakman, E., van Beusechem, V. W., van Krimpen, B. A., Fischer, B., Bolhuis, R. L., and Valerio, D., Genetic correction of cultured T cells from an adenosine deaminase-deficient patient: characteristics of non-transduced and transduced cells, *Eur. J. Immunol.*, 22, 63, 1992.

108. Ferrari, G., Rossini, S., Nobili, N., Maggioni, D., Garofalo, A., Giavazzi, R., Mavilio, F., and Bordignon, C., Transfer of the ADA gene into human ADA-deficient T lymphocytes reconstitutes specific immune functions, *Blood*, 80, 1120, 1992.
109. Demarquoy, J., Retroviral-mediated gene therapy for the treatment of citrullinemia: transfer and expression of argininosuccinate synthetase in human hematopoietic cells, *Experientia*, 49, 345, 1993.
110. Karlsson, S., Correll, P. H., and Xu, L., Gene transfer and bone marrow transplantation with special reference to Gaucher's disease, *Bone Marrow Transplant*, 11 (Suppl. 1), 124, 1993.
111. Mitani, K., Wakamiya, M., and Caskey, C. T., Long-term expression of retroviral-transduced adenosine deaminase in human primitive hematopoietic progenitors, *Hum. Gene Ther.*, 4, 9, 1993.
112. Hughes, P. F., Eaves, C. J., Hogge, D. E., and Humphries, R. K., High efficiency gene transfer to human hematopoietic cells maintained in long-term marrow culture, *Blood*, 74, 1915, 1989.
113. Bordignon, C., Yu, S., Smith, C. A., Hantzopoulos, P., Ungers, G. E., Keever, C. A., O'Reilly, R. J., and Gilboa, E., Retroviral vector-mediated high efficiency expression of adenosine deaminase (ADA) in hematopoietic long-term cultures of ADA-deficient marrow cells, *Proc. Natl. Acad. Sci. U.S.A.*, 86, 6748, 1989.
114. Dick, J. E., Kamel-Reid, S., Murdoch, B., and Doedens, M., Gene transfer into normal human hematopoietic cells using in vitro and in vivo assays, *Blood*, 78, 624, 1991.
115. Cournoyer, D., Scarpa, M., Mitani, K., Moore, K. A., Markowitz, D., Bank, A., Belmont, J. W., and Caskey, C. T., Gene transfer of adenosine deaminase into primitive human hematopoietic progenitor cells, *Hum. Gene Ther.*, 2, 203, 1991.
116. Nolta, J. A. and Kohn, D. B., Comparison of the effects of growth factors on retroviral vector-mediated gene transfer and the proliferative status of human hematopoietic progenitor cells, *Hum. Gene Ther.*, 1, 257, 1990.
117. Moore, K. A., Deisseroth, A. B., Reading, C. L., Williams, D. E., and Belmont, J. W., Stromal support allows efficient cell-free retroviral vector transduction of human bone marrow long-term culture initiating cells, *Blood*, 79, 1393, 1992.
118. Nolta, J. A., Crooks, G. M., Overell, R. W., Williams, D. E., and Kohn, D. B., Retroviral vector-mediated gene transfer into primitive human hematopoietic progenitor cells: effects of mast cell growth factor (MGF) combined with other cytokines, *Exp. Hematol.*, 20, 1065, 1992.
119. Bregni, M., Magni, M., Siena, S., Di Nicola, M., Bonadonna, G., and Gianni, A. M., Human peripheral blood hematopoietic progenitors are optimal targets of retroviral-mediated gene transfer, *Blood*, 80, 1418, 1992.
120. Moritz, T., Keller, D. C., and Williams, D. A., Human cord blood cells as targets for gene transfer: potential use in genetic therapies of severe combined immunodeficiency disease, *J. Exp. Med.*, 178, 529, 1993.
121. Sekhsaria, S., Gallin, J. I., Linton, G. F., Mallory, R. M., Mulligan, R. C., and Malech, H. L., Peripheral blood progenitors as a target for genetic correction of $p47^{phox}$-deficient chronic granulomatous disease, *Proc. Natl. Acad. Sci. U.S.A.*, 90, 7446, 1993.
122. Morgan, R. A., Looney, D. J., Muenchau, D. D., Wong-Stall, F., Gallo, R. C., and Anderson, W. F., Retroviral vectors expressing soluble CD4: a potential gene therapy for AIDS, *AIDS Res. Hum. Retroviruses*, 6, 183, 1990.
123. Caruso, M. and Klatzman, D., Selective killing of $CD4^+$ cells harboring a human immunodeficiency virus-inducible suicide gene prevents viral spread in an infected cell population, *Proc. Natl. Acad. Sci. U.S.A.*, 89, 182, 1992.

124. Harrison, G. S., Long, C. J., Maxwell, F., Glode, L. M., and Maxwell, I. H., Inhibition of HIV production in cells containing an integrated, HIV-regulated diphtheria toxin A chain gene, *AIDS Res. Hum. Retroviruses*, 8, 39, 1992.
125. Buchschacher, G. L. and Panganiban, A. T., Human immunodeficiency virus vector for inducible expression of foreign genes, *J. Virol.*, 66, 2731, 1992.
126. Weerasinghe, M., Liem, S. E., Asad, S., Read, S. E., and Joshi, S., Resistance to human immunodeficiency virus type 1 (HIV-1) infection in human $CD4^+$ lymphocyte-derived cell lines conferred by using retroviral vectors expressing an HIV-1 RNA-specific ribozyme, *J. Virol.*, 65, 5531, 1991.
127. Kasid, A., Morecki, S., Aebersold, P., Cornetta, K., Culver, K., Freeman, S., Director, E., Lotze, M. T., Blaese, R. M., Anderson, W. F., and Rosenberg, S. A., Human gene transfer: characterization of human tumor-infiltrating lymphocytes as vehicles for retroviral-mediated gene transfer in man, *Proc. Natl. Acad. Sci. U.S.A.*, 87, 473, 1990.
128. Rosenberg, S. A., Aehersold, P., Cornetta, K., Kasid, A., Morgan, R. A., Moen, R., Karson, E. M., Lotze, M. T., Yang, J. C., Topalian, S. L., Merino, M. J., Culver, M., Miller, A. D., Blaese, R. M., and Anderson, W. F., Gene transfer into humans: immunotherapy of patients with advanced melanoma using tumor infiltrating lymphocytes modified by retroviral gene transduction, *N. Engl. J. Med.*, 323, 570, 1990.
129. Miller, A. R., Skotzko, M. J., Rhoades, K., Belldegrun, A. S., Tso, C., Kaboo, R., McBride, W. H., Jacobs, E., Kohn, D. B., Moen, R., and Economou, J. S., Simultaneous use of two retroviral vectors in human gene marking trials: feasibility and potential applications, *Hum. Gene Ther.*, 3, 619, 1992.
130. Phase I study of cellular adoptive immunotherapy using genetically modified $CD8^+$ HIV-specific T cells for seropositive patients undergoing allogeneic bone marrow transplant, *Hum. Gene Ther.*, 3, 319, 1992.
131. Claxton, D., Suh, S., Filaccio, M., Ellerson, D., Gaozza, E., Anderson, B., Brenner, M., Reading, C., Feinberg, A., Moen, R., Belmont, J. W., Moore, K., Talpaz, M., Kantarjian, H., and Deisseroth, A., Molecular analysis of retroviral transduction in chronic myelogenous leukemia, *Hum. Gene Ther.*, 2, 317, 1991.
132. Etkin, M., Filaccio, M., Ellerson, D., Suh, S., Claxton, D., Gaozza, E., Brenner, M., Moen, R., Belmont, J. W., Moore, K. A., Moseley, A. M., Reading, C., Khouri, I., Talpaz, M., Kantarjian, H., and Deisseroth, A. B., Use of cell-free supernatants of safety modified viruses for transduction of cells from the marrow of chronic phase and blast crisis CML patients and from normal individuals, *Hum. Gene Ther.*, 3, 137, 1992.
133. Autologous bone marrow transplant for children with AML in first complete remission: use of marker genes to investigate the biology of marrow reconstitution and the mechanism of relapse, *Hum. Gene Ther.*, 2, 137, 1991.
134. Autologous bone marrow transplantation for CML in which retroviral markers are used to discriminate between relapse which arises from systemic disease remaining after preparative therapy versus relapse due to residual leukemia cells in autologous marrow: a pilot trial, *Hum. Gene Ther.*, 2, 359, 1991.
135. Retroviral-mediated gene transfer of bone marrow cells during autologous bone marrow transplantation for acute leukemia, *Hum. Gene Ther.*, 3, 305, 1992.
136. Treatment of severe combined immunodeficiency disease (SCID) due to adenosine deaminase (ADA) deficiency with autologous lymphocytes transduced with a human ADA gene, *Hum. Gene Ther.*, 1, 327, 1990.
137. Culver, K. W., Anderson, W. F., and Blaese, R. M., Lymphocyte gene therapy, *Hum. Gene Ther.*, 2, 107, 1991.

138. Treatment of patients with severe combined immunodeficiency due to adenosine deaminase (ADA) deficiency by autologous transplantation of genetically modified bone marrow cells, *Hum. Gene Ther.*, 3, 553, 1992.
139. Gene therapy of patients with advanced cancer using tumor infiltrating lymphocytes transduced with the gene coding for tumor necrosis factor, *Hum. Gene Ther.*, 1, 441, 1990.
140. Lansdorp, P. M. and Dragowska, W., Maintenance of hematopoiesis in serum-free bone marrow cultures involves sequential recruitment of quiescent progenitors, *Exp. Hematol.*, 21, 1321, 1993.
141. Lansdorp, P. M., Dragowska, W., and Mayani, H., Ontogeny-related changes in proliferative potential of human hematopoietic cells, *J. Exp. Med.*, 178, 787, 1993.

3 Somatic Gene Therapy for Severe Combined Immune Deficiency (SCID)

Kenneth I. Weinberg and Donald B. Kohn

I. BIOLOGY AND CLINICAL MANIFESTATIONS OF SCID

A. Demographic Features

Severe combined immune deficiency (SCID) is a clinical phenotype marked by the congenital absence of T- and B-lymphocyte function.[1,2] The estimated incidence of SCID is approximately 1 per 100,000 live births. Some populations, notably French Canadians, some North American Mennonite communities, and Dine (Navajo), have much higher incidences of SCID, approaching 1 per 500 births. Both autosomal recessive and X-linked modes of inheritance of SCID are observed. There is a male preponderance of SCID cases, suggesting that X-linked SCID (XSCID) may be the most common form of SCID, accounting for as many as 50% of all cases.

B. Clinical Features

The clinical syndrome of SCID is marked by acute and chronic infections due to the severe immunodeficiency. Affected infants present in the first year of life, usually after age 2 months, when protective maternal antibody titers decline. The most common infections observed in 38 children with SCID treated at Children's Hospital Los Angeles (CHLA) between 1984 and 1993 have been recurrent or severe infections due to common encapsulated organisms (otitis media and/or sinusitis), opportunistic fungal infections (oropharyngeal/esophageal candidiasis, cryptococcal meningitis), opportunistic viral infections (respiratory syncytial and parainfluenza virus pneumonia, rotavirus-induced enteritis, and enteroviral meningoencephalitis), and parasitic infections (*pneumocystis carinii* pneumonia). In addition to the above infections, two infants had siblings who had died in infancy from Epstein-Barr virus-induced lymphoproliferative disorder (EBV-LPD). Most infants with SCID develop failure to thrive because of gastrointestinal malabsorption, caused either by chronic viral or bacterial infections, or

0-8493-2440-8/95/$0.00+$.50

maternal graft-vs.-host disease. Untreated SCID is a lethal syndrome, with most infants not surviving more than 2 years.

An unusual clinical feature of SCID is graft-vs.-host disease (GVHD) due to maternal chimerism. SCID patients with maternal GVHD present with chronic weeping and/or scaling erythroderma, malabsorption, and profound hypoalbuminemia. Engraftment of maternal T-lymphocytes in the SCID neonate is thought to occur because the patients lack alloreactive T-lymphocytes that can eliminate maternal T-cells that enter the infant during the perinatal period. Clinically apparent graft-vs.-host disease is observed in approximately 25% of patients. However, maternal chimerism has been observed even in patients who do not have clinical GVHD. The detection of maternal chimerism by HLA-typing is relatively insensitive, and the true incidence of chimerism is probably at least 50%, occurring especially frequently in XSCID.

C. Laboratory Findings

Laboratory analysis of patients with SCID shows the characteristic absence of antigen-specific T- and B-lymphocyte function. While most patients show an absolute T-lymphopenia, especially those with XSCID or ADA deficiency, some patients have normal or near normal numbers of nonfunctional T-lymphocytes. Regardless of the numbers of T-lymphocytes, T-lymphocytes from patients with SCID do not proliferate in vitro to specific antigens, e.g., tetanus toxoid or candida. The proliferative responses to mitogens, such as phytohemagglutinin (PHA) or concanavalin A (Con A), are severely depressed in most patients. Proliferative responses to allogeneic cells are sometimes present, and some investigators have classified patients with alloresponsive cells as having combined immune deficiency (CID). B-lymphocytes are variably present in patients with SCID; XSCID in particular is marked by large numbers of virgin IgM^+ B-lymphocytes. Most patients are hypo- or agammaglobulinemic. No SCID patient can generate specific antibody responses, even those who have B-lymphocytes or normal total immunoglobulin levels.

II. PATHOGENESIS OF SCID

SCID is caused by a variety of mutations that interfere with the differentiation or function of T-lymphocytes or of both T- and B-lymphocytes. In general, the defects in B-lymphocyte function are secondary to the T-lymphoid dysfunction. Theoretically, SCID could be caused by mutations in any gene essential for T-lymphocyte differentiation or function, but which do not also result in lethality from dysfunction of other organs.

A. Adenosine Deaminase (ADA) Deficiency

The most common autosomal recessive form of SCID is due to absence of the enzyme adenosine deaminase (ADA).[3,4] ADA catalyzes the deamination of adenosine and deoxyadenosine metabolites (dAXPs) to the respective inosine analogues. In the absence of ADA activity, dAXP accumulation results in severe derangements of purine metabolism. Feedback inhibition of ribonucleotide reductase occurs, resulting in decreased pools of the deoxynucleotide triphosphates needed for DNA synthesis. The enzyme S-adenosyl homocysteine hydrolase, essential to production of the methyl donor S-adenosyl methionine, is irreversibly inhibited, which prevents critical methylation reactions. Although ADA is expressed by all cells, ADA deficiency produces a phenotype marked predominantly by lymphoid dysfunction, with little evidence of nonlymphoid toxicity. ADA is normally expressed at high levels in lymphoid tissue, and it appears that the high levels of DNA synthesis and purine salvage in lymphopoiesis make lymphocytes particularly susceptible to the toxic effects of ADA deficiency.

B. X-Linked SCID

The X-linked form of SCID is due to the absence of a cell-surface receptor protein, γ_C, which is a component of the receptors for interleukin-2 (IL-2), interleukin-4 (IL-4), and interleukin-7 (IL-7).[5-11] In XSCID, patients have extremely low levels of T-lymphocytes but normal to elevated numbers of virgin B-lymphocytes. Analysis of X-inactivation patterns in the peripheral blood cells of heterozygous women has shown that there is a selective advantage to T-, B-, natural killer, and in some cases myeloid cells, that express the normal γ_C gene.[12-17] The defect in T-lymphopoiesis in X-SCID is probably due to dysfunction of the IL-4 or IL-7 receptors, rather than of IL-2. Although IL-2 is required for proliferation of activated mature T-lymphocytes, it is probably not essential for thymic differentiation.[18-22] In contrast to the profound T-lymphopenia observed in X-linked SCID, both human patients whose cells do not make IL-2 and IL-2 knockout mice have relatively normal production of mature T-lymphocytes, suggesting that IL-2 is not required for thymic differentiation.

C. CD3-γ Deficiency

T-lymphocyte recognition of antigen occurs through the T-cell receptor complex on the cell surface. This complex contains an antigen-specific T-cell receptor (TCR) heterodimer and the invariant proteins collectively termed CD3. Mutations of CD3-γ, which is part of the CD3 complex expressed by cytotoxic T-lymphocytes, have been described in a kindred with SCID.[23] CD3-γ deficiency is the first example of SCID due to mutations in a gene expressed exclusively in T-lymphocytes.

D. ZAP-70 Deficiency

Nonreceptor protein tyrosine kinases (PTKs) are required for both thymic differentiation and activation of mature T-lymphocytes. The PTKs Fyn, Lck, and ZAP-70 have been shown to be critical in TCR-mediated signaling. Mice made deficient in Fyn and Lck by homologous recombination have defective thymic differentiation, but no human patients with Fyn or Lck deficiency have been described thus far.[24-26] Several kindred with an unusual SCID phenotype marked by absence of peripheral $CD8^+$ cells and abnormal signaling in the $CD4^+$ cells have mutations in ZAP-70 (70 kD TCR-ζ–associated protein).[27-29] After TCR engagement with antigen, the ZAP-70 PTK associates with the phosphorylated TCR-ζ protein and itself is phosphorylated. The $CD4^+$ cells from patients lacking ZAP-70 have defective tyrosine phosphorylation, interleukin-2 production, and proliferation. The selection against $CD8^+$ cells in ZAP-70 deficiency occurs at a relatively late stage of thymic differentiation, as thymic biopsies from patients have revealed the presence of $CD4^+CD8^+$ cells, the immediate precursor of $CD4^+$ and $CD8^+$ cells. Like CD3-γ, ZAP-70 expression is lineage-restricted, and is found exclusively in T-cells and natural killer cells.

E. Signaling Defects and IL-2 Deficiency Syndromes

After T-lymphocytes recognize antigen, a signal cascade is triggered, and the T-lymphocyte is activated. Activated T-lymphocytes produce a variety of cytokines and are competent to proliferate. The signal for proliferation is mediated by the interaction of the major T-cell growth factor, IL-2, with its receptor. A variety of cellular defects leading to aberrant activation have been described in some patients with SCID. Some patients have had membrane defects in which signal transduction does not occur.[30] In some patients, activation occurs but IL-2 mRNA is not induced.[18-20] IL-2 deficiency has been described as a specific defect (other cytokines are produced normally) and as part of a generalized defect in cytokine gene activation. The genes responsible for these IL-2 deficiency syndromes are at present unknown, although the generalized cytokine deficiency is probably due to abnormalities of a T-cell-specific transcription factor, NF-AT, which is required for cytokine gene activation.[22]

F. Major Histocompatibility Complex Defects

Recognition of antigens requires formation of a trimolecular complex consisting of the T-cell receptor, antigenic peptide fragments, and major histocompatibility complex proteins. In some patients with SCID, antigen-presenting cells do not express MHC molecules, essentially "blinding" the patients' T-cells. MHC deficiency is also known as "bare lymphocyte syndrome". At least three complementation groups for defective MHC expression are present in mutagenized B-cell lines.[31] SCID patients with bare lymphocyte syndrome have been described with defects corresponding to two of these complementation groups. The first group has absent binding of a family of DNA-binding proteins, RF-X, and abnormal structure of the promoters of the MHC Class II genes.[32-34] No specific gene mutation has been described for this group of patients. A second group of patients has mutations in a Class II transcriptional activator (CIITA), necessary for expression of MHC Class II genes.[31] Unlike the defects in ADA, XSCID, and CD3-γ deficiency, which affect T-lymphocyte progenitors, the primary defect in MHC deficiency is in the antigen-presenting cells (monocytes and B-cells). Therefore, gene therapy for MHC deficiency will require expression of the normal gene in hematopoietic cells, not just in T-lymphocytes.

In summary, SCID is a phenotype that can result from a variety of genetic defects involving T-lymphoid differentiation or function. Gene therapy can only be considered for those forms of SCID in which the mutant gene has been identified. It is likely that in the next few years, the specific molecular defects observed in some forms of SCID, e.g., IL-2 deficiency or the autosomal recessive forms of SCID observed in the Dine or Mennonite communities, will be identified. At the present time, only ADA deficiency, XSCID, CD3-γ deficiency, ZAP-70 deficiency, and the CIITA-deficient form of bare lymphocyte syndrome are candidate diseases for gene therapy.

III. TREATMENTS FOR SCID

A. Bone Marrow Transplantation (BMT)

1. HLA-Matched Sibling BMT

The optimal treatment for SCID continues to be bone marrow transplant (BMT) from a human leukocyte-antigen (HLA) matched sibling donor.[35] Because the patients are immuno-incompetent, no pretransplant chemotherapy (conditioning) is needed. Following intravenous injection of 5×10^7 nucleated donor marrow cells, patients develop normal T- and B-lymphocytes, with resultant normalization of antigen-specific cell-mediated and humoral responses. When histocompatible BMT for SCID is performed without conditioning, only donor T-lymphocytes are found to engraft. The lack of engraftment of other lineages indicates that *only the T-cell progenitors have a selective advantage over the endogenous hematopoietic cells.* In about 50% of patients, the B-lymphocytes following unconditioned BMT are of recipient origin, consistent with the primary defect residing in the T-cell population.

2. Haploidentical BMT

Unfortunately, less than 25% of patients with SCID have a histocompatible sibling donor. For patients lacking a histocompatible donor, several therapeutic approaches have been developed as alternatives. Most commonly, bone marrow from an HLA-haploidentical parental donor is used for transplantation.[36-38] Haploidentical bone marrow must be purged of contaminating donor T-lymphocytes, which, if infused, would cause fatal graft-vs.-host disease. A variety of purging techniques have been developed, using lectin and/or E-rosetting, monoclonal antibodies and complement, or monoclonal antibodies and immunomagnetic beads. Generally, 2 to 3 logs of T-lymphocyte removal are achievable.

A surprising result of the removal of T-lymphocytes has been that some patients reject their graft. Effectors for this graft rejection include maternal T-lymphocytes that engrafted perinatally, natural killer cells, and recipient cytotoxic precursors that mature under the influence of donor cell cytokines, e.g., IL-2. Immunosuppressive chemotherapy, usually cyclophosphamide and anti-thymocyte globulin (ATG), has been given prior to transplant to prevent rejection. In our experience, patients with natural killer cell activity have the best chance of engraftment if marrow-ablative conditioning with busulfan is given along with immunosuppressive therapy. Unlike recipients who do not receive marrow ablative conditioning, patients who receive busulfan pre-BMT show evidence of donor engraftment in all lymphohematopoietic lineages (erythroid, myeloid, megakaryoblastic, T-lymphoid, B-lymphoid). Although pretransplant chemotherapy is clinically important for the engraftment from haploidentical donors, it carries acute risks of infection, bleeding, mucosal irritation, urinary tract bleeding, and liver dysfunction.

3. Problems with Haploidentical BMT

Despite the improvements in haploidentical BMT, several disturbing problems remain. First, there is a consistent delay in reconstitution of T-lymphoid function in recipients of haploidentical BMT compared to those receiving histocompatible transplants. Because haploidentical BMT involves the removal of mature T-lymphocytes from the donor marrow, reconstitution of T-lymphoid function depends on maturation of immature lymphohematopoietic progenitors into T-lymphocytes. The maturation of T-lymphocytes normally takes 16 to 24 weeks *in utero*. T-lymphocyte maturation post-BMT recapitulates ontogeny and follows the same time course. A further confounding factor is the need for post-BMT immune suppression, which contributes to delayed reconstitution to a variable degree. Patients receiving haploidentical BMT have been especially vulnerable to pre-existing viral or fungal infections during this critical time period. Sixty percent of the deaths among the CHLA haploidentical SCID patients were due to infections during the first 6 months post-BMT.

A second problem is defective or delayed B-cell function. Of fourteen surviving recipients of T-depleted BMT at CHLA, thirteen still require monthly intravenous gamma globulin (IVIg) because of hypo- or agammaglobulinemia. The frequency of humoral dysfunction observed among our patients is greater than the 50% incidence reported previously by the Memorial Sloan-Kettering group, probably because most of the surviving patients at CHLA are <2 years post-BMT.[36] The defect in humoral immunity post-BMT is thought to be due to defective cooperation of HLA-disparate T- and B-lymphocytes. The failure to reconstitute full humoral function has not been fatal but has led to significant morbidity as patients must remain on chronic antibiotic prophylaxis and monthly intravenous immunoglobulin.

A third problem observed in survivors of haploidentical BMT has been chronic GVHD. Approximately one-half of our patients transplanted for SCID from haploidentical donors have required immunosuppressive therapy for at least 1 year post-BMT and 25% have required immunosuppression for >1 year.

A final issue involving haploidentical BMT for SCID is the necessity for chemotherapy to ensure engraftment. The late effects of administration of such medications as busulfan and cyclophosphamide in young children are not completely known, as childhood BMT survivors (either SCID or cancer patients) who have received intensive chemotherapy are only now entering their third decade of life. Chronic effects, such as increased incidence of malignancies, sterility, and possibly restrictive lung disease can be anticipated in survivors of BMT for SCID who received intensive pre-transplant conditioning.

4. Matched Unrelated Donor BMT

Filipovich and co-workers have reported the use of marrow from unrelated donors who were phenotypically matched at all HLA loci or who had a single mismatch but were HLA-DR identical.[39] The patients received cytoablative conditioning and post-transplant immune suppression to decrease graft-vs.-host disease. Six of eight SCID patients survived and developed immune reconstitution, with two others succumbing to pre-existing opportunistic infections. These results compare favorably to those achieved with T-cell-depleted parental transplants; further studies will be needed to determine whether either approach is more beneficial.

B. Enzyme Replacement Therapy

Since 1986, enzyme replacement therapy for ADA-deficient SCID has been available as an alternative to haploidentical bone marrow transplantation.[4,40] Clinical experiments had shown that transfusion of normal red blood cells into ADA-deficient children resulted in transient improvement in immune function.[41] The risks of chronic blood transfusion and relatively low levels of ADA delivered by transfusion limited the clinical usefulness of this form of enzyme replacement therapy. Enzyme replacement with purified ADA protein has a short half-life that precluded its use. Modification of purified bovine ADA by covalent attachment of polyethylene glycol to lysine residues results in a biologically active form of ADA with prolonged half-life due to decreased clearance from the blood, resistance to proteolysis, and low antigenicity. Polyethylene glycol modified ADA (PEG-ADA), given by weekly or twice weekly intramuscular injection, results in biochemical normalization of ADA-deficient patients. As ADA levels in the plasma rise, intracellular levels of adenosine metabolites fall to the normal range, and activity of target enzymes such as S-adenosyl homocysteine hydrolase become normal.

Enzyme replacement therapy with PEG-ADA results in immunologic improvement in the majority of treated patients.[4] Analysis of immune development in responding PEG-ADA recipients shows that the reconstitution of T-lymphocytes is similar to that occurring in recipients of haploidentical bone marrow transplant, in that there is a delay due to recapitulation of ontogeny.[42] Patients have a delay in T-cell function but ultimately have normal mitogen and antigen responsiveness. PEG-ADA responders, unlike haploidentical marrow transplant recipients, also have consistently normal B-lymphocyte function.[43] In approximately 30 to 40% of patients who receive PEG-ADA, the response is only partial, e.g., T-lymphocyte numbers remain low or antigen-specific responses are inconsistently present.[4] The basis for these partial responses is unknown. The role of higher or more frequent dosing with PEG-ADA is under investigation.[42] It is possible that extracellular ADA administration is less effective in allowing lymphocyte survival and function[44] than intracellularly produced ADA. If so, then a gene therapy approach might be more effective than PEG-ADA.

The major problem with enzyme replacement with PEG-ADA is that it is noncurative and requires chronic therapy. Repeated intramuscular injections with PEG-ADA are required to maintain normal biochemical and immunologic function. Discontinuation of the PEG-ADA will result in loss of immune function. In general, compliance with chronic injections such as PEG-ADA or insulin is suboptimal, especially in teenagers. Therefore, a curative approach such as that of bone marrow transplant or stem cell gene therapy may offer advantages over PEG-ADA therapy. The cost of PEG-ADA has also been an issue. One year of PEG-ADA therapy costs at least $100,000 and increases with patient size; administration for a lifetime would be more expensive in the long run than either bone marrow transplant or stem cell gene therapy.

IV. GENE THERAPY FOR SCID

Gene therapy has been considered as a potential alternative therapy to allogeneic BMT for SCID for at least a decade.[45] Theoretically, genetic correction of a patient's own T-lymphocytes or their progenitors should achieve the same beneficial results as allogeneic BMT, without the immunologic complications of graft-vs.-host disease or the need for administration of immune suppressive drugs for recipients of non-histocompatible marrow. As with BMT for SCID using a histocompatible donor, autologous gene therapy should not require administration of cytoablative chemotherapy or radiation to eliminate competing endogenous marrow. Because of the high cure rate for SCID patients with histocompatible donors, it is likely that gene therapy will be initially reserved for patients lacking suitable donors.

While PEG-ADA can restore immunity for many patients with this form of SCID, the treatment is palliative, rather than curative.[4,42] Gene therapy may be less costly, as a single treatment, although the specific costs for the reagents that are currently under investigation, such as clinical-grade retroviral supernatants, $CD34^+$ cell isolation kits, recombinant cytokines, etc., are unknown.

A. Theoretical Considerations

1. Choice of Target Cell for Gene Transfer

One obvious choice of target cells for gene therapy of SCID is the mature T-lymphocytes, which are specifically deficient as a result of the various pathologic etiologies and whose absence correlates strongly with the extent of immune deficiency. Among the advantages of using T-lymphocytes as targets for gene therapy of SCID are the ability to readily isolate T-lymphocytes from peripheral blood in nonlymphopenic subjects and to vastly expand the numbers of these T-cells *ex vivo* with IL-2 and a co-mitogen.[46] However, the use of T-lymphocytes may be limited by their ability to confer only the limited repertoire of immune responsiveness contained within the isolated cell population, the unknown life-span of T-cells which have been expanded *ex vivo* with pharmacologic levels of cytokines, and the relatively incomplete transduction (1 to 5%) typically achieved with current retroviral vectors. Most importantly, peripheral blood T-lymphocytes are severely diminished in numbers in many forms of SCID, making this source unavailable.

The hematopoietic stem cells (HSC) represent the other logical target cell for gene therapy of SCID. HSC are long-lived, producing new progeny cells, including T-lymphocytes, for the life of the recipient after transplant. Current techniques allow human HSC to be relatively easily enriched as a portion of the $CD34^+$ cells from bone marrow, from umbilical cord blood, or from the peripheral blood after mobilization by systemic treatment with cytokines, such as G-CSF. Gene transfer into murine hematopoietic stem cells by retroviral vectors has been shown to be highly efficient (see Chapter 2).[47,48] However, the transduction efficiency of marrow stem cells in large animal models and early clinical trials has been relatively low, e.g., 0.1 to 5.0%.[49-52]

2. Selective Survival Advantage of Genetically Normal T-Lymphocytes

A primary reason that SCID has been a major focus for initial attempts at human gene therapy is that, even with the current technical limitations to transducing a high percentage of target cells, a clinical benefit may still be achieved. Results from allogeneic BMT for SCID have shown a selective survival advantage to genetically normal donor T-lymphocytes.[3] Infusion of a relatively low number of bone marrow cells from an HLA-matched sibling into a SCID patient results in repopulation of the recipient immune system with donor-derived lymphocytes, without the need to eliminate the recipient's endogenous hematopoietic/lymphoid system. Additional evidence for a selective advantage for genetically normal T-lymphocytes over those expressing SCID genes comes from observations of the X-chromosome inactivation patterns of women who are carriers of X-linked SCID. Random lyonization, without a selective advantage, would result in equal numbers of circulating cells expressing either the normal X-chromosome or the one bearing the mutant gene.[12] The finding of unbalanced X-inactivation patterns, highly skewed against cells in which the active X-chromosome is the one with the mutant SCID allele, argues for a selective survival advantage of the T-lymphocytes expressing the normal allele of the γ_C gene.

Theoretically, genetic correction of the T-lymphocyte precursors from a SCID patient would also confer a survival advantage. Thus, even with the currently inefficient gene transfer capabilities, sufficient numbers of corrected T-lymphocytes may accumulate over time to restore immune function. Other genetic diseases of hematopoietic cells, such as hemoglobinopathies or lysosomal storage disorders, would not be expected to have a selective advantage over corrected precursors or mature cells.

3. Regulation of Expression by Genes Introduced into T-Lymphocytes or Stem Cells

For each specific form of SCID, the requirements for control of expression of the relevant transduced gene may vary. For a housekeeping enzyme, such as ADA, expression above a minimal threshold (probably 10% of normal) may be sufficient to restore T-lymphocyte survival and overexpression to a high level (>50-fold of normal) may be tolerated. ADA is normally present in all cell types and, therefore, a constitutive promoter that directs expression indiscriminately in stem cell progeny of multiple lineages should be acceptable. In contrast, genes that encode regulatory proteins, for example the IL-2 receptor, intracellular signal transduction molecules, or transcriptional *trans*-acting factors, may require finer control of either the quantitative level or specific cell types in which they are expressed for effective, nontoxic therapy. Potentially, some mutant genes causing SCID may act as dominant negative mutants and counteract the effects of a transduced normal gene. Successful gene therapy for these conditions would either require overexpression of the normal gene or elimination of the mutant allele by a process such as homologous recombination.

B. Preclinical Studies

Introduction of a normal human ADA cDNA in T-lymphocytes from an ADA-deficient SCID patient led to production of ADA enzyme to normal levels and eliminated the hypersensitivity to toxicity from deoxyadenosine, which is the biochemical hallmark of ADA-deficiency.[53] Some of the earliest attempts to express human genes in mouse bone marrow transplant models were performed with the retroviral vectors carrying normal human ADA cDNA. After initial failures, improvements in marrow transduction techniques and vectors design led to successful expression of human ADA in the resultant mature progeny cells, including T-lymphocytes.[54-58] Two groups have demonstrated successful transfer of the human ADA cDNA into primate bone marrow stem cells with sustained *in vivo* expression in progeny hematopoietic cells, although the frequency of transduced cells has been only 0.1 to 2%.[49,50] In the absence of an ADA-deficient animal model, the survival advantage for cells derived from genetically corrected stem cells cannot be tested. Ferrari et al. showed that introduction of a normal human ADA cDNA into ADA-deficient mature PBL from SCID patients restored normal cell survival time when injected into SCID mice.[44,59] Thus, retroviral vectors were shown to be capable of performing the basic functions of ADA gene transfer and expression at levels expected to be clinically beneficial.

C. Clinical Studies of Gene Therapy for ADA-Deficient SCID

As of January 1994, four groups have attempted clinical applications of gene therapy for ADA-deficient SCID patients. Because of the relatively short follow-up time since these trials have begun, preliminary results cited below have been synthesized from oral presentations at meetings and personal communications with the principal investigators, rather than from peer-reviewed reports.

1. National Institutes of Health, U.S.A.

The first use of gene therapy for a genetic disorder was initiated in September of 1990 at the National Institutes of Health, by Michael Blaese and Kenneth Culver of the National Cancer Institute and French Anderson of the National Heart, Lung and Blood Institute.[60] The first patient was a 4-year-old girl with ADA-deficiency, who had only partially responded to PEG-ADA. She was treated by serial monthly leukopheresis to collect peripheral blood mononuclear cells, which were transduced *ex vivo* with the LASN retroviral vector following stimulation with IL-2 and anti-CD3 monoclonal antibody. Following a brief *ex vivo* expansion (typically 10 days), the cells were infused intravenously. Over the course of 2 years of treatment, she had progressive increases in her absolute number of circulating T-lymphocytes, increased leukocyte ADA level to approximately 25% of normal, with an estimated 30 to 50% of her circulating T-lymphocytes containing vector sequences. Concomitantly, she demonstrated significant improvement in many parameters of immune function, such as development of isohemagglutinins (antibodies to red blood cell ABO antigens), cytotoxic T-lymphocytes specific for influenza, and delayed-type hypersensitivity skin test responses.

Mullen et al. recently reported that 1 year after the last infusion of transduced cells, eight of nine T-lymphocyte clones derived from her peripheral blood contained the introduced ADA cDNA and expressed clinically relevant levels of ADA enzymatic activity.[61] This result demonstrates that functional peripheral blood T-lymphocytes can persist for at least 1 year after transduction with the ADA cDNA and infusion into ADA-deficient patients. The long-term durability of immune reconstitution provided by gene transfer into mature T-lymphocytes is presently unknown; this protocol will specifically define the survival time of the genetically corrected T-lymphocytes.

A second child treated by this peripheral blood T-lymphocyte transduction protocol had lower levels of gene transfer into her cells, such that 0.1 to 1% of her circulating T-cells contained the transduced ADA cDNA. In the summer of 1993, she was treated by transduction of G-CSF mobilized peripheral blood $CD34^+$ cells, using a different ADA vector (with ADA under control of an internal SV40 promoter), which could be distinguished from the first by PCR methods. No cells containing the second vector have been detected by PCR analysis of her peripheral blood leukocytes for the first 6 months after therapy.

2. Milan, Italy

Claudio Bordignon treated two ADA-deficient children by gene therapy, beginning in March 1992.[62] A pair of retroviral vectors were used, which each express the human ADA cDNA as part of a minigene with the endogenous ADA promoter, in the "double copy" formation.[63] This protocol combined gene transfer into bone marrow $CD34^+$ cells with one ADA vector and into peripheral blood T-lymphocytes with the second ADA vector. Such dual labeling should allow determination of the relative contribution of these two cell sources to long-lived T-lymphocytes present in the recipients.

The patients currently have approximately 1% transduced hematopoietic progenitors (CFU-GM) in their marrow, and approximately 1% of their circulating T-lymphocytes express ADA. Some cell populations, such as T-lymphocyte clones responsive to tetanus toxoid, show almost 10% gene transduction. Currently, studies are in progress to determine the relative contribution from the initial transduced T-cells and bone marrow to the circulating cells now seen containing vector sequences.

3. University of Leiden, The Netherlands

Dinko Valerio and co-workers performed ADA gene transfer into bone marrow from three ADA-deficient children, two from France and one from England, in the spring of 1993.[64] The vector employed in their studies uses a composite Moloney LTR with a mutant polyoma virus enhancer, which they had previously shown to be effective in both murine and primate gene transfer/BMT studies.[57,65] They cocultured the patients' $CD34^+$ cells directly on the (irradiated) fibroblasts producing the ADA vector and then infused the cells into the patients. Circulating cells containing vector sequences were present for the first few months after treatment but were not detected at 6 months after gene therapy (the latest follow-up point). A bone marrow aspirate from one patient at 6 months after treatment showed vector-containing cells, by PCR analysis.

4. Children's Hospital Los Angeles

Our group at CHLA, in collaboration with Michael Blaese of the NIH, treated three neonates diagnosed *in utero* with ADA-deficiency in May and June of 1993.[66] The infants were born in Los Angeles, San Francisco, and Alberta, Canada, and umbilical cord blood was collected as the source of hematopoietic stem cells. $CD34^+$ cord blood cells were transduced with the LASN vector and infused into each child on the fourth day of life. A six-months follow-up shows the continual presence of peripheral blood mononuclear cells containing the ADA cDNA at levels between 1/1,000 to 1/10,000. The persistence and frequency of gene transduced T-lymphocytes will be assessed over the subsequent time.

5. General Conclusions from Initial Clinical Trials

One common conclusion from all of these studies is that the patients have had no apparent problems from the gene therapy procedures. All remain in good health at the present time.

Each of these protocols avoided the risks of marrow cytoablation from either chemotherapy or total body radiation. Instead, they rely upon the potential selective survival advantage of ADA-corrected cells to allow the small number of gene-corrected cells to overcome the remaining nontransduced cells. Whether this selective advantage will compensate for gene transfer into relatively low percentages of the total pool of stem cells will be revealed by these initial trials.

Additionally, all of the children have been treated with PEG-ADA, which complicates the interpretation of the results. Some portion of the immune restoration seen to date is likely due to the effects of the PEG-ADA, rather than as a direct benefit of the gene therapy. In fact, the PEG-ADA treatment reduces the levels of toxic deoxyadenosine metabolites, which may diminish the selective advantage of the ADA-corrected cells and slow the accumulation of transduced T-lymphocytes. However, the availability of enzyme replacement for the ADA-deficient form of SCID allows one to perform gene therapy ethically, rather than proceeding directly to a haploidentical bone marrow transplant. As administration of PEG-ADA is withdrawn, the effectiveness of the gene transfer may become more apparent.

V. CONSIDERATIONS FOR GENE THERAPY OF OTHER FORMS OF SCID

For patients with non-ADA-deficient forms of SCID, where a palliative treatment such as enzyme replacement therapy is not available, the decision to delay a BMT while assessing the efficacy of gene therapy is more problematic. Certainly, for all SCID patients with an HLA-matched sibling, allogeneic transplant should be performed directly. For patients lacking a matched donor, the choice of treatment options is less clear-cut. SCID patients who present with opportunistic infections have significantly increased morbidity and mortality from haploidentical BMT, due to the need to administer cytoablative chemotherapy and pre- and post-transplant immunosuppressive therapy and the attendant delay in restoration of immune function. If gene therapy were known to be effective, its use would allow these risks from haploidentical BMT to be avoided. However, with the current uncertainty about the efficacy of gene therapy, it may be prudent to reserve this experimental treatment to babies identified prior to development of infections (e.g., neonates or infants identified because of positive family histories). These patients could be maintained in a pathogen-free environment until gene therapy has been beneficial or a BMT is performed.

Knowledge and experience gained from treating patients with SCID have led to key insights into normal human immune function and transplantation biology.[67] Now, SCID continues to lead the way in progress toward wider applications of gene therapy.

ACKNOWLEDGMENT

Thanks to our SCID mentors and to theirs.

REFERENCES

1. Rosen, F. S., Cooper, M. D., and Wedgwood, R. J., The primary immunodeficiencies (two parts), *N. Engl. J. Med.*, 311, 235, 300, 1984.
2. WHO Scientific Group on Immunodeficiency, Primary immunodeficiency diseases, *Immunodeficiency Rev.*, 1, 173, 1989.
3. Parkman, R., Gelfand, E. W., Rosen, F. S., Sanderson, A., and Hirschhorn, R., Severe combined immunodeficiency and adenosine deaminase, *N. Engl. J. Med.*, 292, 714, 1975.
4. Hershfield, M. S. and Chaffee, S., PEG-Enzyme replacement therapy in adenosine deaminase deficiency, in *Treatment of Genetic Diseases*, Desnick, R. J., Ed., Churchill-Livingstone, New York, 1991, 169.
5. Minami, Y., Kono, T., Miyazaki, T., and Taniguchi, T., The IL-2 receptor complex: its structure, function, and target genes, *Annu. Rev. Immunol.*, 11, 245, 1993.
6. Noguchi, M., Yi, H., Rosenblatt, H. M., Filipovich, A. H., Adelstein, S., Modi, W. S., McBride, O. W., and Leonard, W. J., Interleukin 2 receptor γ chain mutation results in X-linked severe combined immunodeficiency in humans, *Cell,* 73, 147, 1993.
7. Puck, J. M., Deschenes, S. M., Porter, J. C., Dutra, A. S., Brown, C. J., Willard, H. F., and Henthorn, P. S., The interleukin-2 receptor gamma chain maps to Xq13.1 and is mutated in X-linked severe combined immune deficiency, SCIDX1, *Hum. Mol. Genet.*, 2, 1099, 1993.
8. Kondo, M., Takeshita, T., Ishii, N., Nakamura, M., Watanabe, S., Arai, K., and Sugamura, K., Sharing of the interleukin-2 (IL-2) receptor γ chain between receptors for IL-2 and IL-4, *Science,* 262, 1874, 1993.
9. Russell, S. M., Keegan, A. D., Harada, N., Nakamura, Y., Noguchi, M., Leland, P., Friedman, M. C., Miyajima, A., Puri, R. K., Paul, W. E., and Leonard, W. J., Interleukin-2 receptor γ chain: a functional component of the interleukin-4 receptor, *Science,* 262, 1880, 1993.

10. Noguchi, M., Nakamura, Y., Russell, S. M., Ziegler, S. F., Tsang, Cao, X., and Leonard, W. J., Interleukin-2 receptor γ chain: a functional component of the interleukin-7 receptor, *Science,* 262, 1877, 1993.
11. Voss, S. D., Hong, R., and Sondel, P. M., Review: severe combined immunodeficiency, interleukin-2 (IL-2), and the IL-2 receptor: experiments of nature continue to point the way, *Blood,* 83, 626, 1994.
12. Puck, J. M., Krauss, C. M., Puck S. M., Buckley, R. H., and Conley, M. E., Prenatal test for X-linked severe combined immunodeficiency by analysis of maternal X-chromosome inactivation and linkage analysis, *N. Engl. J. Med.,* 322, 1063, 1990.
13. Conley, M. E., Buckley, R. H., Hong, R., Guerra-Hanson, C., Roifman, C. M., Brochstein, J. A., Pahwa, S., and Puck, J. M., X-linked severe combined immunodeficiency. Diagnosis in males with sporadic severe combined immunodeficiency and clarification of clinical findings, *J. Clin. Invest.,* 85, 1548, 1990.
14. Conley, M. E., Lavoie, A., Briggs, C., Brown, P., Guerra, C., and Puck, J. M., Nonrandom X chromosome inactivation in B cells from carriers of X chromosome-linked severe combined immunodeficiency, *Proc. Natl. Acad. Sci. U.S.A.,* 85, 3090, 1988.
15. Wengler, G. S., Allen, R. C., Parolini, O., Smith, H., and Conley, M. E., Nonrandom X chromosome inactivation in natural killer cells from obligate carriers of X-linked severe combined immunodeficiency, *J. Immunol.,* 150, 700, 1993.
16. Goodship, J., Malcolm, S., and Levinsky, R. J., Evidence that X-linked severe combined immunodeficiency is not a differentiation defect of T-lymphocytes, *Clin. Exp. Immunol.,* 83, 4, 1991.
17. Kinnon, C. and Levinsky, R., The molecular basis of X-linked immunodeficiency disease, *J. Inher. Metab. Dis.,* 15, 674, 1992.
18. Weinberg, K. and Parkman, R., Severe combined immunodeficiency due to a specific defect in the production of interleukin 2, *N. Engl. J. Med.,* 322, 1718, 1990.
19. Disanto, J. P., Keever, C. A., Small, T. N., Nichols, G. L., O'Reilly, R. J., and Flomenberg, N., Absence of interleukin 2 production in a severe combined immunodeficiency disease syndrome with T cells, *J. Exp. Med.,* 171, 1697, 1990.
20. Chatila, T., Castigli, E., Pahwa, R., Pahwa S., Chirmule, N., Oyaizu, N., Good, R. A., and Geha, R. S., Primary combined immunodeficiency resulting from defective transcription of multiple T-cell lymphokine genes, *Proc. Natl. Acad. Sci. U.S.A.,* 87, 10033, 1990.
21. Schorle, H., Holtschke, T., Hünig, T., Schimpl, A., and Horak, I., Development and function of T cells in mice rendered interleukin-2 deficient by gene targeting, *Nature,* 352, 621, 1991.
22. Castigli, E., Pahwa, R., Good, R. A., Geha, R. S., and Chatila, T. A., Molecular basis of a multiple lymphokine deficiency in a patient with severe combined immunodeficiency, *Proc. Natl. Acad. Sci. U.S.A.,* 90, 4728, 1993.
23. Arnaiz-Villena, A., Timon, M., Corell, A., Perez-Aciego, P., Martin-Villa, J. M, and Regueiro, J. R., Primary immunodeficiency caused by mutations in the gene encoding the CD3-γ subunit of the T-lymphocyte receptor, *N. Engl. J. Med.,* 327, 529, 1992.
24. Molina, T. J., Kishihara, K., Siderovski, D. P., van Ewijk, W., Narendran, A., Timms, E., Wakeham, A., Paige, C. J., Hatmann, K.-U., Veillette, A., Davidson, D., and Mak, T. W., Profound block in thymocyte development in mice lacking $p56^{lck}$, *Nature,* 357, 161, 1992.
25. Appleby, M. W., Gross, J. A., Cook, M. P., Levine, S. D., Qian, X., and Perlmutter, R. M., Defective T cell receptor signaling in mice lacking the thymic isoform of $p59^{fyn}$, *Cell,* 70, 751, 1992.
26. Stein, P. L., Lee, H.-M., Rich, S., and Soriano, P., $pp59^{fyn}$ mutant mice display differential signaling in thymocytes and peripheral T cells, *Cell,* 70, 585, 1992.
27. Arpaia, E., Shahar, M., Dadi, H., Cohen, A., and Roifman, C. M., Defective T cell receptor signaling and CD8+ thymic selection in humans lacking Zap-70 kinase, *Cell,* 76, 947, 1994.

28. Elder, M. E., Lin, D., Clever, J., Chan, A. C., Hope, T. J., Weiss, A., and Parslow, T. G., Human severe combined immunodeficiency due to a defect in ZAP-70, a T cell tyrosine kinase, *Science,* 264, 1596, 1994.
29. Chan, A. C., Kadlecek, T. A., Elder, M. E., Filipovich, A. H., Kuo, W.-L., Iwashima, M., Parslow, T. G., and Weiss, A., ZAP-70 deficiency in an autosomal recessive form of severe combined immunodeficiency, *Science,* 264, 1599, 1994.
30. Chatila, T., Wong, R., Young, M., Miller, R., Terhorst, C., and Geha, R. S., An immunodeficiency characterized by defective signal transduction in T lymphocytes, *N. Engl. J. Med.,* 320, 696, 1989.
31. Steimle, V., Otten, L. A., Zuffery, M., and Mach, B., Complementation cloning of an MHC Class II transactivator mutated in hereditary MHC Class II deficiency (or bare lymphocyte syndrome), *Cell,* 75, 135, 1993.
32. Reith, W., Satola, S., Sanchez, C. H., Amaldi, I., Lisowka-Grospierre, B., Griscelli, C., Hadam, M. R., and Mach, B., Congenital immunodeficiency with a regulatory defect in MHC Class II gene expression lacks a specific HLA-DR promoter binding protein, RF-X, *Cell,* 53, 897, 1988.
33. Reith, W., Barras, E., Satola, S., Kobr, M., Reinhart, D., Sanchez, C. H., and Mach, B., Cloning of the major histocompatibility complex class II promoter binding protein affected in a hereditary defect in class II gene regulation, *Proc. Natl. Acad. Sci. U.S.A.,* 86, 4200, 1989.
34. Herrero Sanchez, C., Reith, W., Silacci, P., and Mach, B., DNA-binding defect observed in major histocompatibility complex class-II regulatory mutants concerns only one member of a family of complexes binding to the X-boxes of class-II promoters, *Mol. Cell Biol.,* 12, 4076, 1992.
35. Parkman, R., The application of bone marrow transplantation to the treatment of genetic diseases, *Science,* 232, 1373, 1986.
36. Moen, R. C., Horowitz, S. D., Sondel, P. M., Borcherding, W. R., Trigg, M. E., Billing, R., and Hong, R., Immunologic reconstitution after haploidentical bone marrow transplantation for immune deficiency disorders: treatment of bone marrow cells with monoclonal antibody CT-2 and complement, *Blood,* 70, 664, 1987.
37. O'Reilly, R. J., Keever, C. A., Small, T. N., and Brochstein, J., The use of HLA-non-identical T-cell depleted marrow transplants for correction of severe combined immunodeficiency disease, *Immunodeficiency Rev.,* 1, 273, 1989.
38. Fischer, A., Landais, P., Friedrich, W., Morgan, G., Gerritsen, B., Fasth, A., Porta, F., Griscelli, C., Goldman, S. F., Levinsky, R., and Vossen, J., European experience of bone-marrow transplantation for severe combined immunodeficiency, *Lancet,* 336, 850, 1990.
39. Filipovich, A. H., Shapiro, R. S., Ramsay, N. K., Kim, T., Blazar, B., Kersey, J., and McGlave, P., Unrelated donor bone marrow transplantation for correction of lethal congenital immunodeficiencies, *Blood,* 80, 270, 1992.
40. Hershfield, M. S., Buckley, R. H., Greenberg, M. L., Melton, A. L., Schiff, R., Hatem, C., Kurtzberg, J., Markert, M. L., Kobayashi, R. H., Kobayashi, A. L., and Abuchowski, A., Treatment of adenosine deaminase deficiency with polyethylene glycol-modified adenosine deaminase, *N. Engl. J. Med.,* 316, 589, 1987.
41. Polmar, S. H.., Stern, R. C., Schwartz, A. L., Wetzler, E. M., Chase, P. A., and Hirschhorn, R., Enzyme replacement therapy for adenosine deaminase deficiency and severe combined immunodeficiency, *N. Engl. J. Med.,* 300, 1337, 1976.
42. Weinberg, K., Hershfield, M. S., Bastian, J., Kohn, D., Sender, L., Parkman, R., and Lenarsky, C., T lymphocyte ontogeny in adenosine deaminase-deficient severe combined immune deficiency after treatment with polyethylene glycol-modified adenosine deaminase, *J. Clin. Invest.,* 92, 596, 1993.

43. Ochs, H. D., Buckley, R. H., Kobayashi, R. H., Kobayashi, A. L., Sorensen, R. U., Douglas, S. D., Hamilton, B. L., and Hershfield, M. S., Antibody responses to bacteriophage ϕX174 in patients with adenosine deaminase deficiency, *Blood*, 80, 1163, 1992.
44. Ferrari, G., Rossini, S., Giavazzi, R., Maggioni, D., Nobili, N., Soldati, M., Ungers, G., Mavilio, F., Gilboa, E., and Bordignon, C., An in vivo model of somatic cell gene therapy for human severe combined immunodeficiency, *Science*, 251, 1363, 1366, 1991.
45. Anderson, W. F., Prospects for human gene therapy, *Science*, 226, 401, 1984.
46. Culver, K., Cornetta, K., Morgan, R., Morecki, S., Aebersold, P., Kasid, A., Lotze, M., Rosenberg, S. A., Anderson, W. F., and Blaese, R. M., Lymphocytes as cellular vehicles for gene therapy in mouse and man, *Proc. Natl. Acad. Sci. U.S.A.*, 88, 3155, 1991.
47. Williams, D.A., Expression of introduced genetic sequences in hematopoietic cells following retroviral-mediated gene transfer, *Hum. Gene Ther.*, 1, 229, 1990.
48. Karlsson, S., Treatment of genetic defects in hematopoietic cell functions by gene transfer, *Blood*, 78, 2481, 1991.
49. Bodine, D. M., Moritz, T., Donahue, R. E., Luskey, B. D., Kessler, S. W., Martin, D. I., Orkin, S. H., Nienhuis, A. W., and Williams, D. A., Long-term in vivo expression of a murine adenosine deaminase gene in rhesus monkey hematopoietic cells of multiple lineages after retroviral mediated gene transfer into CD34+ bone marrow cells, *Blood*, 82, 1975, 1993.
50. Van Beusechem, V. W., Kukler, A., Heidt, P. J., and Valerio, D., Long-term expression of human adenosine deaminase in rhesus monkeys transplanted with retrovirus-infected bone marrow cells, *Proc. Natl. Acad. Sci. U.S.A.*, 89, 7640, 1992.
51. Schuening, F. G., Kawahara, K., Miller, A. D., To, R., Goehle, S., Stewart, D., Mullaly, K., Fisher, L., Graham, T. C., Appelbaum, F. R., et al., Retrovirus-mediated gene transduction into long-term repopulating marrow cells of dogs, *Blood*, 78, 2568, 1991.
52. Brenner, M. K., Rill, D. R., Holladay, M. S., Heslop, H. E., Moen, R. C., Buschle, M., Krance, R. A., Santanta, V. M., Anderson, W. F., and Ihle, J. N., Gene marking to determine whether autologous marrow infusion restores long-term haemopoiesis in cancer patients, *Lancet*, 342, 1134, 1993.
53. Kantoff, P. W., Kohn, D. B., Mitsuya, H., Armentano, D., Sieberg, M., Zweibel, J. A., Eglitis, M. A., McLachlin, J. R., Wiginton, D. A., Hutton, J. J., Horowitz, S. D., Gilboa, E., Blaese, R. M., and Anderson, W. F., Correction of adenosine deaminase deficiency in human T and B cells using retroviral gene transfer, *Proc. Natl. Acad. Sci. U.S.A.*, 83, 6563, 1986.
54. Lim, B., Williams, D. A., and Orkin, S. H., Retrovirus-mediated gene transfer of human adenosine deaminase: expression of functional enzyme in murine hematopoietic stem cells in vivo, *Mol. Cell Biol.*, 7, 3459, 1987.
55. Belmont, J. W., MacGregor, G. R., Wager-Smith, K., Fletcher, F. A., Moore, K. A., Hawkins, D., Villalon, D., Chang, S. M., and Caskey, C. T., Expression of human adenosine deaminase in murine hematopoietic cells, *Mol. Cell Biol.*, 8, 5116, 1988.
56. Wilson, J. M., Danos, O., Grossman, M., Raulet, D. H., and Mulligan, R. C., Expression of human adenosine deaminase in mice reconstituted with retrovirus-transduced hematopoietic stem cells, *Proc. Natl. Acad. Sci. U.S.A.*, 87, 439, 1990.
57. van Beusechem, V. W., Kukler, A., Einerhand, M. P., Bakx, T. A., van der Eb, A. J., van Bekkum, D. W., and Valerio, D., Expression of human adenosine deaminase in mice transplanted with hematopoietic stem cells infected with amphotropic retroviruses, *J. Exp. Med.*, 172, 729, 1990.
58. Kaleko, M., Garcia, J. V., Osborne, W. R., and Miller, A. D., Expression of human adenosine deaminase in mice after transplantation of genetically-modified bone marrow, *Blood*, 75, 1733, 1990.

59. Ferrari, G., Rossini, S., Nobili, N., Maggioni, D., Garofalo, A., Giavazzi, R., Mavilio, F., and Bordignon, C., Transfer of the ADA gene into human ADA-deficient T lymphocytes reconstitutes specific immune functions, *Blood,* 80, 1120, 1992.
60. Culver, K. W., Anderson, W. F., and Blaese, R. M., Lymphocyte gene therapy, *Hum. Gene Ther.,* 2, 107, 1991.
61. Mullen, C. A., Snitzer, K., and Blaese, R. M., Long term in vivo expression of retrovirally transferred ADA genes in lymphocyte clones from and ADA-SCID patient treated with gene therapy, *J. Cell Biochem.,* Suppl. 18A, 240, 1994.
62. Ferrari, G., Notarangelo, L., Servida, P., Casorati, G., Rossini, S., Ugazio, A., Mavilio, A., and Bordignon, C., The role of peripheral blood lymphocytes and bone marrow cells in the development of a functional immune system after gene therapy for ADA deficient SCID, *J. Cell Biochem.,* Suppl. 17E, 187, 1992.
63. Hantzopoulos, P. A., Sullenger, B. A., Ungers, G., and Gilboa, E., Improved gene expression upon transfer of the adenosine deaminase minigene outside the transcriptional unit of a retroviral vector, *Proc. Natl. Acad. Sci. U.S.A.,* 86, 3519, 1989.
64. Hoogerbrugge, P. M., van Beusechem, V., Valerio, D., Moseley, A., Harvey, M., Fischer, A., Debree, M., Gaspar, B., Morgan, G., and Levinsky, R., Gene therapy of 3 children with adenosine deaminase deficiency, *Blood,* 82 (Suppl. 1), 315a, 1993.
65. van Beusechem, V. W., Bakx, T. A., Kaptein, L. C., Bart-Baumeister, J. A., Kukler, A., Braakman, E., and Valerio, D., Retrovirus-mediated gene transfer into rhesus monkey hematopoietic stem cells: effect of viral titers on transduction efficiency, *Hum. Gene Ther.,* 4, 239, 1993.
66. Kohn, D. B., Weinberg, K. I., Parkman, R., Lenarsky, C., Crooks, G. M., Shaw, K., Hanley, M. E., Lawrence, K., Annett, G., Brooks, J. S., Wara, D., Elder, M., Bowen, T., Hershfield, M. S., Berenson, R. I., Moen, R. C., Mullen, C. A., and Blaese, R. M., Gene therapy for neonates with ADA-deficient SCID by retroviral-mediated transfer of the human ADA cDNA into umbilical cord CD34+ cells, *Blood,* 82 (Suppl. 1), 315a, 1993.
67. Gelfand, E. W., SCID continues to point the way, *N. Engl. J. Med.,* 322, 1741, 1990.

4 Fibroblast Cell Biology and Gene Therapy

Douglas A. Treco, Michael W. Heartlein, and Richard F Selden

I. INTRODUCTION

Over the past two decades, a wide variety of cell types have been studied in the context of gene therapy. Of the work performed with myoblasts, leukocytes, fibroblasts, hepatocytes, and endothelial cells, at least one unifying conclusion can be drawn: no single cell type is appropriate for all cell-based gene therapies. Nevertheless, human fibroblasts do exhibit a number of characteristics and properties that make them appropriate for a broad range of gene therapy applications. The purpose of this review is to summarize the types of studies that have employed fibroblasts and to draw some broad conclusions about the future utility of fibroblasts in gene therapy.

Although it is not generally considered to be the case, the practice of cell-based gene therapy falls squarely within the discipline of cell biology. (Only molecular biologists would consider a therapy in which the exogenous DNA added to a cell relative to total cellular DNA amounts to a few parts per million to fall primarily within the discipline of molecular biology.) The basic steps required in a cell-based gene therapy system — cell harvest, propagation, transfection, and implantation — are fundamentally issues of *in vitro* cell biology, and the ability of the implanted cells to survive, to express (and, if desired, secrete) the product of therapeutic interest, and to retain normal morphologic and growth characteristics are fundamentally issues of *in vivo* cell biology.

The *in vitro* and *in vivo* properties of cells determine their therapeutic roles and effectiveness. With few cell types is the relationship between cell biology and utility in gene therapy so well studied as with the human fibroblast. The advantages of fibroblasts in gene therapy are based in part on knowledge gained from *in vitro* studies performed over the past 40 years — the first clone deposited at the ATCC was a mouse fibroblast line (ATCC CCL 1),[1,2] and one of the first human cell types to be cultured successfully was the human fibroblast.[3,4] The major advantages can be summarized as follows.

Fibroblasts are readily obtained by punch biopsy of the skin and can be cultured free of other cell types. A 4 mm punch yields large numbers of dermal fibroblasts, and it is straightforward to obtain a pure population free from keratinocytes, the other major cell type present in skin. This can be achieved by separating the epidermis (containing the

0-8493-2440-8/95/$0.00+$.50

keratinocytes) from the dermis and by growing the fibroblasts under conditions that are not compatible with keratinocyte survival.[5] As a punch biopsy is a minimally invasive procedure, it is generally performed on an outpatient basis, certainly a benefit for the gene therapy of a number of diseases. Alternatively, fibroblasts can be obtained from most tissues (including lung, liver, and bone marrow) and, accordingly, can be obtained from surgical specimens when appropriate.

Fibroblasts can be administered simply and safely. Implantation of fibroblasts can be performed intradermally, subcutaneously, and intramuscularly, three minimally invasive approaches that lend themselves to treatment in an outpatient setting. Since fibroblasts are normally present throughout the body, they may also be implanted and expected to survive in somewhat less accessible sites, including intrahepatically, under the renal subcapsule, intraperitoneally, intracranially, and intrathecally. Although all of these sites pose significantly greater risks, discomforts, and inconveniences to the patient as compared to the subcutaneous, intradermal, and intramuscular sites, there are certainly clinical conditions that justify their use. Intracranial implantation, for example, has the major advantage of circumventing the blood-brain barrier. Similarly, the site of implantation may be chosen to allow high concentrations of the transgene product to be delivered locally.

Fibroblasts can be retrieved following implantation into most sites, and, accordingly, fibroblast-based gene therapies have the advantage of being reversible in the event the patient's underlying condition improves or otherwise warrants their removal. For example, in the case of fibroblast-mediated erythropoietin delivery to a patient suffering from renal disease, the engineered fibroblasts could be removed following a successful renal transplant. The reversibility of the therapy is possible because, with the notable exception of wound healing (reviewed in Reference 6), fibroblasts are not migratory cells (and even in the case of wound healing, the cells only migrate short distances to participate in scar formation).

Fibroblasts possess significant capacity for proliferation *in vitro*, allowing the cells to be transfected, propagated, and, in some cases, characterized prior to implantation into the patient. This proliferative capacity is the *in vitro* equivalent of wound healing. In fact, the stimulus for *in vitro* and *in vivo* fibroblast growth is identical — serum factors — and it is likely that the early successes with fibroblast growth in the 1950s were due in part to the sensible decision, made at the beginning of this century, to include blood components in the growth medium.[7] Although somewhat diminished, this proliferative capacity is preserved even in older individuals.[8]

Fibroblasts are capable of expressing a wide variety of exogenous genes at high levels, of efficiently performing a diverse array of protein processing reactions, and of secreting the resulting products. This flexibility has had a profound impact on the utility of fibroblasts in gene therapy and will be discussed below.

Fibroblasts, in general, exhibit a slow rate of growth *in vivo*.[9] In the adult dermis, for example, dividing fibroblasts and dying fibroblasts are rarely observed, and the absolute number of dermal fibroblasts remains relatively constant during aging. This constancy is most likely related to the structural role the fibroblast plays in supporting the tissues and organs of the body and the body as a whole. The limited rate of fibroblast growth *in vivo* has a critical implication for gene therapy — implanted fibroblasts should exhibit minimal growth following implantation and may have the potential to deliver relatively constant levels of a given therapeutic protein over extended periods of time. Fibroblasts do exhibit significant growth *in vivo* in two settings, during growth and development and wound healing. In theory, even these properties may have fundamental applications in gene therapy. For example, if transfected fibroblasts grow at the same rate as untransfected fibroblasts in children, the implanted fibroblasts may, in theory, "dose themselves" as the patient grows.

Human fibroblasts do not spontaneously immortalize in culture, a property that has significant implications for the safety of fibroblast-mediated gene therapy (discussed below). Although nonimmortalized human fibroblasts will be the focus of this article, immortalized fibroblasts and nonhuman fibroblasts will be discussed as appropriate. Finally, although human fibroblasts can be transfected *in vivo* (by injection of DNA, DNA-lipid complexes, or infections with viruses, for example), the scope of this review concerns *in vitro* approaches to fibroblast engineering and gene therapy.

II. GROWTH OF FIBROBLASTS IN VITRO

The fibroblast has been the best-studied mammalian cell type over the past 40 years, and much of its value is due to its well-characterized ability to be propagated in culture. That fibroblasts were successfully cultured before other cell types was somewhat fortuitous. Since the late 1800s, a variety of growth media were considered for *in vitro* culture of mammalian cells, tissue, and organs; ultimately, serum was found to be a valuable source of undefined "factors" that were required to support cell growth.[10] One of the stimuli to fibroblast growth *in vivo* is the presence of serum components in an otherwise hostile milieu, and to a very real extent, fibroblast growth *in vitro* was so effective in that it mirrored the environment in which fibroblast growth is stimulated, namely wound healing.

The growth of fibroblasts *in vitro* has significant implications for gene therapy in general and clonal gene therapy in particular, and an understanding of their growth patterns has come from a comparison of fibroblasts derived from a number of species. In order to discuss these studies, it is useful to define a few key terms. A "primary cell culture" refers to any nonimmortalized cell type that has been removed from the body and placed on an *in vitro* substrate for the first time. A "secondary cell strain" (or simply a "cell strain") refers to a nonimmortalized cell type that has been passaged at least once following primary cell culture (and therefore has been propagated on at least two *in vitro* substrates). Both primary and secondary cell strains have defined replicative life spans *in vitro*. In sharp contrast, an established cell line refers to an immortalized cell type that is capable of essentially unlimited growth *in vitro*; such lines are, under the appropriate circumstances, tumorigenic.

Murine fibroblasts are representative of one of two major fibroblast growth patterns.[11] Primary and secondary strains of murine fibroblasts are capable of undergoing approximately 10 to 15 cumulative population doublings (cpd) *in vitro*. At this point, the growth rate of the cells declines dramatically, and, depending on the their density, they either die off (a phenomenon known as "crisis") or survive to form rapidly growing foci of transformed cells. These surviving cells have undergone massive genotypic and phenotypic changes and are established cell lines.

Human fibroblasts, in contrast, exhibit a remarkably different *in vitro* growth pattern.[4] They do not become cell lines at all; instead, they attain over 60 cpd in culture as primary and secondary cells and then cease growing in culture. As the cells approach 60 cpd, they undergo a series of extremely well-characterized changes associated with aging, termed "senescence": the cells take longer to double, their volume increases, and their saturation density decreases. The tendency of a fibroblast to become transformed is a property of species type and the fibroblasts of certain species (human fibroblasts in particular) are remarkably resistant to spontaneous transformation in culture. That human fibroblasts senesce rather than spontaneously immortalize was one of the primary reasons leading to the suggestion that they may be useful in somatic cell gene therapy.[12]

III. GENETIC ENGINEERING OF HUMAN FIBROBLASTS

Fibroblast transfection was first accomplished in immortalized cells and was based on the use of the calcium phosphate coprecipitation technique.[13] This method is generally viewed as appropriate for use in immortalized cells but has never gained widespread acceptance for use in gene therapy. This is presumably because transfection by calcium phosphate coprecipitation is viewed as an inefficient methodology, with transfection efficiencies generally in the range of 10^{-3} to 10^{-7} per treated cell. As a result, the bulk of the work in the field has focused on a "high efficiency" transfection methodology, namely retroviral-mediated gene transfer.

Retroviral gene transfer is based on the replacement of retroviral genes with genes of therapeutic interest, and the technical aspects of the system have been extensively reviewed.[14] Recombinant retroviruses have been utilized to express a variety of genes in cultured fibroblasts, and a listing of representative studies is shown in Table 1. Of note, all of these studies were focused on rare genetic disorders (primarily those involving intermediary metabolism), which is consistent with the field of gene therapy as a whole in the late 1980s and early 1990s. In addition, the use of nonimmortalized fibroblasts has been favored (based on the obvious considerations of the relationship between immortalization *in vitro* and tumorigenicity *in vivo*).

Though genes of therapeutic interest have been introduced successfully into fibroblasts *in vitro*, the *in vivo* extension of these experiments has proven somewhat disappointing. In general, it has proven quite difficult to obtain long-term expression of the therapeutic gene following implantation of retrovirally-infected cells, a phenomenon first noted in 1991[26] and now referred to as "retroviral shutdown." The phenomenon was named based on the observation that fibroblasts expressing a given retrovirally-encoded gene would cease such expression soon after implantation. On removal from the animals, the recovered fibroblasts had irreversibly lost the ability to express the gene of interest (although the retrovirus itself was present in the cells). It is now quite clear that one approach to minimizing the effects of this shutdown involves the employment of

Table 1

Summary of *In Vitro* Gene Therapy Studies in Human Fibroblasts

Disease	Gene	Cells[a]	Gene Transfer	Reference
Citrullinemia	Arginosuccinate synthetase	Primary	Retroviral	15
Fucosidosis	α-L-fucosidase	Primary	Retroviral	16
Gyrate atrophy	Ornithine aminotransferase	Primary	Retroviral[b]	17
Hemophilia B	Factor IX	Primary/immortalized	Retroviral	18
Methylmalonic acidemia	Methylmalonyl-CoA mutase	Primary	Retroviral	19, 20
Mucopolysaccharidosis type I	α-L-iduronidase	Primary	Retroviral	21
Niemann-Pick	Acid sphingomyelinase	Primary	Retroviral	22, 23
Propionic acidemia	Propionyl CoA carboxylase α	Primary	Electroporation	24
PNP deficiency	Purine nucleoside phosphorylase	Primary	Retroviral	25

[a] Primary refers to primary cells or secondary cell strains.
[b] Transient expression only.

appropriate promoters.[27] Nonetheless, it has been difficult to obtain long-term expression *in vivo* using retroviral infection-based systems.

There are a number of nonviral approaches to the transfection of mammalian cells in culture, including fusion methodologies, physical methodologies such as microinjection and electroporation, and chemical methodologies such as calcium phosphate coprecipitation and polybrene coprecipitation. It is clear that these have not been widely applied to the transfection of normal human fibroblasts (for an exception, refer to Reference 24), and this is most likely due to their lower efficiencies of transfection as compared to viral methodologies. Of course, it is possible that essentially any method of transfection could be optimized to obtain high efficiency of stably transfected fibroblasts. This optimization was clearly performed in the case of the retroviruses, but the very success of this approach may have dampened interest in optimizing other approaches.

Although the use of clonal lines of transfected immortalized cells was of major importance in molecular biology from the late 1970s on, the use of transfected nonimmortalizing cells to generate a genetically-engineered clonal cell strain has not been widespread. It is critical to note that transfected clonal cell strains have certain advantages as compared to heterogeneous populations of transfected or infected cells — a uniform population can be chosen that has a higher per cell expression level than a typical heterogeneous population, which will, by definition, contain a mixture of high- and low-expressing cells, which differ with respect to site of integration of the transgene. Similarly, the cells of a clonal population have the same integration site, and inasmuch as the number of integration sites in a population and the introduction of deleterious mutations are directly related,[28] their use in gene therapy may minimize the possibility of producing cells that may have undesired properties. Finally, the use of clonal cell strains obviates the need for high-efficiency transfection or infection, in that only a single parental transfected cell is required to generate the clonal strain. This approach to gene therapy has been termed "transkaryotic therapy,"[12,29] and "clonal gene therapy."[30]

IV. DISEASE MODELS: IMPLANTATION OF GENETICALLY-MODIFIED FIBROBLASTS

Gene therapy has the potential to treat and perhaps cure diseases in almost every medical specialty, and it should come as no surprise that fibroblasts have been studied for the gene therapy of hematologic, oncologic, metabolic, endocrinologic, and CNS diseases (Table 2). Much of the work in the field has focused on hematologic and endocrinologic diseases, possibly because some of the first human genes isolated fall into these categories and because the disorders are amenable to treatment by the systemic delivery of proteins.

A number of groups have developed *in vivo* model systems for the treatment of hemophilia A and B. Of the two, work in hemophilia B appears to be somewhat more advanced — *in vivo* expression of factor VIII has yet to be reported in a fibroblast-based gene therapy system. In the hemophilias (as with most deficiencies of a systemic protein), a rate-limiting step in the development of a working gene therapy concerns the

Table 2

Summary of *In Vivo* Fibroblast Gene Therapy Studies

Gene	Cells	Gene Transfer	Expression	Reference
Hematologic				
Factor VIII	Primary human	Retroviral	None[a]	31
Factor IX, β-gal	Primary mouse	Retroviral	60 days	27
Factor IX	Primary mouse	Retroviral	10–12 days	32
Factor IX	Primary human/	Retroviral	7–14 days	52
	Immortalized mouse	Retroviral	28 days	
G-CSF	Immortalized mouse	Calcium phosphate	28 days	33
Endocrinologic				
Growth hormone	Immortalized rat	Calcium phosphate	41 days	34
Growth hormone	Immortalized mouse	Calcium phosphate	94 days	12
Growth hormone	Primary rabbit	Electroporation	550 days	35
Insulin	Immortalized mouse	Calcium phosphate	46 days	36
Insulin	Immortalized mouse	Calcium phosphate	50 days	29
Central nervous system				
Nerve Growth Factor	Immortalized rat	Retroviral	2 weeks	37
Oncologic				
α-interferon	Immortalized mouse	Calcium phosphate	NA[b]	38
Thymidine kinase	Immortalized mouse	Retroviral	NA[c]	39, 40
Other				
Adenosine deaminase	Primary dog	Retroviral	4 weeks	42
β-glucuronidase	Primary mouse	Retroviral	155 days	43

[a] No expression detected *in vivo* but expression detected in cells recovered two months post-implantation.
[b] Experiments designed to kill tumor cells.
[c] *In vivo* transduction experiments designed to kill tumor cells.

level of expression of the protein of interest; factor IX expression is apparently more straightforward than that of factor VIII. The required level of expression can be estimated in a variety of ways, and one approach to estimating the required amount of factor IX expression per cell follows (based on References 44 and 45):

$$C_{SS} = K_O T_{1/2}/0.693\ V_d$$

For maintenance of 5% normal activity in 70 kg patient: C_{SS} = steady state concentration (250 ng/ml), K_O = rate of production, $T_{1/2}$ = factor IX plasma half-life, and V_d = body distribution volume.

$$250\ \text{ng/ml} = K_O\ 23\ \text{hrs}/(0.693)(4900\ \text{ml})$$

$$K_O = 885\ \mu\text{g/day}$$

Therefore, assuming an implant consisting of 2×10^8 fibroblasts, the factor IX production level must be 4.4 μg factor IX produced per 10^6 cells per 24 hr. Similarly, the required amount of factor VIII expression can be calculated. For example, for maintenance of 10% normal activity in a 70-kg patient, with C_{SS} = steady state concentration (10 ng/ml),

$$10\ \text{ng/ml} = K_O\ 12\ \text{hr}/(0.693)(3000\ \text{ml})$$

$$K_O = 41.6\ \mu\text{g/day}$$

Again assuming an implant of 2×10^8 fibroblasts, the factor VIII production levels must be 208 ng factor VIII produced per 10^6 cells per 24 hr. It is critical to note that this analysis must be modified for estimates concerning animal model systems. In particular, the half-life of a given protein may vary substantially between a laboratory animal and a human. (The half-life of human growth hormone is 4 min in a mouse as compared to 20 min in a human, for example.)

The delivery of growth hormone and insulin via genetically-modified fibroblasts has been demonstrated by a number of groups. The studies of insulin delivery are particularly informative, in that they illustrate some critical points regarding the use of fibroblasts in gene therapy. When immortalized fibroblasts expressing insulin (or proinsulin) are implanted into animals that allow growth of the cells, hyperinsulinemia or hyperproinsulinemia results, a situation incompatible with long-term survival.[29,36]

At the most simplistic level, such experiments clearly demonstrate the problems inherent in utilizing immortalized cells as vehicles for protein delivery. Just as importantly, however, the studies can be generalized to reveal the potential of dosing problems in any cell type (including nonimmortalized cells) that experiences a fluctuation in numbers *in vivo*. Even in a setting in which cell number is constant *in vivo*, certain diseases (including diabetes) may require regulated expression of the gene of therapeutic interest. The ability to utilize fibroblasts to perform regulated expression and secretion for a given disease will depend on the ability of fibroblasts to respond naturally to the stimulus as required (e.g. glucose or certain amino acids in diabetes), to be engineered to respond to the stimulus (e.g. by introducing glucose-sensitive transcription units or secretory pathways), or to be engineered to respond to an alternative stimulus that accomplishes the same purpose (e.g., to secrete insulin in response to gastrin). It seems likely that all three approaches will have a place in the fibroblast gene therapist's armamentarium.

V. IMPLANTATION OF FIBROBLASTS IN TANDEM WITH SUBSTRATES

On implantation, fibroblasts generate the organized tissue to which they are accustomed by producing a collagen matrix. Several groups have experimented with the creation of matrices of fibroblasts and collagen *in vitro*.[32,46,47] Potential advantages of this approach are that the fibroblasts are implanted as organized tissue and may become stabilized *in vivo* more rapidly than naked fibroblasts and that the matrices may be handled somewhat more easily *in vitro*. Potential disadvantages include the use of nonautologous collagen, whether human or bovine, which increases the potential for transmission of adventitious agents and neither of which may prove stable for extended periods of time. A related technique involves the generation of fibroblast matrices based on polymers such as polytetrafluoroethylene coated with components of the extracellular matrix.[43,48]

The second major approach to the development of artificial substrates for use in fibroblast gene therapy concerns microencapsulation of the genetically-engineered cells. The approach is based on the concept that engineered cells can be contained physically within a membrane that serves two basic functions: it prevents the cells of the immune system from entering the structure, protecting the cells from rejection, and it allows the therapeutic protein secreted by the cells to freely diffuse out of the structure. One of the best-studied membrane materials is alginate-poly-L-lysine alginate, which has been used to encapsulate cells secreting growth hormone[49] and factor IX.[50] Related techniques are based on the use of other materials for forming the microcapsules and, for example, have been applied to the delivery of proinsulin from fibroblasts encapsulated in agarose.[51]

Encapsulation techniques have the potential to allow nonpatient-specific (nonautologous) gene therapy. In theory, a single individual's cells could be modified to produce a given protein, and any patient requiring this protein would receive encapsulated cells generated from the same individual. It is also possible that the type of cell to be encapsulated could be xenogeneic, which has been a major focus of islet transplantation research, or immortalized. While both categories of cells may be useful, it should be noted that xenogeneic cells may produce secreted antigens that could provoke a host-immune response, and immortalized cells (xeno- or allogeneic) demand an encapsulation system or device of the highest integrity because their escape could result in tumor formation. An additional issue that must be considered with encapsulation approaches is that the substrate material may be considered to be a foreign body by the host. As foreign bodies tend to be destroyed or sequestered by the host, it may be difficult to obtain expression of the gene of therapeutic interest beyond some number of months.

Based on the significant efforts currently being devoted to new types of microcapsules and to new types of membranes, however, this limitation may be overcome in the future.

VI. CONCLUSION: ROLE OF FIBROBLASTS IN GENE THERAPY

As stated above, no single cell type is appropriate for all cell-based gene therapies. Nevertheless, it is clear from the properties of fibroblasts and from the gene therapy studies performed to date that fibroblasts have the potential to play a prominent role in the clinical application of gene therapy. Fibroblasts lend themselves to applications in several classes of disease. In particular, implanted fibroblasts can secrete proteins that will ultimately find their way into the systemic circulation (applicable in diabetes, short stature, and hemophilia). The cells can also be implanted to allow local secretion (applicable in Parkinson's disease and Alzheimer's disease using intracranial implantation to circumvent the blood-brain barrier). Finally, the cells can serve as filters for toxic products (applicable in a host of disorders of intermediary metabolism) or to secrete toxic products (applicable in the treatment of certain cancers).

Fibroblast gene therapy has contributed in significant ways to the field of gene therapy, and, to this day, the fibroblast represents one of the best (if not the best) characterized and understood cell types *in vitro*. The fundamental relationship between the cell biology of the fibroblast and its utility in gene therapy has provided the rationale for continued research in these areas and bodes well for the successful application of genetically-engineered fibroblasts to the treatment and cure of a broad array of diseases in the future.

REFERENCES

1. *Catalogue of Cell Lines and Hybridomas*, 7th ed., American Type Culture Collection, Rockville, MD, 1992, 1.
2. Sanford, K. et al., The growth *in vitro* of single isolated tissue cells, *J. Natl. Cancer Inst.*, 9, 229, 1948.
3. Swim, H. E. and Parker, R. F., Culture characteristics of human fibroblasts propagated serially, *Am. J. Hyg.*, 66, 235, 1957.
4. Hayflick, L. and Moorhead, P. S., The serial cultivation of human diploid cells strains, *Exp. Cell Res.*, 25, 585, 1961.
5. Freshney, R. I., *Culture of Animal Cells,* 2nd ed., Alan R. Liss, New York, 1987.
6. Clark, R. A. F., *Physiology, Biochemistry, and Molecular Biology of the Skin*, 2nd ed., Oxford University Press, Oxford, 1991, 576.
7. Burrows, M. T., A method of furnishing a continuous supply of new medium to a tissue culture *in vitro*, *Anat. Rec.*, 6, 141, 1912.
8. Schneider, E. L. and Mitsui, Y., The relationship between *in vitro* cellular aging and *in vivo* human age, *Proc. Natl. Acad. Sci. U.S.A.*, 73, 3584, 1976.
9. Lapiere, C. M., The aging dermis: the main cause for the appearance of 'old' skin, *Br. J. Dermatol.*, 122 (Suppl. 35), 5, 1990.

10. Eagle, H., Nutrition needs of mammalian cells in tissue culture, *Science,* 122, 501, 1955.
11. Todaro, G. J. and Green, H., Quantitative studies of the growth of mouse embryo cells in culture and their development into established lines, *J. Cell Biol.*, 17, 299, 1963.
12. Selden, R. F. et al., Implantation of genetically engineered fibroblasts into mice: Implications for gene therapy, *Science*, 236, 714, 1987.
13. Graham, F. L. and van der Eb, A. J., Transformation of rat cells by DNA of human adenovirus 5, *Virology*, 54, 536, 1973.
14. Gilboa, E. et al., Transfer and expression of cloned genes using retroviral vectors, *Biotechniques*, 4, 504, 1986.
15. Demarquoy, J., Retroviral-mediated gene therapy for the treatment of citrullinemia. Transfer and expression of argininosuccinate synthetase in human hematopoietic cells, *Experientia*, 49, 345, 1993.
16. Occhiodoro, T. et al., Correction of α-L-fucosidase deficiency in fucosidosis fibroblasts by retroviral vector-mediated gene transfer, *Hum. Gene Ther.,* 3, 365, 1992.
17. Hotta, Y. and Inana, G., Expression of human ornithine aminotransferase (OAT) in OAT-deficient Chinese hamster ovary cells and fibroblasts of gyrate atrophy patient, *Jpn. J. Ophthalmol.*, 36, 28, 1992.
18. Dai, Y.-F. et al., High efficient transfer and expression of human clotting factor IX cDNA in cultured human primary skin fibroblasts from hemophilia B patient by retroviral vectors, *Sci. China (Ser. B)*, 35, 2, 1992.
19. Sawada, T. and Ledley, F. D., Correction of methylmalonyl-Co-A mutase deficiency in Mut^o fibroblasts and constitution of gene expression in primary human hepatocytes by retroviral-mediated gene transfer, *Somatic Cell Molec. Genet.*, 18, 6, 1992.
20. Wilkemeyer, M., Propionate metabolism in cultured human cells after overexpression of recombinant methylmalonyl CoA mutase: implications for somatic gene therapy, *Somatic Cell Molec. Genet.,* 18, 6, 1992.
21. Anson, D. S. et al., Correction of mucopolysaccharidosis type I fibroblasts by retroviral-mediated transfer of the human α-L-iduronidase gene, *Hum. Gene Ther.*, 3, 371, 1992.
22. Dinur, T. et al., Toward gene therapy for Niemann-Pick disease (NPD); separation of retrovirally corrected and noncorrected NPD fibroblasts using a novel fluorescent sphingomyelin, *Hum. Gene Ther.*, 3, 633, 1992.
23. Suchi, M. et al., Retroviral-mediated transfer of the human acid sphingomyelinase cDNA: correction of the metabolic defect in cultured Niemann-Pick disease cells, *Proc. Natl. Acad. Sci. U.S.A.*, 89, 3227, 1992.
24. Stankovics, J. and Ledley, F. D., Cloning of functional alpha propionyl CoA carboxylase and correction of enzyme deficiency in pccA fibroblasts, *Am. J. Hum. Genet.*, 52, 144, 1993.
25. Osborne, W. R. A. and Miller, A. D., Design of vectors for efficient expression of human purine nucleoside phosphorylase in skin fibroblasts from enzyme-deficient humans, *Proc. Natl. Acad. Sci. U.S.A.*, 85, 6851, 1988.
26. Palmer, T. D. et al., Genetically modified skin fibroblasts persist long after transplantation but gradually inactivate introduced genes, *Proc. Natl. Acad. Sci. U.S.A.*, 88, 1330, 1991.
27. Scharfmann, R. et al., Long-term *in vivo* expression of retrovirus-mediated gene transfer in mouse fibroblast implants, *Proc. Natl. Acad. Sci. U.S.A.,* 88, 4626, 1991.
28. Bardwell, L., The mutagenic and carcinogenic effects of gene transfer, *Mutagenesis*, 2, 245, 1989.
29. Selden, R. F et al., Regulation of insulin gene expression: implications for gene therapy, *N. Engl. J. Med.*, 317, 1067, 1987.
30. Garver, R. I. et al., Clonal gene therapy: transplanted mouse fibroblast clones express human α1-antitrypsin gene in vivo, *Science*, 236, 762, 1987.

31. Hoebens, R. C. et al., Toward gene therapy for hemophilia A: long-term persistence of Factor VIII-secreting fibroblasts after transplantation into immunodeficient mice, *Hum. Gene Ther.*, 4, 179, 1993.
32. St. Louis, D. and Verma, I. M., An alternative approach to somatic cell gene therapy, *Proc. Natl. Acad. Sci. U.S.A.*, 85, 3150, 1988.
33. Tani, K. et al., Implantation of fibroblasts transfected with human granulocyte colony-stimulating factor cDNA into mice as a model of cytokine-supplemented gene therapy, *Blood*, 74, 1274, 1989.
34. Behara, A. M. P. et al., Intrathymic implants of genetically modified fibroblasts, *FASEB J.*, 6, 2853, 1992.
35. Heartlein, M. et al., Long-term production and delivery of human growth hormone *in vivo*, in press.
36. Kawakami, Y. et al., Somatic gene therapy for diabetes with an immunological safety system for complete removal of transplanted cells, *Diabetes*, 41, 956, 1992.
37. Rosenberg, M. F. et al., Grafting genetically modified cells to the damaged brain: restorative effects of NGF expression, *Science*, 242, 1575, 1988.
38. Ogura, H. et al., Implantation of genetically manipulated fibroblasts into mice as antitumor α-interferon therapy, *Cancer Res.*, 50, 5102, 1990.
39. Zvi, R. et al., *In situ* retroviral-mediated gene transfer for the treatment of brain tumors in rats, *Cancer Res.*, 53, 83, 1993.
40. Takamiya, Y. et al., An experimental model of retrovirus gene therapy for malignant brain tumors, *J. Neurosurg.*, 79, 104, 1993.
41. Gage, F. H. et al., Grafting genetically modified cells to the brain: possibilities for the future, *Neuroscience*, 23(3), 795, 1987.
42. Ramesh, N. et al., High-level human adenosine deaminase expression in dog skin fibroblasts is not sustained following transplantation, *Hum. Gene Ther.*, 4, 3, 1993.
43. Moullier, P. et al., Correction of lysosomal storage in the liver and spleen of MPS VII mice by implantation of genetically modified skin fibroblasts, *Nat. Genet.*, 4, 154, 1993.
44. Braunwald, E. et al., in *Harrison's Principles of Internal Medicine*, McGraw-Hill, New York, 342.
45. Yao, S.-N. and Kurachi, K., Expression of human factor IX in mice after injection of genetically modified myoblasts, *Proc. Natl. Acad. Sci. U.S.A.*, 89, 3357, 1992.
46. Elsdale, T. and Bard, J. Collagen substrata for studies of cell behavior, *J. Cell Biol.*, 54(3), 626, 1972.
47. Bell, E. et al., Production of a tissue-like structure by contraction of collagen lattices by human fibroblasts of different proliferative potential in vitro, *Proc. Natl. Acad. Sci. U.S.A.*, 76, 1274, 1979.
48. Thompson, J. et al., Heparin-binding growth factor 1 induces the formation of organoid neovascular structures in vivo, *Proc. Natl. Acad. Sci. U.S.A.*, 86, 7928, 1989.
49. Tai, I. T. and Sun, A. M., Microencapsulation of recombinant cells: a new delivery system for gene therapy, *FASEB J.*, 7, 1061, 1993.
50. Liu, H.-W. et al., Expression of human factor IX by microencapsulated recombinant fibroblasts, *Hum. Gene Ther.*, 4, 291, 1993.
51. Taniguchi, H. et al., Treatment of diabetic mice with encapsulated fibroblasts producing human proinsulin, *Transplant. Proc.*, 24, 2977, 1992.
52. Palmer, T. D. et al., Production of human factor IX by genetically-modified skin fibroblasts: potential therapy for hemophilia B, *Blood*, 73, 438, 1989.

5 Hepatic Gene Therapy

Fred D. Ledley

I. INTRODUCTION

The liver, site of many essential metabolic and secretory functions, will be an important target for somatic gene therapy. For example, gene therapy for inherited metabolic diseases, such as phenylketonuria, familial hypercholesterolemia, hemophilia, organic acidemia, and urea cycle disorders, may require gene replacement in hepatocytes. These cells have the capacity to provide complementary subunits of heterologous enzymes, cofactors, regulatory factors, and metabolites which may be involved in constituting effective biologic activity of the recombinant gene product.[1] The liver may also be the preferred target for gene therapies that require secretion of products into the blood since many serum proteins require specific post-translational modifications or are effective only at high concentrations, which may not be achieved by production of recombinant gene products from smaller organs. Gene therapy may also be used for various liver-specific diseases including infectious hepatitis, autoimmune hepatitis, or cirrhosis.

Hepatic gene therapy may take advantage of many unique anatomic, histologic, cellular, and molecular characteristics of the liver. These include (1) a blood supply (the portal system) which directly and selectively perfuses the liver, (2) a sinusoidal structure in which interstitial cells are not separated from the vascular space by a continuous endothelial layer or basement membrane, and (3) a constituent cell type (the hepatocyte) that will proliferate *in vivo*, which can be selectively maintained and cultivated in a differentiated state *in vitro*, which has well characterized regulatory mechanisms, and which has unique cell surface receptors.

Early methods for hepatic gene therapy have not been based primarily on the unique biologic features of the liver, but rather have employed general strategies for gene delivery which have been successful in other target organs. These include *ex vivo* transduction of cells with retroviral vectors, direct infection of organs *in vivo* with retroviral or adenoviral vectors, and the use of nonspecific, viral promoter elements.[2,3] More recent methods are beginning to take advantage of the unique characteristics of the liver for targeting genes selectively to the liver and achieving specific expression of recombinant genes within hepatocytes. The utility of several approaches for hepatic gene delivery has now been demonstrated in *in vitro* models, animal experiments, and clinical trials. There are significant differences between these methods in terms of their

0-8493-2440-8/95/$0.00+$.50

invasiveness, the number of cells that can be transformed with the recombinant gene, the persistence of therapeutic gene expression, and the suitability of repetitive dosing regiments. While many of these methods are elegant, the acceptance of gene therapy in clinical practice will not be based solely on its effectiveness but also on the relative safety, invasiveness, toxicity, compliance, and cost. Moreover, gene therapies will have to be evaluated relative to other therapeutic approaches, including surgical procedures, liver assist devices, pharmaceutical products, and biologic products that may be developed for the same clinical indications. This review describes various approaches and methods which may be used for hepatic gene therapy and the state of preclinical and clinical studies to critically evaluate the effectiveness, efficacy, and clinical utility of these methods.

II. METHODS FOR GENE DELIVERY TO THE LIVER

A. Ex Vivo Gene Delivery to Hepatocytes

Initial strategies for hepatic gene therapy, like initial efforts toward gene therapy in general, focused on *ex vivo* methods for gene delivery. This involved performing a partial hepatectomy from patients, cultivating primary hepatocytes *in vitro*, transducing these cells with defective retroviral vectors that carried a therapeutic gene, and then transplanting these cells into the autologous host by infusion into the portal venous system. Studies in experimental animals demonstrated that primary hepatocytes could be transduced efficiently with retroviral vectors[4-6] and that hepatocytes transplanted into the portal venous system would engraft within the normal architecture of the liver and persist for the lifetime of the animal.[7,8] In model systems, expression from incorporated viruses was demonstrated as long as 6 months after gene delivery.[9] Further preclinical studies demonstrated the feasibility of transducing human hepatocytes,[10,11] transplanting human hepatocytes into immunodeficient mice,[11] and performing hepatocellular transplantation in nonhuman primates.[11,12] Two clinical trials involving gene transfer into the liver were proposed and approved by the RAC and FDA using this *ex vivo* approach.[13,14] In the work of Wilson and his colleagues, several patients with familial hypercholesterolemia due to LDL-receptor deficiency have been treated using a protocol in which hepatocytes were harvested, transduced with a retrovirus capable of expressing the LDL-receptor, and then reintroduced by infusion into the portal vascular system. The results reported for one patient indicate that hepatic engraftment of transduced hepatocytes has been achieved and that some biological function can be detected.[15]

The clinical utility of *ex vivo* hepatic gene therapy is inherently limited by several factors. First, *ex vivo* gene therapy requires an invasive surgical procedure (5 to 20% partial hepatectomy), with the attendant risks of any major abdominal surgical procedure. Second, the number of cells in the liver that will contain the recombinant gene is severely limited. This number is determined by the efficiency by which primary human hepatocytes in culture can be transduced with retroviral vectors[16] and the fraction of hepatocytes in a patient that can be reconstituted by hepatocellular transplantation.[11] In preclinical experiments as well as in the initial clinical trials, existing methods have been shown to introduce the gene into approximately 1/1000 hepatocytes in the recipient, a fraction that may be suitable for treating certain disease states, depending upon the biology of the disease process. Various methods have been proposed to increase this fraction by increasing the efficiency of transduction,[16] selecting transduced cells *in vitro*,[5] inducing selective proliferation of transplanted cells *in vivo*,[17] or isolating the elusive hepatic stem cell.[18] With these methods it is theoretically possible to establish the recombinant gene in up to 1 to 2% of the hepatocyte population of the host. Third, *ex vivo* gene therapy is a costly procedure requiring extensive surgical procedures, hospitalization, and large scale *in vitro* manipulation of cells which could limit the widespread application of such a procedure to common diseases.

B. In Vivo Transduction with Retroviral Vectors

Recent studies have raised the possibility of performing retroviral gene transfer *in vivo* by direct infusion of retroviral vectors into the portal vein.[19-23] Since retroviral integration into infected cells is dependent upon cell division, *in vivo* delivery requires methods to induce hepatocyte proliferation, such as a 70 to 80% (subtotal) hepatectomy. *In vivo* transduction with retroviral vectors has been most efficient in postnatal, rather than adult, animals where >10% of hepatocytes have been transduced.[23] Sustained levels of recombinant gene products, including α_1-antitrypsin and factor IX, have been established by *in vivo* retroviral infusion.[21,24]

While *in vivo* delivery under select conditions in animal models is more efficient than *ex vivo* gene therapy, continuing studies are needed to establish whether these methods will be applicable in humans. Particular attention will need to be paid to the ability of human complements to lyse retroviral vectors of various trophism[25,26] and the level of expression of the appropriate receptor for the retrovirus on human hepatocytes. The failure of older studies to demonstrate transduction of rodent hepatocytes *in vivo* with ecotropic virus has recently been explained by the observation that the ecotropic receptor is not expressed by hepatocytes *in vivo* even though it is expressed under normal culture conditions *in vitro*.[27-30] While the amphotropic receptor has not yet been identified or cloned and there is no reason to believe *a priori* that it will be regulated like the ecotropic receptor, studies with primary human hepatocytes *in vitro* suggest that under certain conditions the expression of the amphotropic receptor may be inadequate to allow efficient transduction with amphotropic vectors (Adams and Ledley, manuscript in preparation). These studies also suggest that this limitation may be circumvented with the use of xenotropic vectors that have a distinct host-range[16] or with development of formulations that enable retroviral transduction of cells across host-range barriers in the absence of a cognate receptor.[31] Another important issue in developing clinical applications for *in vivo* retroviral transduction is to develop alternative methods for stimulating hepatocyte cell division without 70 to 80% hepatectomy, a procedure that has considerable operative short-term and long-term complications. It is possible that the hepatocellular regeneration associated with diseases such as infectious hepatitis or certain inborn errors of metabolism may be sufficient to enable *in vivo* retroviral gene delivery without hepatectomy, though this has yet to be demonstrated in animal models. Alternatively, methods for "chemical hepatectomy" or methods for pharmacologically stimulating cell division may be useful but may also embody considerable risks. Such methods, however, may enable retroviral delivery to be performed using relatively noninvasive procedures, such as direct, transhepatic, or transjugular catheterization of the portal vein.

C. In Vivo Gene Transfer Using Recombinant Adenoviral Vectors

Adenoviral vectors have been shown to be extremely effective agents for introducing recombinant genes into hepatocytes of several different animal models. Using high-titer adenoviral vectors, it is possible to introduce recombinant genes into virtually all cells within the liver by infusion into the portal vein or systemic circulation.[32-34] Adenoviral infection of the hepatocytes is feasible because of the sinusoidal structure of the liver, which provides these viral particles with direct access to heptocytes without the barrier of a continuous epithelium or a basement membrane. Adenoviral vectors have been used to transduce hepatocytes with reporter genes such as β-galactosidase and to constitute potentially therapeutic levels of gene products such as ornithine transcarbamylase, factor IX, and apoA. Unlike retrovirus vectors, adenoviral vectors are not designed to integrate recombinant genes stably into infected cells. Rather, the vector appears to persist in the liver in an episomal state for a period of days to weeks before the foreign DNA is eliminated from the cell and expression of the therapeutic gene product is lost.

The major limitation of the current generation of adenoviral vectors relates to the fact that these vectors are immunogenic and, at high concentrations, cytopathic. There is a relatively narrow therapeutic index between the concentrations of virus that are required to transduce the majority of cells in the liver and concentrations of virus that induce substantial cytopathicity. Furthermore, a single administration of adenoviral vectors produces neutralizing antibodies that render subsequent doses of the same vector, or vectors expressing other genes, completely ineffective. This could preclude the use of adenoviral vectors for gene therapy in patients who had natural exposure to adenovirus having certain serotypes as well as the application of repetitive dosing strategies for chronic diseases. Because of the enormous efficiency of adenoviral gene transfer *in vitro* and *in vivo*, efforts are being made to develop new forms of recombinant adenoviral vectors to achieve either stable transformation or to eliminate the immunogenicity of the vectors. Until such efforts bear fruit, however, the clinical applications of adenovirus for hepatic gene therapy appear limited. Nevertheless, adenoviral vectors will continue to be useful experimental tools for studying the biologic effect of recombinant gene expression for short periods of time in experimental animals.

D. Hepatic Delivery of DNA-Based Vectors

While early efforts at gene therapy arising from research in molecular biology were based on the use of viral vectors for gene transfer, more recent efforts have focused on exploiting methods for delivery of DNA vectors to the liver using methods developed for conventional drug delivery. Initial interest has focused on the use of ligands for the asialoglycoprotein receptor, which is known to be effective in delivering various drugs to the liver. Important success has been achieved with the use of asialo-orosomucoid,[26,35,36,38-42] which is covalently coupled to polylysine and then complexed noncovalently to DNA by an ionic interaction between the negatively charged DNA and positively charged polylysine. This complex remains capable of binding to the asialoglycoprotein receptor, and the subsequent receptor-mediated endocytosis of this complex leads to the uptake and expression of DNA from the complex. This method has been used to express several genes in the liver, including reporter genes such as chloramphenicol acetyltransferase as well as therapeutic genes such as the LDL-receptor, methylmalonyl CoA mutase, or albumin.[35-42] In each case, gene delivery provides expression of the recombinant gene for several days. Pharmacokinetic studies demonstrate that the half-life of DNA within the liver is only several hours. Commonly, the gene product can be detected in animals for several days due to the longer half-lives of the mRNA transcript or the recombinant product itself.[42] Even with the relatively short duration of action of the gene therapy, repetitive dosing could make this method of gene delivery useful for treating acute disease or establishing steady-state levels of recombinant products to treat chronic diseases.

The major limitation of this method is the antigenicity of the covalent asialoglycoprotein-polylysine complex and the high titers of antibodies that can be formed after repetitive administration of this complex. Such antibodies would complicate the ability to have a reproducible dose-response, which would be required for repetitive dosing regiments. Various alternative ligands for the asialoglycoprotein receptor have been described, which may not be antigenic, including synthetic ligands containing tris-galactose[43] or asialolactoferrin, which is a DNA-binding protein in its native form and capable of forming gene delivery complexes without antigen-producing covalent modifications.[44] Efforts have also been made to use liposomes,[45,46] cationic lipids,[47] and cyclic amphipathic peptides.[48] Cationic lipids and other synthetic formulations are highly effective in many *in vitro* systems. Variable success has been achieved with these methods in *in vivo* studies. While each of these approaches has considerable promise, further research is required to obtain predictable results *in vivo*.

E. Future Directions for Gene Delivery to the Liver

The focus of future research in gene delivery will be on achieving controlled delivery of DNA vectors to the liver, controlled intracellular trafficking of DNA within the targeted cell, and controlled persistence of expression of the therapeutic gene product. It is likely that future methods will increasingly be based on the application of methods for advanced drug delivery[49] and exploitation of the unique histologic, cellular, and molecular characteristics of hepatocytes to achieve targeted delivery and selective gene expression. The liver is an attractive and achievable target for gene delivery due to its sinusoidal structure, which provides access to particulate material in the blood, as well as the presence of a variety of hepatocyte-specific receptors. Moreover, selective expression of recombinant genes in hepatocytes may be achieved by combining methods for delivery that may provide efficient but not completely specific targeting of DNA to hepatocytes, with the use of hepatocyte-specific promoters that will not direct expression of recombinant genes in cells other than hepatocytes.

A particularly important issue in DNA delivery to the liver is to control the intracellular trafficking of DNA after uptake into hepatocytes. Most methods for targeted delivery of DNA to the liver involve endocytosis of the vector by the targeted cell. Studies have shown that most of the DNA is destroyed in the endosome soon after the initial uptake. *In vitro* studies have demonstrated that combining ligands that target DNA to cells with viral particles or viral-derived proteins that are capable of lysing the endosome improves the efficiency of gene transfer and the levels of gene expression by several orders of magnitude.[50-54] Replication-defective adenovirus has been most effective, though promising results have also been obtained with the influenza virus hemagglutinin peptide as well as synthetic peptides. While these experiments demonstrate the importance of controlling the intracellular trafficking of DNA, these methods have yet to be used effectively in *in vivo* experiments targeted to the liver. It is likely that effective gene transfer to the liver will require the incorporation of methods for enhancing the release of DNA from the endosomal compartment using methods which are nonantigenic and still permit the administered particle to enter hepatocytes from the sinusoids by endocytosis.

The final issue is to achieve control over gene expression to ensure that the level of the therapeutic product remains within the therapeutic index of the disease being treated. This will involve optimizing the dose and schedule of administration of the gene to achieve steady-state levels within the therapeutic profile, the use of promoters that are regulated by physiologic feedback mechanisms, or the use of strategies for regulating the expression of recombinant genes in the body by exogenously administered pharmaceutical agents.

III. APPROACHING CLINICAL APPLICATIONS OF HEPATIC GENE THERAPY

In the decade that has passed since the first efforts to assess the feasibility of hepatic gene therapy, considerable progress has been made. Effective gene delivery to the liver has been demonstrated in several animal models, *in vitro* or *in vivo* studies have begun to assess the feasibility of gene therapy for a large number of genetic and acquired diseases of the liver, and the first patients have been enrolled in clinical trials. Deficiencies of serum proteins such as clotting factors (factor VIII or factor IX) or α_1-antitrypsin, metabolic disorders including those of organic acid metabolism, the urea cycle, and amino acid metabolism appear to be likely targets for gene therapy. Monogenic and multifactorial disorders of cholesterol metabolism are also important targets for hepatic gene therapy as is infectious or autoimmune hepatitis and cirrhosis.

Despite considerable optimism resulting from the initial animal and clinical trials, a great deal of analysis remains to be done to determine which strategies for gene therapy will be most efficacious in clinical practice, which methods for gene delivery will be most effective, safe, economical, and acceptable to physicians and patients, and how gene therapy will come to be employed in real clinical practice. One critical issue is whether a proposed gene therapy will involve the permanent introduction of genes into the liver with stable expression of the therapeutic gene product, or whether gene therapy will involve repetitive delivery of genes to the liver to achieve controlled persistence of the DNA and the gene product. Permanent gene replacement strategies would not be suitable for treating acute, limited diseases such as hepatitis, but are attractive for life-long conditions such as inherited metabolic disease. Since such strategies for gene therapy envision performing only a single procedure during the patients's lifetime, a certain degree of invasiveness, toxicity, risk, and cost may be acceptable. Permanent gene therapy, however, requires a great deal of confidence in the accuracy of diagnosis, in the reproducibility of the procedure used to administer gene therapy, and in the ability to control the level of gene expression within the therapeutic index through normal growth, maturity, ageing, and the inevitable intercurrent illness. The development of strategies for providing long-term follow-up care and surveillance is particularly important for patients who have recombinant genes permanently integrated into their cells.[55]

In contrast, strategies that provide limited persistence of the gene and gene product will be applicable to acute diseases as well as chronic diseases that might be managed by repetitive administration of the gene therapy. In this strategy, the gene will be administered like a conventional drug or medicine; the dose and schedule of administration will be adjusted to meet the patient's clinical needs or even stopped if the patient's condition improves or alternative therapies are available. Preliminary experimental and theoretical work has begun to address the pharmacokinetics of such therapy, which will be governed not only by conventional issues such as the distribution and fate of the administered gene and gene product but also the processes of intracellular compartmentalization, transcription, and translation.[56] For repetitive dosing strategies to be clinically acceptable, such therapies must be relatively noninvasive, have a low degree of toxicity and risk, and have a reproducible dose-response over time. Appropriate therapies for different diseases may require different degrees of persistence to achieve adequate compliance and cost. Frequent injections (i.e., daily) might be acceptable for many life-threatening acute or chronic diseases as well as diseases currently treated by transplantation. Less frequent administration would be required for gene therapy of less severe disease. Prophylactic therapy for disorders such as hypercholesterolemia may require very infrequent administration using noninvasive approaches with an extremely low level of risk.

Another important issue is cost. The application of gene therapy to many rare diseases may be limited by the cost of developing, validating, manufacturing, and providing the products required for gene therapy. While academic trials of gene therapy for many orphan diseases are likely, the problems of providing continuous availability of such therapies has not been adequately addressed. This is particularly true for the many inborn errors of metabolism, which might be treated by hepatic gene therapy. Thus, gene therapy may not come to be the treatment of choice for many rare inherited diseases if general methods such as hepatocellular transplantation can be developed into equally efficacious therapies.

A final issue is the importance of developing accessory clinical technologies that may be required for gene delivery. One of the early problems facing *ex vivo* strategies for hepatic gene therapy was the need to establish a fundamentally new surgical procedure (hepatocellular transplantation) in addition to novel methods for molecular and cellular manipulation of hepatocytes. The lack of any clinical precedent or acceptance of hepatocellular transplantation,[57] and the need for basic surgical research to study this procedure and its effectiveness remain a significant impediment to further development of *ex vivo* gene delivery technologies to solid organs. Gene therapies requiring intraportal administration of materials may be based upon methods developed for invasive radiologic procedures, such as ultrasound-directed transhepatic or transjugular portal catheterization of the portal vein. Gene therapies involving infection of patients with defective or attenuated virus can take advantage of clinical experience with vaccines based on similar vectors. Gene therapies involving delivery of DNA to liver may take advantage of materials and methods for advanced drug delivery that have already been validated in *in vitro*, animal, or even clinical experiments. The clinical application and acceptance of gene therapy may be greatly facilitated by developing therapies that are based on materials, methods, and modes of delivery that are familiar to physicians and patients.

Despite the progress that has been made in developing and validating methods for gene therapy, the initial clinical trials of gene delivery to the liver and other organs have only begun to address the feasibility of gene delivery in human subjects. These trials are not designed to obtain statistically significant and controlled data on either the effectiveness or safety of gene transfer.[58,59] The critical evaluation of gene therapy will require large, controlled clinical trials in various patient populations to obtain statistically significant data on the effectiveness, safety, and appropriate clinical indications for gene therapy.[59] These trials, which will for the first time assess gene therapy in the context of the vicissitudes and constraints of clinical medicine, will provide the next important indication of how gene therapy should be further developed to fulfill its considerable promise.

IV. CONCLUSION

Several different methods for gene delivery to the liver have been shown to be feasible in animal experiments and in initial clinical trial. The next stage in the development of hepatic gene therapy will involve not only further refinement of these methods but critical preclinical assessment of their utility, safety, acceptability, and cost, as well as controlled clinical analysis of how these methods may come to be employed in clinical practice.

ACKNOWLEDGMENTS

The author's work has been supported by the Howard Hughes Medical Institute, NIH grants PO1-DK-44989, the ACTA Foundation, and a grant from the Mathers Foundation. Dr. Ledley is a founder with equity interest in GeneMedicine, Inc.

REFERENCES

1. Ledley, F. D., Clinical application of somatic gene therapy in inborn errors of metabolism, *J. Inher. Metab. Dis.*, 13, 597, 1990.
2. Ledley, F. D., Are contemporary methods for somatic gene therapy suitable for clinical applications?, *Clin. Invest. Med.*, 16, 78, 1993.
3. Ledley, F. D., Hepatic gene therapy: present and future, *Hepatology*, 18, 1263, 1993.
4. Wolff, J. A., Yee, J. K., and Skelly, H. F., Expression of retrovirally transduced genes in primary cultures of adult rat hepatocytes, *Proc. Natl. Acad. Sci. U.S.A.*, 84, 3344, 1987.
5. Ledley, F. D., Darlington, G. J., Hahn, T., and Woo, S. L., Retroviral gene transfer into primary hepatocytes: implications for genetic therapy of liver specific functions, *Proc. Natl. Acad. Sci. U.S.A.*, 84, 5335, 1987.
6. Wilson, J. M., Jefferson, D. M., Chowdhury, J., Novikoff, P. M., Johnston, D. E., and Mulligan, R. C., Retrovirus-mediated transduction of adult hepatocytes, *Proc. Natl. Acad. Sci. U.S.A.*, 85, 3014, 1988.
7. Ponder, K. P., Gupta, S., Leland, F., Darlington, G., Finegold, M., DeMayo, J., Ledley, F. D., Chowdhury, J. R., and Woo, S. L. C., Mouse hepatocytes migrate to liver parenchyma and function indefinitely after intrasplenic transplantation, *Proc. Natl. Acad. Sci. U.S.A.*, 88, 1217, 1991.
8. Gupta, S., Aragona, E., Vemuru, R. P., Bhagarva, K. K., Burk, R. D., and Chowdhury, J., Permanent engraftment and function of hepatocytes delivered to the liver: implications for gene therapy and liver repopulation, *Hepatology*, 14, 144, 1990.
9. Chowdhury, J., Grossman, M., Gupta, S. J., Chowdhury, N., Baker, J. R., Jr., and Wilson, J. M., Long-term improvement of hypercholesterolemia after ex vivo gene therapy in LDLR-deficient rabbits, *Science*, 254, 1802, 1991.
10. Grossman, M., Raper, S., and Wilson, J. M., Towards liver-directed gene therapy: retrovirus-mediated gene transfer into human hepatocytes, *Somat. Cell Mol. Genet.*, 17, 601, 1992.
11. Ledley, F. D., Adams, R. M., Soriano, H. E., Darlington, G. J., Finegold, M., Lanford, R., Carey, D., Lewis, D., Baley, P., Rothenberg, S., Kay, M., Brandt, M., Moen, R., Anderson, W. F., Whittington, P., Pokorny, W., and Woo, S. L. C., Development of a clinical protocol for hepatic gene transfer: lessons learned in pre-clinical studies, *Pediatr. Res.*, 33, 313, 1993.
12. Grossman, M., Raper, S. E., and Wilson, J. M., Transplantation of genetically modified autologous hepatocytes into nonhuman primates: feasibility and short-term toxicity, *Hum. Gene Ther.*, 5, 501, 1992.
13. Ledley, F. D., Woo, S. L., Ferry, G. D., Whisennand, H. H., Brandt, M. L., Darlington, G. J., Demmler, G. J., Finegold, M. J., Pokorny, W. J., Rosenblatt, H., Schwartz, P., Moen, R. C., and Anderson, W. F., Hepatocellular transplantation in acute hepatic failure and targeting genetic markers to hepatic cells, *Hum. Gene Ther.*, 2, 331, 1991.
14. Wilson, J. M., Grossman, M., Raper, S. E., Baker, J. R., Jr., Newton, R. S., and Thoene, J. G., Ex vivo gene therapy of familial hypercholesterolemia, *Hum. Gene Ther.*, 3, 179, 1992.
15. Grossman, M., Raper, S. E., Kozarsky, K., Stein, E. A., Englehardt, J. F., Muller, D., Lupien, P. J., and Wilson, J. M., Successful ex vivo gene therapy directed to the liver in a patient with familial hypercholesterolemia, *Nat. Genet.*, 1994.
16. Adams, R. M., Soriano, H., Wang, M., Darlington, G., Steffen, D., and Ledley, F. D., Transduction of primary human hepatocytes with amphotropic and xenotropic retroviral vectors, *Proc. Natl. Acad. Sci. U.S.A.*, 89, 9981, 1992.
17. Gupta, S., Yerneni, P. R., Vemuru, R. P., Lee, C. D., Yellin, E. L., and Bhargava, K. K., Studies on the safety of intrasplenic hepatocyte transplantation: relevance to ex vivo gene therapy and liver repopulation in acute hepatic failure, *Hum. Gene Ther.*, 4, 249, 1993.

18. Brill, S., Holst, P., Sigal, S., Zvibel, I., Fiorino, A., Ochs, A., Somasundaran, U., and Reid, L. M., Hepatic progenitor populations in embryonic, neonatal, and adult liver, *Proc. Soc. Exp. Biol. Med.*, 204, 261, 1993.
19. Kaleko, M., Garcia, J. V., and Miller, A. D., Persistent gene expression after retroviral gene transfer into liver cells in vivo, *Hum. Gene Ther.*, 2, 27, 1991.
20. Ferry, N., Duplessis, O., Houssin, D., Danos, O., and Heard, J. M., Retroviral mediated gene transfer into hepatocytes in vivo, *Proc. Natl. Acad. Sci. U.S.A.*, 88, 8377, 1991.
21. Kay, M. A., Li, Q., Liu, T. J., Leland, F., Toman, C., Finegold, M., and Woo, S. L., Hepatic gene therapy: persistent expression of human $alpha_1$-antitrypsin in mice after direct gene delivery in vivo, *Hum. Gene Ther.*, 3, 541, 1992.
22. Cardoso, J. E., Branchereau, S., Jeyaraj, P. R., Houssin, D., Danos, O., and Heard, J. M., *In situ* retrovirus mediated gene tranfser into dog liver, *Hum. Gene Ther.*, 4, 411, 1993.
23. Kolodka, T. M., Finegold, M., and Woo, S. L. C., Hepatic gene therapy: efficient retroviral-mediated gene transfer into rat hepatocytes *in vivo*, *Somatic Cell. Molec. Genet.*, 19, 491, 1993.
24. Kay, M. A., Rothenberg, S., Landen, C. N., Bellinger, D. A., Leland, F., Toman, C., Finegold, M., Thompson, A. R., Read, M. S., Brinkhous, K. M. et al., In vivo gene therapy of hemophilia B: sustained partial correction in factor IX-deficient dogs, *Science*, 262, 117, 1993.
25. Weiss, R., The search for human RNA tumor viruses, in *RNA Tumor Viruses*, Weiss, R., Teich, N., Varmus, H., and Coffin, J., Eds., Cold Spring Harbor Laboratory, 1984, 1205.
26. Welsh, R. M., Cooper, N. R., Jensen, F. C., and Oldstone, M. B., *Nature*, 257, 612, 1975.
27. Jaenisch, R. and Hoffmann, E., Transcription of endogenous C-type viruses in resting and proliferating tissues of BALB/Mo mice, *Virology*, 98, 289, 1979.
28. Jaenisch, R., Fan, H., and Croker, B., Infection of preimplantation mouse embryos and of newborn mice with leukemia virus: tissue distribution of viral DNA and RNA and leukemogenesis in the adult animal, *Proc. Natl. Acad. Sci. U.S.A.*, 72, 4008, 1975.
29. Albritton, L. M., Tseng, L., Scadden, D., and Cunningham, J. M., A putative murine ecotropic retrovirus receptor gene encodes a multiple membrane-spanning protein and confers susceptibility to virus infection, *Cell*, 57, 659, 1989.
30. Albritton, L. M., Kim, J. W., Tseng, L., and Cunningham, J. M., Envelope-binding domain the cationic amino acid transporter determines the host range of ecotropic murine retroviruses, *J. Virol.*, 67, 2091, 1993.
31. Adams, R. M., Wang, M., Steffen, D., and Ledley, F. D., Replication defective adenovirus enables transduction by retroviral vectors of cells outside of their host range, submitted.
32. Stratford-Perricaudet, L. D., Levrero, M., Chasse, J. F., Perricaudet, M., and Briand, P., Evaluation of the transfer and expression in mice of an enzyme-encoding gene using a human adenovirus vector, *Hum. Gene Ther.*, 1, 241, 1990.
33. Jaffe, H. A., Danel, C., Longenecker, G., Metzger, M., Setoguchi, Y., Rosenfeld, M. A., Gant, T. W., Thorgeirssen, S. S., Stratford-Perricaudet, L. D., Pericaudet, M., Pavirani, A., Lecocq, J. P., and Crystal, R. G., Adenovirus-mediated in vivo gene transfer and expression in normal rat liver, *Nat. Genet.*, 1, 372, 1992.
34. Li, Q., Kay, M. A., Finegold, M., Stratford-Perricaudet, L. D., and Woo, S. L. C., Assessment of recombinant adenoviral vectors for hepatic gene therapy, *Hum. Gene Ther.*, 4, 403, 1993.
35. Wu, G. Y., Wilson, J. M., and Wu, C. H., Targeting genes: delivery and persistent expression of a foreign gene driven by mammalian regulatory elements in vivo, *J. Biol. Chem.*, 264, 16985, 1989.
36. Wu, G. Y. and Wu, C. H., Receptor-mediated gene delivery and expression in vivo, *J. Biol. Chem.*, 263, 14621, 1988.

37. Wu, G. Y. and Wu, C. H., Evidence for targeted gene delivery to hepG2 hepatoma cells in vitro, *Biochemistry*, 27, 887, 1988.
38. Wu, G. Y. and Wu, C. H., Receptor-mediated in vitro gene transformation by a soluble DNA carrier system, *J. Biol. Chem.*, 262, 4429, 1987.
39. Wu, G. Y., Wilson, J. M., Shalaby, F., Grossman, M., Shafritz, D. A., and Wu, C. H., Receptor-mediated gene delivery in vivo. Partial correction of genetic analbuminemia in Nagase rats, *J. Biol. Chem.*, 266, 14338, 1991.
40. Wilson, J. M., Wu, C. H., and Wu, G. Y., Targeting genes: delivery and persistent expression of a foreign gene driven by mammalian regulatory elements in vivo, *J. Biol. Chem.*, 264, 16985, 1989.
41. Wilson, J. M., Grossman, M., Cabrera, J. A., Wu, C. H., and Wu, G. Y., A novel mechanism for achieving transgene persistence in vivo after somatic gene transfer into hepatocytes, *J. Biol. Chem.*, 267, 11483, 1992.
42. Stankovics, J., Andrews, E., Wu, G., Wu, C. T., and Ledley, F. D., Overexpression of human methylmalonyl CoA mutase (MCM) in mouse liver after *in vivo* gene delivery using asialoglycoprotein complexes, *Hum. Gene Ther.*, in press.
43. Haensler, J. and Szoka, F. C., Jr., Synthesis and characterization of a trigalactosylated bisacridine compound to target DNA to hepatocytes, *Bioconj. Chem.*, 4, 85, 1993.
44. Yovandich, J., Stankovics, J., and Ledley, F. D., Receptor mediated gene transfer using a DNA-binding ligand, unpublished data.
45. Zhou, X. and Huang, L., Targeted delivery of DNA by liposomes and polymers, *J. Control. Rel.*, 19, 269, 1991.
46. Legendre, J. Y. and Szoka, F. C., Jr., Delivery of plasmid DNA into mammalian cell lines using pH sensitive liposomes: comparison with cationic liposomes, *Pharmaceutical Res.*, 10, 1235, 1991.
47. Zhu, N., Liggitt, D., Liu, Y., and Debs, R., Systemic gene expression after intravenous DNA delivery into adult mice, *Science*, 261, 209, 1993.
48. Legendre, J. Y. and Szoka, F. C., Jr., Cyclic amphipathic peptide-DNA complexes mediate high-efficiency transfection of adherent mammalian cells, *Proc. Natl. Acad. Sci. U.S.A.*, 90, 893, 1993.
49. Tomlinson, E. T., Theory and practice of site-specific drug delivery, *Adv. Drug Del. Rev.*, 1, 87, 1987.
50. Wagner, E., Plank, C., Zatloukal, K., Cotten, M., and Birnstiel, M. L., Influenza virus hemagglutinin HA-2 N-terminal fusogenic peptides augment gene transfer by transferrin-polylysine-DNA complexes: toward a synthetic virus-like gene-transfer vehicle, *Proc. Natl. Acad. Sci. U.S.A.*, 89, 7934, 1992.
51. Wagner, E., Zatloukal, K., Cotten, M., Kirlappos, H., Mechtler, K., Curiel, D. T., and Birnstiel, M. L., Coupling of adenovirus to transferrin-polylysine/DNA complexes greatly enhances receptor-mediated gene delivery and expression of transfected genes, *Proc. Natl. Acad. Sci. U.S.A.*, 89, 6099, 1992.
52. Curiel, D. T., Agarwal, S., Wagner, E., and Cotten, M., Adenovirus enhancement of transferrin-polylysine-mediated gene delivery, *Proc. Natl. Acad. Sci. U.S.A.*, 88, 8850, 1991.
53. Cristiano, R. J., Smith, L. C., and Woo, S. L. C., Hepatic Gene Therapy: adenovirus enhancement of receptor-mediated gene delivery and expression in primary hepatocytes, *Proc. Natl. Acad. Sci. U.S.A.*, 90, 2122, 1993.
54. Cristiano, R. J., Smith, L. C., Kay, M. A., Brinkley, B. R., and Woo, S. L., Hepatic gene therapy: efficient gene delivery and expression in primary hepatocytes utilizing a conjugated adenovirus-DNA complex, *Proc. Natl. Acad. Sci. U.S.A.*, 90, 11548, 1993.

55. Ledley, F. D., Brody, B., Kozinetz, C., and Mize, M., The challenge of follow-up for clinical trials of somatic gene therapy, *Hum. Gene Ther.*, 3, 657, 1992.
56. Ledley, T. S. and Ledley, F. D., A multicompartment, numerical model of the pharmacokinetics of gene medicines, *Hum. Gene Ther.*, 5, 679, 1994.
57. Gupta, S. and Chowdhury, R. J., Hepatocyte transplantation: back to the future, *Hepatology*, 15, 156, 1992.
58. Ledley, F. D., Clinical considerations in the design of protocols for somatic gene therapy, *Hum. Gene Ther.*, 2, 77, 1991.
59. Ledley, F. D., Issues in the design of clinical trials of somatic gene therapy, *Proc. N.Y. Acad. Sci.*, 716, 283, 1994.

6 Epidermal Keratinocytes: Opportunities and Applications in Somatic Cell Gene Therapy

Brian J. Aneskievich, Anthony G. Letai, and Elaine Fuchs

Recently, two fields of intense research efforts, namely gene transfer and epidermal keratinocyte biology, have been united to provide a new avenue of somatic cell gene therapy. In this review, we will examine the choice of epidermal keratinocytes as recipient cells in gene therapy and their application in current research directed toward nonepidermal genetic diseases. Additional approaches for gene therapy utilizing keratinocytes in treating genetically-based epidermal diseases will also be discussed.

I. EPIDERMAL KERATINOCYTES: TARGET CELLS FOR GENE THERAPY

A. In Vivo Characteristics

Mammalian skin is separated histologically into two compartments.[1] First is an underlying dermis composed of extracellular matrix proteins studded with fibroblasts and containing the supporting vasculature. Separating the two compartments is a basement membrane, and above this is the surface or epidermal component composed largely (>95%) of keratinocytes.[1] Although epidermal thickness varies depending on body site, its overall organization of histologically and biochemically distinct layers (Figure 1) is maintained.[1]

Each layer of the epidermis expresses characteristic keratin and nonkeratin proteins,[2,3] whose expression is coordinately linked to differentiation.[4,5] For example, transcription of keratin 5 (K5) and keratin 14 (K14) is largely restricted to the innermost basal layer. Additionally, mitotic activity is confined to the innermost layer, apposed to the basement membrane.[4,5] This renewing population of basal keratinocytes is the ultimate target of gene delivery. In response to as yet incompletely defined cues, basal keratinocytes cease replication and undergo a strictly regulated program of terminal

0-8493-2440-8/95/$0.00+$.50

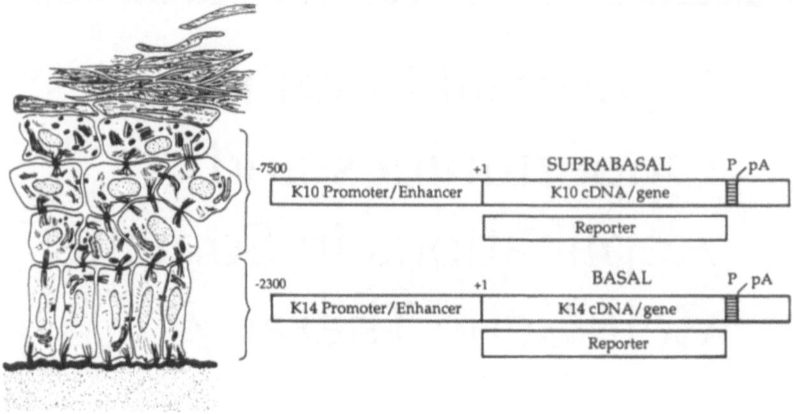

FIGURE 1. Targeting gene expression to the epidermis. This schematic representation of a histologic section of epidermis shows the layered keratinocytes resting on a basement membrane (heavy black line), beneath which is the dermis and supporting vasculature (not shown). Optimal expression of therapeutic genes in the epidermis may come from the use of differentiation-specific promoters such as K14 or K10. To this end, reporter genes or the keratin cDNAs themselves with specific tags for immunologic recognition (P, neurosubstance P) have been expressed in the epidermis of transgenic mice. Promoter size is expressed in base pairs starting at the transcription initiation site and is not drawn to scale. pA, polyadenylation signal.

differentiation involving characteristic morphologic and biochemical changes. For instance, K5 and K14 transcription is down-regulated, and a new suprabasal-specific keratin pair, K1 and K10, is transcribed abundantly (Figure 1). While transitting to the skin surface, keratinocytes remain transcriptionally and metabolically active through the spinous and granular layers. Typifying this is the expression of an intermediate-filament-associated protein (IFAP), filaggrin, implicated in the formation of keratohyalin granules and keratin bundling. The uppermost or cornified layer is physiologically inert but structurally important for the mechanical and chemical resiliency of the skin. This last layer is shed to the environment. Both the predictable gene expression schedule and the high metabolic activity of keratinocytes can be exploited for gene therapy.

The selection of epidermal keratinocytes as the cellular vehicle for gene therapy is supported when certain salient features of these cells and their tissue of origin are considered. One of their foremost advantages is body site location. Skin keratinocytes offer easy accessibility for initial cell acquisition for culture and subsequent return to the patient. These are distinct benefits as compared to other cells, such as hepatocytes, myoblasts, and hematopoietic cells, that are often considered for gene therapy manipulations (for review see References 6 and 7). After transplantation to the patient, the surface location offers the ongoing asset of easy surveillance to monitor gene product synthesis and the health and clinical progression of the graft itself.

B. Keratinocyte Culture Systems

Before a therapeutic benefit from gene insertion may be realized, the cell of choice must be amenable to considerable laboratory and clinical demands. In this respect, there is substantial experience regarding *in vitro* growth and expansion of epidermal keratinocytes and their clinical application. The success of such previous endeavors is in part due to well-characterized tissue culture systems. Basic research interests in the balance and control of proliferation vs. differentiation have spurred the development of *in vitro* conditions favoring populations of either replicating, basal-like cells or mixed cultures with some features of terminal differentiation. Such versatility is of paramount importance when technical requirements vary from expansion of cell clones positive for the therapeutic gene to *in vitro* generation of an epithelium with sufficient structural integrity for transplant.

In general, two distinct culture systems have been established for *in vitro* growth of keratinocytes. The first was developed by Rheinwald and Green and is characterized by cocultivation of human epidermal keratinocytes and mitotically-inactivated fibroblasts.[8,9] Cell division in the fibroblasts is halted by pretreatment with either gamma irradiation or DNA synthesis-inhibiting agents such as mitomycin C. Although the culture system was originally developed with an established line of mouse fibroblasts, human dermal fibroblasts may be substituted.[10,11] Fibroblasts improve keratinocyte plating efficiency and overall growth presumably because they contribute soluble growth factors and extracellular matrix proteins to the culture environment. If desired, the fibroblasts may be removed selectively from the culture by treatment with EDTA. Culture performance may be further augmented by addition of purified growth factors (epidermal growth factor, keratinocyte growth factor), hormones (insulin, triiodothyronine), and trace elements (selenium) to the serum-containing media.[12] This system permits the long-term serial passage of epidermal keratinocytes necessary for generating large numbers of epithelial cells and a correspondingly large surface area. Under optimal conditions, a 1-cm^2 biopsy may be expected to produce one square meter of cultured epithelia in about 3 weeks.[13] This prolific ability, though tempered by the quality of the surgical specimen and donor age, is critical to provide sufficient numbers of keratinocytes following genetic manipulations.

An alternative culture regimen is available.[14,15] Ham and co-workers have developed serum-free, low-calcium media for the culture of human keratinocytes in the absence of fibroblast feeder cells. The lack of fibroblast feeder cells and serum proteins facilitates preparation of keratinocytes grown *in vitro* for transplant. It should be noted that long-term serial cultivation does not seem as efficient in this system as in the Rheinwald and Green method, although it may be improved by the addition of bovine pituitary extract. However, for applications where chemically defined media are important, this method may be preferred.

Finally, *in vitro* systems have been developed that allow for better recapitulation of *in vivo* skin differentiation characteristics.[16-19] Unlike the above methods where keratinocytes are grown on tissue-culture plastic submerged in media, culturing keratinocytes at the air-liquid interface and supporting them by various dermal substitutes triggers a differentiation cascade more reminiscent of native epidermis. Such techniques would produce a more resilient epidermis for transplant but would require the use of nonimmunogenic synthetic matrix polymers and the patient's dermal fibroblasts to avoid potential problems with graft rejection. Some work with such artificial matrices, although not of a transplant-suitable nature, has been reported for epidermal keratinocytes.[20]

Practical application of epithelium generated *in vitro* on plastic has already been realized in treatment of patients suffering from large-scale burns and chronic skin ulcers.[21-23] Preparation of the epidermal graft involves growth of the keratinocyte culture to confluence (removal of the fibroblast feeder cells, if used) and exhaustive rinsing in serum-free, growth factor-free media. The epithelium is then removed from the culture vessel by the use of dispase, a proteolytic enzyme that degrades basilar contacts with the underlying matrix proteins, leaving the cell-cell junctions unaffected and thus the epidermal sheet intact. Additional rinsing removes the enzyme. Surgical gauze is applied to the apical surface to support the epidermal sheet and to aid in application to the patient. Patient studies over a 5-year period have revealed a normal, healthy morphology in regions that have received a cultured epidermal cell autograft.[21]

In principle, keratinocyte-mediated gene therapy could utilize many of the same techniques that have already been applied to burn treatment. Since diagnoses of certain epidermal (and nonepidermal) genetic diseases can be made in a newborn child or even in a fetus,[24] gene therapy could be initiated at an extremely early age. A young patient would provide keratinocytes with a greater generative capacity, a feature that could be useful when extreme expansion of a keratinocyte culture is required. In the case of epidermal diseases, he/she would also have a smaller surface area of pathologic epidermis to replace. Indeed, youth of the patient is likely to be an important factor in the feasibility of any gene therapy using somatic cells. Below we discuss recent studies that explore the feasibility of gene therapy using keratinocytes as a model system.

II. TREATMENT OF NONEPIDERMAL GENETIC DISEASE WITH KERATINOCYTES

A. Experimental Systems Suggest In Vivo Success

Given that gene delivery or uptake processes are dependent on DNA synthesis and/or mitotic activity of the recipient cell,[7] the use of appropriately cycling or synchronized keratinocytes could greatly improve overall gene transfer efficiency. Extensive studies have been performed detailing the biochemical and proliferative nature of epidermal cells.[2,25,26] Both kinetic and clonogenic variation have been described in the replicating population of cultured keratinocytes.[27,28] Barrandon and Green have identified three such groups.[28,29] Keratinocyte paraclones withdraw from the cell cycle after a very limited number of divisions. Meroclones have mixed features and may represent a transition phase between a proliferative phase and a terminally differentiated phenotype. Holoclones produce rapidly growing colonies where proliferation is vastly favored over differentiation. After ~23 doublings, these cells still transfer well, with less than 5% of the population converting to a poorly dividing, terminally-differentiated phenotype. The production of single cell-derived cultures with such differing proliferating potential lends credence to the proposal that a stem cell compartment exists in the epidermis.[30,31] Presumably, in culture, this population of cells would be close to or one step removed from the holoclones. Similarly, the meroclones might be a transit-amplifying compartment that would go through limited cell cycles to produce mostly differentiated cells.

In vitro manipulations required for successful gene therapy make it preferable to isolate keratinocytes with a high proliferative potential. Building on earlier studies,[32,33] a recent report has demonstrated that stem cell-like cultured keratinocytes with high proliferative potential could be isolated based on expression of integrins[34] on their cell surface. Specifically, those keratinocytes with a high expression of the β1 integrin subunit were most efficient at forming rapidly growing colonies reminiscent of holoclones.[34] Using fluorescence-activated cell sorting and an anti-β1 integrin antibody, it might be possible to develop a population of keratinocytes more amenable to the rigors of selection and expansion. Enrichment for holoclones could be done either directly from the skin biopsy, soon after culturing, or when DNA recombinant clones are being functionally screened.

While these advances hold promise for augmenting general keratinocyte culture, certain questions specific to epidermis-based gene therapy remain to be examined. One immediate concern is the efficiency of DNA uptake via conventional transfection methods. Extensive growth potential will be of little value if holoclones prove refractory to gene transfer. This said, DNA transfection methods rely on mitotic activity for efficient integration of a transgene into chromosomal DNA (see Reference 7). In this regard, holoclones should have an advantage since they are highly proliferative in culture, a feature atypical of the stem cells of most cell types.[35-38] Thus, while further studies will be necessary to assess the extent to which holoclones will be the optimal keratinocyte for epidermis-based gene transfer technology, there is every indication to suggest that these cells will be suitable for this purpose.

B. Selection and Delivery of Therapeutic Genes

Several criteria will guide an informed selection of which genetic disease(s) might most appropriately be treated by keratinocyte-based gene therapy. Diseases where the levels of circulating factors need to be completely or partially compensated are particularly attractive candidates. Insulin-dependent diabetes or clotting disorders like hemophilia A (factor VII) and hemophilia B (factor IX), now commonly treated by hypodermic injection of animal-derived or bacterially-produced recombinant protein, might be controlled alternatively through production of the relevant protein by grafted epidermal keratinocytes genetically engineered to produce the needed factor.

A crucial prerequisite in determining whether keratinocytes might be suitable for gene therapy of systemic substances is verifying that keratinocytes can secrete factors that can transit the basement membrane and find their way to the bloodstream. The first demonstration of an exogenous gene expressed by primary epidermal keratinocytes for the purpose of gene therapy of systemic agents was reported by Morgan and coworkers,[13] who used retroviral vectors to confer expression of human growth hormone (HGH). In culture, biologically active hormone of the appropriate molecular mass was found secreted into the media, and these transduced cells were capable of reforming a histologically normal stratified squamous epithelium upon grafting to athymic mice.

Although hormone could be detected in engrafted epidermis, the question of systemic delivery was not resolved initially. Subsequent studies aimed at improving serum levels of the transgene product have focused on alternative grafting techniques[39,40] and promoter choice for the transgene.[41] Importantly, it was discovered that efficient expression in culture is not necessarily predictive of expression in the grafted epidermis: secreted HGH was undetectable by 4 weeks after transplant of cells bearing a herpes simplex virus (HSV) thymidine kinase (tk) promoter-expressed construct. In contrast, hormone could be detected from expression derived via the metallothionein promoter.[41] Thus, selective promoter function could provide insights toward keratinocyte gene expression and differentiation processes.

The plausibility of systemic gene therapy via recombinant skin keratinocytes has also been addressed by Taichman and co-workers. These researchers grafted keratinocytes onto a mouse and determined the degree to which a naturally secreted keratinocyte product, apolipoprotein E (apoE),[42] makes its way to the blood stream. A species-specific antibody to human apoE confirmed that athymic mice bearing grafts of cultured human epidermal primary keratinocytes delivered apoE (33,000 kD) to the mouse's circulation soon after application of the graft (about 4 days) and then for as long as the graft was present (at least 12 weeks).[43] These results demonstrate unequivocally that systemic delivery of certain factors can continue after the reconstitution of an intact basement membrane.

Subsequently, a keratinocyte cell line bearing an introduced apoE gene construct expressed via a retroviral long-terminal repeat (LTR) has been tested for systemic delivery by intradermal injection of the transduced cells.[44] Human apoE was detected in sera from mice injected with the transduced but not parental line.

One recent study has focused on the use of keratinocytes for gene therapy of hemophilia B.[45] Primary epidermal cells successfully transduced to express factor IX in culture were transplanted to athymic mice using an alternative grafting protocol that includes preparation of a collagen bed[40] at the graft site. Although human factor IX was delivered to the mouse circulatory system, one problem still to be overcome is the relatively limited amount of factor IX detected in the sera. Long-term studies regarding continued promoter activity also remain to be conducted. Nevertheless, such efforts serve to advance the reality of somatic cell gene therapy via epidermal keratinocytes.

In all of these examples, selection of keratinocytes positive for gene transfer relied on selection with the aminoglycoside G418. However at least two of these reports[13,41] raised the concern that the proliferative nature of the keratinocytes was negatively affected by exposure to the drug. In this respect it is noteworthy that two selectable resistance markers, neomycin phosphotransferase (*neo*) and histidinol dehydrogenase (*hisD*), have been compared in keratinocytes.[46] In these studies, canine epidermal keratinocytes were transduced with the human gene for adenosine deaminase, which encodes a protein of potential use in genetic therapy of severe combined immunodeficiency.[46] Selection with neomycin produced colonies with a reduced proliferative potential and morphologic changes suggestive of the onset of terminal differentiation. Such negative features were not detected in the histidinol-selected cultures and may reflect the innate drug-sensitivity differences in holoclone and paraclone cells where the same level of drug resistance does not equally protect all cells. Given the demands placed on mitotic potential of positively selected cells, further examination of this and other selectable markers should be pursued.

Thus far, most keratinocyte gene therapy studies have utilized retroviruses for gene delivery despite their inherent short-comings with promoter-silencing (see below) and possible contamination with replication-competent virus.[7,47] This has been especially true of work done with primary keratinocytes,[13,45] due to their relatively low efficiency of transfection. In fact, Teumer et al.[41] report that they were unsuccessful in selecting for stable gene transfer in primary keratinocyte cultures following standard calcium phosphate-plasmid DNA transfections, although others[48] have reported success in transient transfection assays. With the recent introduction of improved methods, such as lipid and lipid-polyamine-mediated DNA delivery,[49-51] dependency on retroviral transduction may no longer be necessary.

Electroporation may also provide a superior method for delivery of biologic macromolecules.[52,53] This method has been used with success for both transient and stable gene expression.[54,55] Unlike calcium phosphate and liposomal transfection methods that can result in multiple integrations of the construct, clones derived from electroporated cells usually have a single copy of the transgene,[53,54] a desired trait for the purposes of gene therapy via homologous recombination (see below). Electroporation can allow efficient colony formation (>1/1000 cells survive the electroporation shock),[56] although such efficiency depends on the construct, locus, and cell line. As there is little information in the literature regarding the efficient transformation of keratinocytes by electroporation, development of this protocol would be an early priority in epidermal-specific gene therapy. While a large number of clones may need to be screened no matter what the DNA delivery technique, the final selection of positive clones can be considerably facilitated by the design of unambiguous PCR and/or Southern assays.

C. Promoter Choices for Epidermal Keratinocyte-Based Gene Therapy

Another issue important for gene therapy entails the selection of a gene promoter to drive expression of therapeutic agents. Although studies are limited, there is some suspicion that foreign promoters, such as the virally-derived sequences HSV tk and retroviral LTR, may be subject to a progressive decay or complete loss of expression after epithelium transfer to an animal host (see above).[13,41] Alternatives such as retroviral vectors containing secondary promoters[46] remain as yet poorly developed. In addition, it may be valuable to explore the possibility of using promoters based on ubiquitously expressed genes, such as those responsible for actin or phosphoglycerate kinase expression.

The promoters best-suited to epidermis-based gene therapy are likely to be those directing tissue-specific proteins, particularly keratins. In this respect, it is noteworthy that basal keratinocytes transcribe at high rates[57] the genes encoding K5 and K14, whose products account for approximately 30% of the total protein of these cells.[2,58] To date, differentiation specificity in transgenic mouse epidermis has been achieved by capitalizing on basal, K5 and K14,[59,60] vs. suprabasal, K1 and K10,[61-63] keratin promoters. This may prove beneficial should it be desirable to restrict expression of the therapeutic transgene to either mitotic or postmitotic cell populations. The approach has already been successful in designing transgenic mouse models to study epidermal disease[64-67] and the role of growth factors and growth regulators[61,63,68-71] in epidermal growth and differentiation. Although the mode of gene delivery for transgenic mice is quite different from keratinocyte transfections or transductions, long-term expression from these native promoters has been achieved in transgenic mouse epidermis.

Of particular interest to the matter at hand is the direct demonstration in at least one study of transgene product present both locally in the epidermis and distant in the serum.[68]

This example, and that cited above for apoE, of systemic delivery of a transgene product (range ~17,000 to 33,000 mol wt) from site of production in the skin further strengthens the proposed use of epidermal keratinocytes as secretory cells for use in gene therapy. Moreover, it underscores the potentially powerful use of the human K5 and K14 gene promoters as effective agents to drive the expression of foreign genes in the epidermis.

III. EPIDERMAL-SPECIFIC DISEASE: TREATMENT VIA HOMOLOGOUS RECOMBINATION

A. Candidate Epidermal Diseases

While transfected keratinocytes have a wide range of potential uses, from drug delivery to the treatment of recessive epidermal diseases, there are special problems to address when treatment of autosomal dominant diseases is considered. In the treatment of such diseases, selective suppression of the mutant gene product might be desirable, particularly if one copy of the wild-type gene can compensate to supply 100% of the wild-type product required. The healthy corrected keratinocytes would then be expanded to produce an epithelium suitable for transplantation back to the patient's skin in a manner analogous to the treatment of burns with cultured autografts.[22,23,72] In the ideal situation, all of the patient's affected epidermis could be replaced by genetically modified autografts, either directly by a series of transplants, or indirectly by *in situ* expansion of the modified autografts following transplantation. The latter situation would be possible as long the autograft would have a selective advantage over the pathologic skin. This strategy presents two main technical problems. First, one must isolate the successfully modified clonal populations of keratinocytes in which the pathologic conditions have been eliminated. Second, these clonal populations must possess sufficient proliferative potential to provide grafts large enough to cover affected parts of the patient's body. These concerns are parallel to those of producing a graft large enough for sufficient systemic delivery of a soluble therapeutic gene product (see above).

To outline an approach for genetic therapy of an autosomal dominant epidermal disease, we will consider the skin disorders epidermolysis bullosa simplex (EBS) and epidermolytic hyperkeratosis (EH). These diseases are characterized by skin blistering occurring upon mild physical stress or trauma. The physical defect occurs as a result of keratinocyte lysis.[73-76] In the case of EBS, the lysing cells are in the basal layer, while for EH, the involved cells are primarily in the suprabasal layers. These diseases have as their genetic bases single point mutations in the epidermal keratin genes (for review see References 77 and 78). For EBS, the mutations are in the genes encoding the basally expressed keratins, K5 and K14 (EBS),[64,65,79-81] whereas for EH, mutations occur in the suprabasally expressed keratins, K1 and K10.[67,82-84]

The keratin network disruption typical of these diseases is duplicated *in vitro* by expression of deletion or point mutant keratins in cultured keratinocytes[85-87] and in transgenic mice.[62,65,66] Thus, it seems likely that if the mutant keratin allele were exchanged for a wild-type version, a normal keratin network phenotype, and thus normal function, could be restored to the keratinocyte. This might be accomplished either by (1) knocking out the mutant gene using homologous recombination, (2) replacing the mutant gene with a wild-type allele using homologous recombination, or (3) using an anti-sense vector which could selectively suppress translation of the mutant, but not the wild-type, keratin transcript.

A priori, the ablation of a K14 or K5 allele in correcting an EBS keratinocyte defect might be expected to pose a gene dosage problem. However, epidermal keratins are remarkably stable in the presence of their partners ($t_{1/2}$ >48 h), while they are rapidly turned over in the absence of their partner ($t_{1/2}$ <30 min; R. Lersch, Ph.D. thesis, University of Chicago, 1991). Thus, it seems likely that a single wild-type allele would be capable of providing enough of one keratin to pair with the keratin expressed from the two wild-type alleles of its partner.

These potential techniques of gene therapy described below may be generally applicable to other epidermal disorders that are genetic in origin; however, EBS and EH may be ideal targets for such therapies, given that there is a natural selection for the successfully treated keratinocyte due to their increased resistance to cytolysis compared to the untreated diseased keratinocytes.

B. Gene Targeting: Designing Knockout Vectors and Protocols

Since the first demonstration of targeted homologous recombination in a mammalian genome,[88] considerable work has been performed in an effort to increase the efficiency of such a recombination event. Optimal vectors for gene ablation contain genomic DNA sequence encompassing the promoter and the beginning portions of the relevant gene. This gene segment is interrupted by a selectable marker, such as neomycin resistance, allowing for positive selection of cells that have integrated the construct into their genome. Often, a negatively selectable gene, such as the HSV tk gene (selected against by ganciclovir) or the bacterial gene for diphtheria toxin[89-91] (which requires no drug treatment to cause cell death), is placed outside the region of homology. If the construct is integrated via a desirable homologous recombination event, these sequences outside the regions of homology will not be incorporated into the chromosome, and thus will not affect selection (Figure 2A). If the construct is incorporated by an undesirable nonhomologous recombination event, however, the negatively selectable gene will likely be incorporated into the chromosomal DNA (Figure 2B) and cause cell death when expressed in the appropriate culture conditions.[92,93] It is important that the positive and negative selection be separated by the shorter arm of homology to maximize the chances that both genes will be simultaneously incorporated into the genome in a nonhomologous recombination event, a requirement for enrichment of homologous recombination events by negative selection. Estimates for enrichment obtained from using a negatively selectable marker vary from 3-fold[94] to 2000-fold.[95] Depending on the strategies used and alleles and cell lines involved, efficiencies of homologous recombination in stably transfected mammalian cells range from approximately 10^{-5} to $1–5 \times 10^{-2}$.

Homologous recombination vectors have been the construct of choice for the creation of null alleles.[96-100] Recombination events are genetically stable and can be used

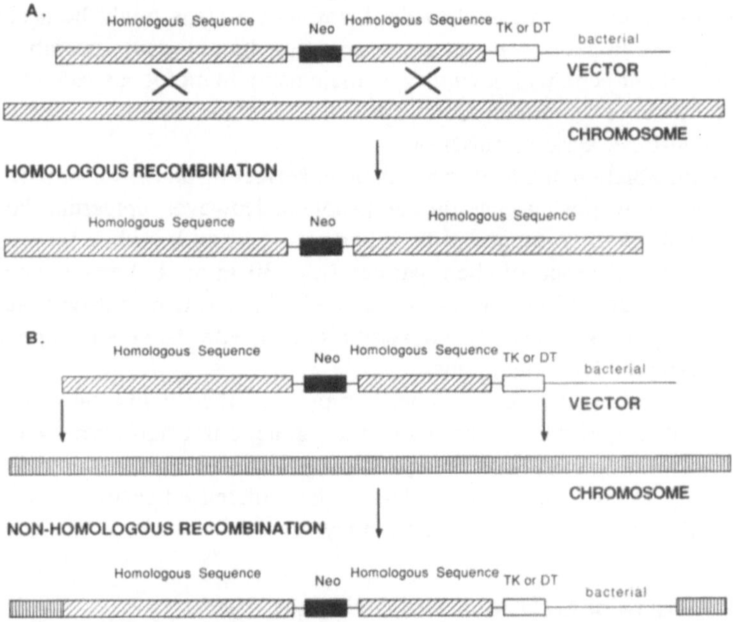

FIGURE 2. A schematic representation of homologous (A) and nonhomologous (B) recombination events using a construct known as a replacement vector. (A) Homologous recombination by a double reciprocal crossover event. Hatched boxes represent chromosomal sequence of the relevant gene, contained either in the chromosome or in the replacement vector; black boxes represent the Neo gene; gray box represents the negatively selectable element (TK: herpes simplex virus thymidine kinase; DT: diphtheria toxin genes), which is not incorporated into the chromosome in this homologous recombination event; thin line represents vector bacterial sequences. (B) Nonhomologous recombination. Hatched boxes represent chromosomal sequence of the relevant gene; vertically hatched boxes represent chromosomal sequence unrelated to and distant from the gene of interest; black boxes represent the Neo gene; gray box represents the negatively selectable element (TK of DT genes), which is incorporated into the chromosome by this nonhomologous recombination event; thin line represents vector bacterial sequences. Note that treatment with G418 and ganciclovir (if TK used) or with G418 alone (if DT used) will select for cells containing the final chromosomal arrangement in (A) but will select against cells containing the final chromosomal arrangement in (B).

either to ablate a mutant gene, if the *neo* gene is inserted as described above, or to replace a mutant coding sequence with wild-type sequence, if the *neo* gene is placed outside the coding region.[101]

There are many parameters to be considered in designing a replacement vector, each of which affects the efficiency of homologous recombination. The efficiency of gene targeting can depend critically upon length of homology used,[98,102-104] with a 100-fold increase in targeting efficiency observed when the total length of homology was increased from 2.8 to 14.6 kb.[102] However, this parameter can often be limited by practical considerations, such as the amount of genomic sequence available and the presence of convenient restriction sites. Particular to skin-specific genetic disease, ample genomic sequence information for certain relevant genes, such as those that encode the epidermal keratins, is already available either as complete nucleotide sequence or detailed restriction maps.[105-109] This information makes it an accessible candidate for development of *in vitro* gene therapy protocols.

Another consideration is whether to use isogenic or nonisogenic DNA in the targeting construct. In mice, constructs containing isogenic DNA are 5 to 20 times more efficient at gene targeting than those containing nonisogenic DNA.[100,102] However, for gene therapy of pathologic keratinocytes, the use of isogenic DNA constructs would require cloning the relevant genes from each individual patient. Hence, it would likely save time, materials, and labor if nonisogenic DNA could be used to provide sufficient efficiency of homologous recombination to be feasible for such types of gene therapy.

IV. PREPARATION OF GENETICALLY-ALTERED KERATINOCYTE CLONES FOR TRANSFER

A. Expansion of the Clonal Population

Subsequent to the isolation of clones of recombinant keratinocytes, adequate numbers of cells must be generated for either sufficient delivery of the systemic gene product or suitable area for replacement of the diseased epidermis. In the case of a genetic therapy by homologous recombination, this could pose difficulties given the low frequency of such recombination events. Moreover, positive selection methods may not be desirable, since they leave a foreign gene in the human genome. In the case of EBS and EH, this caveat could be circumvented, since there should be a natural selective advantage for the survival of the corrected keratinocyte.

Since 3×10^6 epidermal cells can be obtained from 1 cm^2 biopsy,[72] and since the surface area of the body of an adult is approximately 2 m^2, we estimate that there are approximately 10^{11} epidermal cells in the body. Since the epidermis is at least five cell layers thick, $\sim 2 \times 10^{10}$ basal cells from a monolayer culture would be required to cover the body. Therefore, if one starts with one to ten genetically targeted cells, an expansion of at least 10^9-fold would be required for replacement of the entire epidermis. This corresponds to approximately 30 doublings. Is there sufficient proliferative potential in a keratinocyte stem cell to accomplish these doublings *in vitro*? Recent studies suggest that there is. Gallico and co-workers[22] have reported that expansion beyond 20 doublings of epidermal cultures from burn patients "is not limited by the performance of the culture system." In our own laboratory, epidermal cells from newborn foreskin may be cultured through >100 doublings. Thus, it seems likely that the expansion required for the repopulation of the epidermis by genetically altered cells may be possible under optimal culture and experimental conditions. If so, homologous recombination as a means of gene therapy may be feasible in the keratinocyte.

B. Grafting In Vitro-Generated Epithelia

It is not known how long a greatly expanded clonal population could survive as an autograft, since it is possible that most of the reproductive capacity of the population might have been used up in culture. The expansion of a small number of clones, such as would be isolated for homologous recombination-based gene therapy, would require 10^5 to 10^7 greater amplification. This may be especially difficult for keratinocytes cultured from an older patient. This said, it is notable that burn victims with >85% of their surface epidermis obliterated can still be rescued by expansion and grafting of keratinocytes taken from the few regions of unaffected skin. Thus, the remarkable proliferative capacity of the cultured keratinocyte lends hope for the prospects of methods in gene therapy that might be unthinkable for other cell types.

It is important to emphasize that, *in vivo*, there should be a major selective advantage of corrected keratinocytes over the more fragile original EBS or EH keratinocytes. Thus, it may be possible for grafted keratinocytes to spread considerably beyond the margins of their graft. Evidence for the selective advantage of wild-type keratin-expressing keratinocytes over mutant keratin-expressing keratinocytes comes from a transgenic mouse model. In transgenic mice with keratinocytes expressing a mutant keratin in a mosaic distribution, the phenotype of the mice improved over time, and there was eventually little evidence of mutant protein in the skin (Vassar and Fuchs, unpublished results). If multiple clones could therefore be seeded initially at several sites around the body, it is hoped that this serial grafting method might allow for the repopulation of the epidermis with keratinocytes containing normal keratin networks. Even if the serial grafting were unsuccessful, it is hoped that the primary grafts would provide relief from the disease in the most severely affected parts of the body.

A final note worthy of mention is the possibility that in the next decade, problems of graft-vs.-host rejection will be overcome. If so, then it would be possible to graft normal keratinocytes from a healthy individual onto a patient with EBS or EH. This would circumvent the need for homologous recombination of EBS or EH keratinocytes and greatly improve the prospects for keratinocyte-based therapy of these diseases.

V. CONCLUSIONS AND FUTURE PROSPECTS

Two recent reviews have succinctly related both the great promise and great hurdles remaining in the clinical application of somatic cell gene therapy. Anderson[6] cites the encouraging results regarding the treatment of severe combined immunodeficiency disease by expression of a recombinant adenosine deaminase gene via retroviral transfer. Although representing an extremely limited number of cases, it nevertheless tests the utility of treatment of genetic disease by gene transfer. As presented by Mulligan,[7] therein lies the next important challenge: the melding of advances in molecular biology and *in vitro* cell manipulations with the demands of clinical medicine. While gene therapy using genetically engineered keratinocytes has not yet advanced to a clinical trial stage, the *in vitro* accomplishments of gene delivery to epidermal keratinocytes with their subsequent transplant to animal hosts suggest a real possibility of success in human patients. Regarding the transfer of cultured keratinocytes to the patient, much of the relevant groundwork has already been established through the use of autologous grafts of cultured keratinocytes for burn therapy. While more work is required to assure optimal and prolonged expression of transgenes, these are problems not necessarily unique to keratinocyte-based therapies. Advances in other cell systems may prove applicable to keratinocyte gene therapy. Finally, although gene therapy protocols have not yet included homologous recombination, diseases such as EBS and EH offer excellent opportunities to develop and explore this alternative for correction of autosomal dominant disorders of keratinocytes and other cell types. The continued advances in molecular techniques coupled with an increased understanding of the mechanisms underlying epidermal growth and differentiation lend great hope to the future prospects of keratinocyte-based gene therapy.

REFERENCES

1. Fitzpatrick, T. B., Eisen, A. Z., Wolff, K., Freedberg, I. M., and Austen, K. F., *Dermatology in General Medicine*, 4th ed., McGraw-Hill, New York, 1993.
2. Fuchs, E. and Green, H., Changes in keratin gene expression during terminal differentiation of the keratinocyte, *Cell*, 10, 1033, 1980.
3. Fuchs, E., Tyner, A. L., Giudice, G. J., Marchuk, D., RayChaudhury, A., and Rosenberg, M., The human keratin genes and their differential expression, *Curr. Top. Dev. Biol.*, 22, 5, 1987.
4. Watt, F. M., Terminal differentiation in epidermal keratinocytes, *Curr. Opin. Cell Biol.*, 1, 1107, 1989.
5. Fuchs, E., Epidermal differentiation, *Curr. Opin. Cell Biol.*, 2, 1028, 1990.
6. Anderson, W. F., Human gene therapy, *Science*, 256, 808, 1992.
7. Mulligan, R. C., The basic science of gene therapy, *Science*, 260, 926, 1993.
8. Rheinwald, J. G. and Green, H., Serial cultivation of strains of human epidermal keratinocytes: the formation of keratinizing colonies from single cells, *Cell*, 6, 331, 1975.
9. Rheinwald, J. G. and Green, H., Epidermal growth factor and the multiplication of cultured human keratinocytes, *Nature*, 265, 421, 1977.
10. Limat, A., Hunziker, T., Boillat, C., Bayreuther, K., and Noser, F., Post-mitotic human dermal fibroblasts efficiently support the growth of human follicular keratinocytes, *J. Invest. Dermatol.*, 92, 758, 1989.
11. Limat, A., Hunziker, T., Boillat, C., Noser, F., and Weismann, U., Post-mitotic human dermal fibroblasts preserve intact feeder properties for epithelial cell growth after long-term cryopreservation, *In Vitro Cell Dev. Biol.*, 26, 709, 1990.
12. Wu, Y. J., Parker, L. M., Binder, N. E., Beckett, M. A., Sinand, J. H., Griffithes, C. T., and Rheinwald, J. G., The mesothelial keratins: a new family of cytoskeletal proteins identified in cultured mesothelial cells and nonkeratinizing epithelia, *Cell*, 31, 693, 1982.
13. Morgan, J. R., Barrandon, Y., Green, H., and Mulligan, R. C., Expression of an exogenous growth hormone gene by transplantable human epidermal cells, *Science*, 237, 1476, 1987.
14. Boyce, S. T. and Ham, R. G., Calcium-regulated differentiation of normal human epidermal keratinocytes in chemically defined clonal culture and serum-free culture, *J. Invest. Dermatol.*, 81, 33, 1983.
15. Tsao, M. C., Walthall, B. J., and Ham, R. G., Clonal growth of normal human epidermal keratinocytes in a defined medium, *J. Cell Physiol*, 110, 219, 1982.
16. Bell, E., Ivarssen, B., and Merrill, C., A living tissue formed in vitro and accepted as a full thickness skin equivalent, *Science*, 211, 1042, 1979.
17. Bell, E., Sher, S., Huss, B., Merrill, C., Rosen, S., Chamson, A., Asselineau, D., Dubertret, L., Coulomb, B., Lapiere, C., Nusgens, G., and Neveux, Y., The reconstitution of living skin, *J. Invest. Dermatol.*, 81S, 2, 1983.
18. Prunieras, M., Regneir, M., and Woodley, D., Growth and differentiation of adult human epidermal cells on substrates, *Front. Matrix Biol.*, 9, 4, 1981.

19. Prunieras, M., Regneir, M., and Woodley, D., Methods for cultivation of keratinocytes with an air-liquid interface, *J. Invest. Dermatol.*, 81s, 28, 1983.
20. Vaughan, F., Gray, R. H., and Bernstein, I. A., Growth and differentiation of rat keratinocytes on synthetic membranes, *In Vitro Cell. Dev. Biol.*, 22, 141, 1986.
21. Compton, C. C., Gill, J. M., Bradford, D. A., Regauer, S., Gallico, G. G., and O'Connor, N. E., Skin regenerated from cultured epithelial autografts on full-thickness burn wounds from 6 days to 5 years after grafting: a light, electron microscopic and immunohistochemical study, *Lab. Invest.*, 60, 600, 1989.
22. Gallico, G. G., O'Connor, N. E., Compton, C. C., Kehinde, O., and Green, H., Permanent coverage of large burn wounds with autologous cultured human epithelium, *N. Engl. J. Med.*, 311, 448, 1984.
23. O'Connor, N. E., Mulliken, J. B., Banks-Schlegel, S., Kehinde, O., and Green, H., Grafting of burns with cultured epithelium prepared from autologous epidermal cells, *Lancet*, 1, 75, 1981.
24. Holbrook, K. A., Wapner, R., Jackson, L., and Zaeri, N., Diagnosis and prenatal diagnosis of epidermolysis bullosa herpetiformis (Dowling-Meara) in a mother, two affected children, and an affected fetus, *Prenat. Diagn.*, 12, 725, 1992.
25. Dover, R. and Potten, C. S., Heterogeneity and cell cycle analyses from time-lapse-studies of human keratinocytes *in vitro*, *J. Cell Sci.*, 89, 359, 1988.
26. Watt, F. M., Jordan, P. W., and O'Neill, C. H., Cell shape controls terminal differentiation of human epidermal keratinocytes, *Proc. Natl. Acad. Sci. U.S.A.*, 85, 5576, 1988.
27. Albers, K. M., Greif, F., Setzer, R. W., and Taichman, L. B., Cell-cycle withdrawal in cultured keratinocytes, *Differentiation*, 34, 236, 1987.
28. Barrandon, Y. and Green, H., Three clonal types of keratinocytes with different capacities for multiplication, *Proc. Natl. Acad. Sci. U.S.A.*, 84, 2302, 1987.
29. Barrandon, Y. and Green, H., Cell size as a determinant of the clone-forming ability of human keratinocytes, *Proc. Natl. Acad. Sci. U.S.A.*, 82, 5390, 1985.
30. Potten, C. S. and Hendry, J. H., Clonogenic cells and stem cells in the epidermis, *Int. J. Radiat. Biol.*, 24, 537, 1973.
31. Potten, C. S. and Morris, R. J., Epithelial stem cells *in vivo*, *J. Cell Sci.*, 10s, 45, 1988.
32. Adams, J. C. and Watt, F. M., Changes in keratinocyte adhesion during terminal differentiation: reduction in fibronectin binding precedes $\alpha_5\beta_1$ integrin loss from the cell surface, *Cell*, 63, 425, 1990.
33. Adams, J. C. and Watt, F. M., Expression of β_1, β_3, β_4, and β_5 integrins by human epidermal keratinocytes and non-differentiating keratinocytes, *J. Cell Biol.*, 115, 829, 1991.
34. Jones, P. H. and Watt, F. M., Separation of human epidermal stem cells from transit amplifying cells on the basis of differences in integrin function and expression, *Cell*, 73, 713, 1993.
35. Lavker, R. M. and Sun, T.-T., Heterogeneity in epidermal basal keratinocytes: morphological and functional correlations, *Science*, 215, 1239, 1982.
36. Lavker, R. M. and Sun, T.-T., Epidermal stem cells, *J. Invest. Dermatol.*, 81S, 121, 1983.
37. Cotsarelis, G., Chenh, S.-Z., Dong, G., Sun, T.-T., and Lavker, R. M., Existence of slow-cycling limbal epithelial cells that can be preferentially stimulated to proliferate: implications on epidermal stem cells, *Cell*, 57, 201, 1989.
38. Cotsarelis, G., Sun, T.-T., and Lavker, R. M., Label-retaining cells reside in the bulge area of pilosebaceous unit: implications for follicular stem cells, hair cycle, and skin carcinogenesis, *Cell*, 61, 1329, 1990.
39. Barrandon, Y., Li, V., and Green, H., New techniques for the grafting of cultured human epidermal cells onto athymic animals, *J. Invest. Dermatol.*, 91, 315, 1988.

40. Flowers, M. E. D., Stockschlaeder, M. A. R., Schuening, F. G., Niederwieser, D., Hackman, R., Miller, A. D., and Storb, R., Long-term transplantation of canine keratinocytes made resistant to G418 through retrovirus-mediated gene transfer, *Proc. Natl. Acad. Sci. U.S.A.*, 87, 2349, 1990.
41. Teumer, J., Lindahl, A., and Green, H., Human growth hormone in the blood of athymic mice grafted with cultures of hormone-secreting human keratinocytes, *FASEB J.*, 4, 3245, 1990.
42. Gordon, D. A., Fenjves, E. S., Williams, D. L., and Taichman, L. B., Synthesis and secretion of apolipoprotein E by cultured human keratinocytes, *J. Invest. Dermatol.*, 92, 96, 1989.
43. Fenjves, E. S., Gordon, D. A., Pershing, L. K., Williams, D. L., and Taichman, L. B., Systemic distribution of apolipoprotein E secreted by grafts of epidermal keratinocytes: implications for epidermal function and gene therapy, *Proc. Natl. Acad. Sci. U.S.A.*, 86, 8803, 1989.
44. Fenjves, E. S., Lee, J. I., Garlick, J. A., Williams, D. L., and Taichman, L. B., *Prospects for Epithelial Gene Therapy*, Plenum Press, New York, 1990.
45. Gerrard, A. J., Hudson, D. L., Brownlee, G. G., and Watt, F. M., Towards gene therapy for haemophilia using primary human keratinocytes, *Nat. Genet.*, 3, 180, 1993.
46. Stockschlaeder, M. A., Storb, R., Osborne, W. R., and Miller, A. D., L-histidinol provides effective selection of retrovirus-vector-transduced keratinocytes without impairing their proliferative potential, *Hum. Gene Ther.*, 2, 33, 1991.
47. Goff, S. P. and Shenk, T., Sleeping with the enemy: viruses as gene transfer agents, *Curr. Opin. Gen. Dev.*, 3, 71, 1993.
48. Lee, J. I. and Taichman, L. B., Transient expression of a transfected gene in cultured epidermal keratinocytes: implications for future studies, *J. Invest. Dermatol.*, 92, 267, 1989.
49. Gregoriadis, G., *Incorporation of Drugs, Proteins and Genetic Materials*, CRC Press, Boca Raton, FL, 1984.
50. Li, L., Margolis, L. B., Lishko, V. K., and Hoffman, R. M., Product-delivery liposomes specifically target hair follicles in histocultured intact skin, *In Vitro Cell. Dev. Biol.*, 28A, 679, 1992.
51. Li, L., Lishko, V., and Hoffman, B. M., Liposome targeting of high molecular weight DNA to the hair follicles of histocultured skin, *In Vitro Cell. Dev. Biol.*, 29A, 258, 1993.
52. Potter, H., Weir, L., and Leder, P., Enhancer dependent expression of human κ immunoglobulin genes introduced into mouse pre-B lymphocytes by electroporation, *Proc. Natl. Acad. Sci. U.S.A.*, 81, 7161, 1984.
53. Potter, H., Electroporation in biology: methods, applications and instrumentation, *Anal. Biochem.*, 174, 361, 1984.
54. Chu, G., Hayakawa, H., and Berg, P., Electroporation for the efficient transfection of mammalian cells, *Nucl. Acids Res.*, 15, 1311, 1987.
55. Chu, D. C., Cell poration and cell fusion using an oscillating electric field, *Biophys. J.*, 56, 641, 1989.
56. Thomas, K. and Capecchi, M., Site-directed mutagenesis by gene targeting in mouse embryo-derived stem cells, *Cell*, 51, 503, 1987.
57. Stellmach, V., Leask, A., and Fuchs, E., Retinoid-mediated transcriptional regulation of keratin genes in human epidermal and squamous cell carcinoma cells, *Proc. Natl. Acad. Sci. U.S.A.*, 88, 4582, 1991.
58. Sun, T.-T. and Green, H., Keratin filaments of cultured human epidermal cells, *J. Biol. Chem.*, 253, 2053, 1978.
59. Vassar, R., Rosenberg, M., Ross, S., Tyner, A., and Fuchs, E., Tissue-specific and differentiation-specific expression of a human K14 keratin gene in transgenic mice, *Proc. Natl. Acad. Sci. U.S.A.*, 86, 1563, 1989.
60. Byrne, C. and Fuchs, E., Probing keratinocyte and differentiation specificity of the human K5 promoter *in vitro* and in transgenic mice, *Mol. Cell. Biol.*, 13, 3176, 1993.

61. Bailleul, B., Surani, M. A., White, S., Barton, S. C., Brown, K., Blessing, M., Jorcano, J., and Balmain, A., Skin hyperkeratosis and papilloma formation in transgenic mice expressing a *ras* oncogene from a suprabasal keratin promoter, *Cell*, 62, 697, 1990.
62. Fuchs, E., Esteves, R. A., and Coulombe, P. A., Transgenic mice expressing a mutant keratin 10 gene reveal the likely genetic basis for epidermolytic hyperkeratosis, *Proc. Natl. Acad. Sci. U.S.A.*, 89, 6906, 1992.
63. Sellheyer, K., Bickenbach, J. R., Rothnagel, J. A., Bundman, D., Longley, M. A., Kreig, T., Roche, N. S., Roberts, A. B., and Roop, D. R., Inhibition of skin development by overexpression of transforming growth factor β_1 in the epidermis of transgeneic mice, *Proc. Natl. Acad. Sci. U.S.A.*, 90, 5237, 1993.
64. Coulombe, P. A., Hutton, M. E., Letai, A., Hebert, A., Paller, A. S., and Fuchs, E., Point mutations in human keratin 14 genes of epidermolysis bullosa simplex patients: genetic and functional analyses, *Cell*, 66, 1301, 1991.
65. Coulombe, P. A., Hutton, M. E., Vassar, R., and Fuchs, E., A function for keratins and a common thread among different types of epidermolysis bullosa simplex diseases, *J. Cell Biol.*, 115, 1661, 1991.
66. Vassar, R., Coulombe, P. A., Degenstein, L., Albers, K., and Fuchs, E., Mutant keratin expression in transgenic mice causes marked abnormalities resembling a human genetic skin disease, *Cell*, 64, 365, 1991.
67. Cheng, J., Syder, A. J., Yu, Q.-C., Letai, A., Paller, A. S., and Fuchs, E., The genetic basis of epidermolytic hyperkeratosis: a disorder of differentiation-specific keratin genes, *Cell*, 70, 811, 1992.
68. Cheng, J., Turksen, K., Yu, Q.-C., Schreiber, H., Teng, M., and Fuchs, E., Cachexia and graft-vs.-host-disease-type skin changes in keratin promoter-driven TNFα transgenic mice, *Genes Dev.*, 6, 1444, 1992.
69. Guo, L., Yu, Q.-C., and Fuchs, E., Targeting expression of keratinocyte growth factor to keratinocytes elicits striking changes in epithelial differentiation in transgenic mice, *EMBO J.*, 12, 973, 1993.
70. Turksen, K., Kupper, T., Degenstein, L., Williams, I., and Fuchs, E., IL-6: insights to its function in skin by overexpression in transgenic mice, *Proc. Natl. Acad. Sci. U.S.A.*, 89, 5068, 1992.
71. Vassar, R. and Fuchs, E., Transgenic mice provide new insights into the role of TGF-alpha during epidermal development and differentiation, *Genes Dev.*, 5, 714, 1991.
72. Green, H., Kehinde, O., and Thomas, J., Growth of cultured human epidermal cells into multiple epithelia suitable for grafting, *Proc. Natl. Acad. Sci. U.S.A.*, 76, 5665, 1979.
73. Fine, J.-D., Johnson, L., Wright, T., and Horiguchi, Y., Epidermolysis bullosa simplex: identification of a kindred with autosomal recessive transmission of the Weber-Cockayne variety, *Pediatr. Dermatol.*, 6, 1, 1989.
74. Fine, J.-D., Bauer, E. A., Briggaman, R. A., Carter, D. M., Eady, R. A. J., Esterly, N. B., Holbrook, K. A., Hurwitz, S., Johnson, L., Lin, A., Pearson, R., and Sybert, V., Revised clinical and laboratory criteria for subtypes of inherited epidermolysis bullosa, *J. Am. Acad. Dermatol.*, 24, 119, 1991.
75. Haneke, E. and Anton-Lamprecht, I., Ultrastructure of blister formation in epidermolysis bullosa hereditaria: V. epidermolysis bullosa simplex localista type Weber-Cockayne, *J. Invest. Dermatol.*, 78, 219, 1982.
76. Leigh, I. M., Tidman, M. J., and Eady, R. A. J., Epidermolysis bullosa: preliminary observations of blister formation in keratinocyte cultures, *Br. J. Dermatol.*, 111, 527, 1984.
77. Fuchs, E. and Coulombe, P. A., Of mice and men: genetic skin diseases of keratin, *Cell*, 69, 899, 1992.

78. Epstein, E. H., Molecular genetics of epidermolysis bullosa, *Science*, 256, 799, 1992.
79. Bonifas, J. M., Rothman, A. L., and Epstein, E. H., Epidermolysis bullosa simplex: evidence in two families for keratin gene abnormalities, *Science*, 254, 1202, 1991.
80. Lane, E. B., Rugg, E. L., Navsaria, H., Leigh, I. M., Heagerty, A. H. M., Ishida-Yamamoto, A., and Eady, R. A. J., A mutation in the conserved helix termination peptide of keratin 5 in hereditary skin blistering, *Nature*, 356, 244, 1992.
81. Chan, Y.-M., Yu, Q.-C., Fine, J.-D., and Fuchs, E., The genetic basis of Weber-Cockayne epidermolysis bullosa simplex, *Proc. Natl. Acad. Sci. U.S.A.*, 90, 7414, 1993.
82. Compton, J. G., DiGiovanna, J. J., Santucci, S. K., Kearns, K. S., Amos, C. I., Abangan, D. L., Korge, B. P., McBride, O. W., Steinert, P. M., and Bale, S. J., Linkage of epidermolytic hyperkeratosis to the type II keratin gene cluster on chromosome 12q, *Nat. Genet.*, 1, 301, 1992.
83. Rothnagel, J. A., Dominey, A. M., Dempsey, L. D., Longley, M. A., Greenhalgh, D. A., Gagne, T. A., Huber, M., Frenk, E., Hohl, D., and Roop, D. R., Mutations in the rod domains of keratins 1 and 10 in epidermolytic hyperkeratosis, *Science*, 257, 1128, 1992.
84. Chipev, C. C., Korge, B. P., Markova, N., Bale, S. J., DiGiovanna, J. J., Compton, J. G., and Steinert, P. M., A leucine-proline mutation in the H1 subdomain of keratin 1 causes epidermolytic hyperkeratosis, *Cell*, 70, 821, 1992.
85. Albers, K. and Fuchs, E., The expression of mutant epidermal keratin cDNAs transfected in simple epithelial and squamous cell carcinoma lines, *J. Cell Biol.*, 105, 791, 1987.
86. Albers, K. and Fuchs, E., Expression of mutant keratin cDNAs in epithelial cells reveals possible mechanisms for initiation and assembly of intermediate filaments, *J. Cell Biol.*, 108, 1477, 1989.
87. Letai, A., Coulombe, P. A., McCormick, M. B., Yu, Q.-C., Hutton, E., and Fuchs, E., Disease severity correlates with position of keratin point mutations in patients with epidermolysis bullosa simplex, *Proc. Natl. Acad. Sci. U.S.A.*, 90, 3197, 1993.
88. Smithies, O., Gregg, R. G., Boggs, S. S., Koralewski, M. A., and Kucherlapati, R. S., Insertion of DNA sequences into the human chromosomal β-globin locus by homologous recombination, *Nature*, 317, 230, 1985.
89. Breitman, M. L., Clapoff, S., Rossant, J., Tsui, L. C., Glode, L. M., Maxwell, I. H., and Bernstein, A., Genetic ablation: progammed lineage suicide by tissue-specific expression of the diphtheria toxin A gene in transgenic mice, *Science*, 238, 1563, 1987.
90. Maxwell, I. H., Maxwell, F., and Glode, L. M., Regulated expression of a diphteria toxin A-chain gene transfected into human cells: possible strategy for inducing cancer cell suicide, *Cancer Res.*, 46, 4660, 1986.
91. Maxwell, I. H., Glode, L. M., and Maxwell, F., Expression of the diphtheria toxin A-chain coding sequence under the control of promoters and enhancers from immunoglobulin genes as a means of directing toxicity to B-lymphoid cells, *Cancer Res.*, 51, 4299, 1991.
92. Fung-Leung, W.-P. and Mak, T. W., Embryonic stem cells and homologous recombination, *Curr. Opin. Immunol.*, 4, 189, 1992.
93. Hasty, P., Ramirez-Solis, R., Krumlauf, R., and Bradley, A., Introduction of a subtle mutation into the Hox-2.6 locus in embryonic stem cells, *Nature*, 350, 243, 1991.
94. Koller, B. H., Kim, H.-S., Latour, A. M., Brigman, K., Boucher, R. C., Scambler, P., Wainwright, B., and Smithies, O., Toward an animal model of cystic fibrosis: targeted interruption of exon 10 of the cystic fibrosis transmembrane regulator gene in embryonic stem cells, *Proc. Natl. Acad. Sci. U.S.A.*, 88, 10730, 1991.
95. Mansour, S. L., Thomas, K. R., and Capecchi, M. R., Disruption of the proto-oncogene int-2 in mouse embryo-derived stem cells: a general strategy for targeting mutations to non-selectable genes, *Nature*, 336, 348, 1988.

96. Chisaka, O. and Capecchi, M. R., Regionally restricted developmental defects resulting from targeted disruption of the mouse homeobox gene hox-1.5, *Nature*, 350, 473, 1991.
97. Soriano, P., Montgomery, C., Geske, R., and Bradley, A., Targeted disruption of the c-src proto-oncogene leads to osteopetrosis in mice, *Cell*, 64, 693, 1991.
98. Thomas, K. R., Deng, C., and Capecchi, M. R., High-fidelity gene targeting in embryonic stem cells by using sequence replacement vectors, *Mol. Cell. Biol.*, 12, 2912, 1992.
99. Kim, C. G., Epner, E. M., Forrester, W. C., and Groudine, M., Inactivation of the human β-globin gene by targeted insertion into the β-globin locus control region, *Genes Dev.*, 6, 928, 1992.
100. te Riele, H., Maandag, E. R., and Berns, A., Highly efficient gene targeting in embryonic stem cells through homologous recombination with isogenic DNA constructs, *Proc. Natl. Acad. Sci. U.S.A.*, 89, 5128, 1992.
101. Shesely, E. G., Kim, H.-S., Shehee, W. R., Papayannopoulou, T., Smithies, O., and Popovich, B. W., Correction of a human βs-globin gene by gene targeting, *Proc. Natl. Acad. Sci. U.S.A.*, 88, 4294, 1991.
102. Deng, C. and Capecchi, M., Reexamination of gene targeting frequency as a function of the extent of homology between the targeting vector and the target locus, *Mol. Cell. Biol.*, 12, 3365, 1992.
103. Hasty, P., Rivera-Perez, J., and Bradley, A., The length of homology required for gene targeting in embryonic stem cells, *Mol. Cell. Biol.*, 11, 5586, 1991.
104. Hasty, P., Rivera-Perez, J., and Bradley, A., The role and fate of DNA ends for homologous recombination in embryonic stem cells, *Mol. Cell. Biol.*, 12, 2464, 1992.
105. Marchuk, D., McCrohon, S., and Fuchs, E., Complete sequence of a gene encoding a human type I keratin: sequences homologous to enhancer elements in the regulatory region of the gene, *Proc. Natl. Acad. Sci. U.S.A.*, 82, 1609, 1985.
106. Marchuk, D., McCrohon, S., and Fuchs, E., Remarkable conservation of structure among intermediate filament genes, *Cell*, 39, 491, 1984.
107. Rieger, M. and Franke, W. W., Identification of an orthologous mammalian cytokeratin gene: high degree of intron sequence conservation during evolution of human cytokeratin 10, *J. Mol. Biol.*, 204, 841, 1988.
108. Lersch, R., Stellmach, V., Stocks, C., Giudice, G., and Fuchs, E., Isolation, sequence, and expression of a human keratin K5 gene: transcriptional regulation of keratins and insights into pairwise control, *Mol. Cell. Biol.*, 9, 3685, 1989.
109. Johnson, L., Idler, W., Zhou, X.-M., Roop, D., and Steinert, P., Structure of a gene for the human 67-kDa keratin, *Proc. Natl. Acad. Sci. U.S.A.*, 82, 1896, 1985.

7 Endothelial Cells: Endothelium as a Target Organ for Somatic Gene Therapy

Louis M. Messina and James C. Stanley

I. INTRODUCTION

Endothelial cells form a monolayer that covers the inner surfaces of vessels throughout the arterial, venous, and capillary circulation. Endothelial cells are not simply passive cells that serve only as a lining of blood vessels. Rather, endothelial cells serve important regulatory roles in (1) angiogenesis, (2) blood coagulation and fibrinolysis, (3) vasomotor tone, (4) solute and protein exchange between the blood tissues, as well as (5) in the systemic immune and inflammatory responses through constitutive and inducible expression of various adhesion receptors.[1-4]

Endothelial cells have assumed a prominent place in the early development of somatic gene therapy technology.[5-12] A number of endothelial cell-specific characteristics account for this. First, these are durable cells, whose half-life is measured in years. Second, endothelial cells are pluripotent, their specific phenotypic expression and thus their functional capabilities being regulated in part by their local environment. This enhances their capacity for successful transplantation to different locations within the circulation. Third and most important, endothelial cells are located at the blood-surface interface. This provides a logical site for recombinant protein secretion to achieve a local paracrine effect as well as for delivery of recombinant proteins into the circulating blood to achieve a systemic effect.

Somatic gene therapy targeted to the endothelium has application to the treatment of both vascular and nonvascular diseases. The purpose of this chapter will be to review the existing body of experimental work that has explored the use of endothelial cells for somatic gene therapy, including (1) techniques of gene transfer to endothelial cells *in vitro*, (2) transplantation of genetically modified endothelium *in vivo*, (3) techniques of direct gene transfer *in vivo*, and (4) potential applications of this technology to treatment of human diseases.

0-8493-2440-8/95/$0.00+$.50

II. GENE TRANSFER TO ENDOTHELIAL CELLS IN VITRO

Somatic gene therapy involves transfer of a new gene into nongerm cells of a patient to treat a specific disease. A major goal of such therapy is to achieve efficient gene transfer and durable recombinant gene expression in specific target cells of the host. There are two principle methods of gene transfer: transfer by plasmids or transfer by replication-incompetent viral vectors containing the gene of interest.[13,14] A critical factor in assessing techniques of gene transfer is whether incorporation of the transgene into the chromosomal DNA of the host cell occurs and thereby persists after host cell proliferation. To date, the techniques of gene transfer most likely to ensure persistence of the transgene after host cell proliferation are retroviral or adeno-associated viral vectors.

Gene transfer with plasmids has been accomplished under a variety of conditions. Cells in culture exposed to naked plasmid DNA will take up the plasmid, but usually at a very low rate of efficiency.[14] Microinjection of DNA or electroporation, the formation of pores in the cell membrane induced by high voltage electric shock, are physical methods used to enhance gene transfer.[15] Chemical techniques to enhance plasmid-mediated gene transfer include the use of calcium phosphate precipitation, receptor-mediated transfer in which the DNA is complexed to a polypeptide targeted to a receptor in the target cell wall, and lipofection.[16-18] Lipofection is accomplished by encapsidation of the DNA within liposomes. The positively charged polycationic liposomes have a high affinity for the negatively charged surface membrane of target cells, such as endothelial cells, and their fusion allows gene introduction into the host cell's cytoplasm.[18]

Many replication-incompetent viral vectors have been developed to facilitate gene transfer to target cells.[19-26] The most sophisticated and widely used are the retroviral vectors.[19-21] The other commonly used viral vectors are the adenoviruses.[22-26] The principle advantage of retroviral vectors is their potential to achieve stable incorporation of an unrearranged copy of a gene into the host chromosomal DNA. A disadvantage of retroviral vectors is that they can carry only a limited amount of DNA, up to 7 kilobases (kb). Furthermore, retroviral vectors require a receptor-mediated mechanism of transfection, and the host cell must be dividing for genomic incorporation of the transgene to occur. Although unlikely, retroviral vector-mediated gene transfer carries the risk of generation of a replication-competent virus during proviral production in the host cell. A second potential hazard is that the vector's incorporation into the host genome may result in insertional mutagenesis. Insertional mutagenesis may occur if the viral vector with its mammalian promoter inserts near a proto-oncogene. Last, because retroviral vectors are generated in live packaging cell lines, unwanted contamination with other genes may occur.

Adenoviral vectors offer a number of advantages as a technique of gene transfer.[22-26] Adenoviral vectors infect a large range of host cells, including human cells, usually at a low degree of pathogenicity. Adenoviral vectors have a large capacity to carry DNA, up to 7.5 kb. Adenoviral vectors can be generated in very high titers, which is particularly advantageous for *in vivo* gene therapy. The major disadvantage of adenoviral vectors is that the gene of interest is rarely incorporated into the host chromosomal DNA. Rather, the new DNA persists in the nucleus in an unintegrated form.[13] The fate of this unintegrated DNA is largely unknown. Nonetheless, there have been reports of persistence of transgene expression *in vivo* for up to 12 months following transfection using adenoviral vectors.[23] Less experience exists with the use of adeno-associated viruses, herpes virus, vaccinia virus, and polio virus as vectors to facilitate gene transfer.

A. Gene Transfer by Retroviral Vectors

The first reports of gene transfer to endothelial cells did not involve experimental studies directed toward gene therapy, but focused on mechanisms that regulate endothelial cell gene expression. For example, Faller and associates transduced endothelial cells with an amphotrophic murine leukemia virus containing the *ras* or *mos* oncogenes to permit the introduction of the rous sarcoma virus into human umbilical vein endothelial cells. This resulted in an immortalized cell line that was morphologically and phenotypically normal.[24] Other early *in vitro* experiments involved retroviral-mediated gene transfer of the promoters for tissue plasminogen activator inhibitor 1 (PAI-1) and the endothelin-I receptor.[28-30] These studies were undertaken to identify the *cis*-acting regulatory factors and the tissue specificity of PAI-1, as well as to identify the *trans*-acting factor of the endothelin-I receptor that results in endothelial cell-specific expression of this gene.[30] These studies established the fact that endothelial cells could be transduced using retroviral vectors to achieve durable recombinant gene expression.

Zweibel and associates reported the first successful retroviral-mediated gene transfer to human endothelial cells for the purpose of gene therapy in 1989.[31,32] These investigators used three retroviral vectors expressing three different recombinant genes. The first contained the neomycin resistance gene that encodes for neomycin phosphotransferase and is commonly used in the construction of retroviral and nonretroviral vectors. This allows selection of transduced cells on the basis of the conferred resistance to the neomycin analogue G418. The two other vectors studied contained the gene for either human adenosine deaminase or rat growth hormone. These constructs represented examples for both secreted and nonsecreted gene products. The initial efficiency of gene transfer in these experiments was low, but transduced cells were successfully selected in G418. The potential of using these cells for gene therapy was evaluated by growing them successfully on the surface of synthetic vascular grafts *in vitro*, where they remained viable for 32 days. These cells exhibited persistent secretion of growth hormone, although at levels insufficient to achieve detectable systemic concentrations. Wilson and colleagues, using a different retroviral vector, achieved secretion of recombinant parathyroid hormone (PTH), as well as expression of the receptor for low-density lipoprotein.[33] It is noteworthy that PTH requires extensive post-translational modification, an event accomplished successfully by these transduced endothelial cells. The efficiency of gene transfer ranged from 5 to 60% in these latter experiments.

The gene for human tissue plasminogen activator (tPA) has considerable potential application to somatic gene therapy for vascular disease. Augmentation of tPA secretion by human endothelial cells *in vivo* could reduce the frequency of acute thrombosis after angioplasty of atherosclerotic coronary or peripheral arterial lesions, as well as reduce the thrombogenicity of small caliber synthetic grafts seeded with autologous tPA-

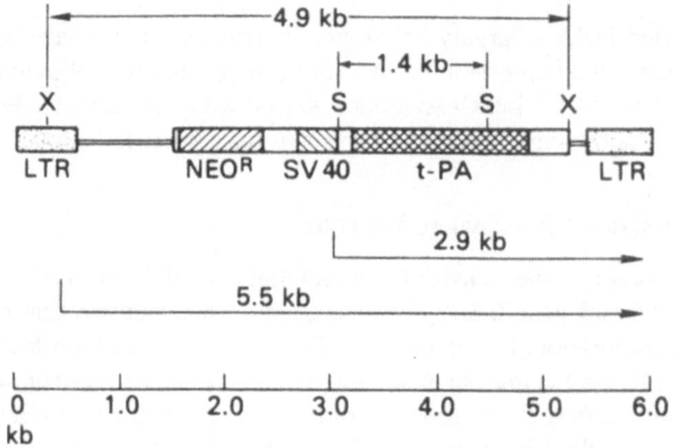

FIGURE 1. B2NSt, a retroviral vector containing the human tPA gene. The vector consists of the Maloney murine leukemia virus long-terminal repeats (LTR) and flanking sequences (small unshaded bars), the neomycin phosphotransferase gene (*neo*ʳ), the SV40 early promoter, and the human tPA cDNA. The 5′ and 3′ untranslated regions of both the *neo*ʳ gene and the tPA cDNA are left unshaded. Important restriction sites are indicated as follows: S, *Sac1*, X, *XBA1* sites 3′ to the tPA sequence and three *Sac1* sites outside the tPA cDNA are not shown. (With permission from Reference 34.)

transduced endothelial cells. Considerable work has been accomplished by a number of groups to determine the feasibility and consequences of transferring the tPA gene to endothelial cells.[34-38]

Elegant studies of retroviral vector-mediated gene transfer and expression of the human tPA gene in endothelial cells have been performed by Dichek and associates at the National Institutes of Health.[34-37] These investigators enhanced the fibrinolytic activity of sheep endothelial cells transfected with a retroviral vector constructed with the cDNA for human tPA, driven by a SV-40 early promoter (Figure 1). This B2NSt vector was subsequently produced by PA317 cells, an amphotrophic packaging cell line. Transfection of sheep arterial endothelial cells with the B2NSt vector resulted in a 30-fold increase in tPA secretion over that from control endothelial cells. Secretion in arterial endothelium was durable, lasting up to 11 weeks, averaging 100 ng/24 hr (Figure 2). Secretion in venous endothelium was not stable, declining to 50% of the initial secretion rate by 6 weeks. The secreted tPA was functionally active. However, the activity relative to the amount of protein present was less than anticipated. Casein zymography, an *in situ* enzymatic assay of different protein moieties separated by electrophoresis, documented that very little of the recombinant tPA was free, most being bound to its inhibitor, PA-I. In spite of this, levels of tPA secretion were higher than any reported previously. An additional finding was that the total activity of endogenous sheep plasminogen activators was increased. In the absence of species-specific assays, the investigators were unable to determine if this was due to an increase in the total amount of the activator or to a reduction in sheep PAI-1 available for binding as a result of its interaction with the human tPA. The important finding of this study was that retroviral-mediated tPA gene transfer resulted in a net increase in endothelial cell

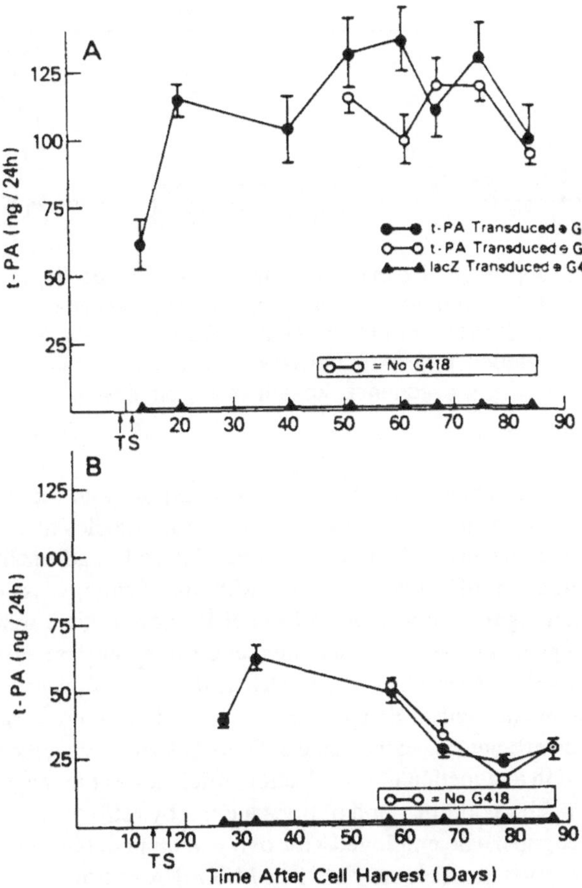

FIGURE 2. (A) Expression of human tPA by sheep carotid artery endothelial cells in culture. (B) Secretion of human tPA by jugular vein endothelial cells of sheep in culture. Data points represent mean ± SD of two or more ELISA wells from each of duplicate or triplicate tissue culture wells. (With permission from Reference 34.)

fibrinolytic activity. An issue unanswered by these studies was whether human endothelium transduced with this vector would also achieve a substantial increase in fibrinolytic activity or whether the level of human PAI-1 would increase and negate the augmented secretion of tPA.

In subsequent experiments, Dichek and colleagues transduced human umbilical vein endothelial cells with the same B2NSt vector.[8] The level of the tPA transcript was increased in these studies by 100 to 200% over that of the endogenous tPA transcript. The tPA antigen levels were two to three times higher than that in control transduced cells. However, this amount of tPA was not considered sufficient to produce a functional *in vivo* biologic effect in humans.

Dichek and colleagues also studied a wide range of techniques to optimize transfection efficiencies using retroviral vectors expressing either the β-galactosidase or

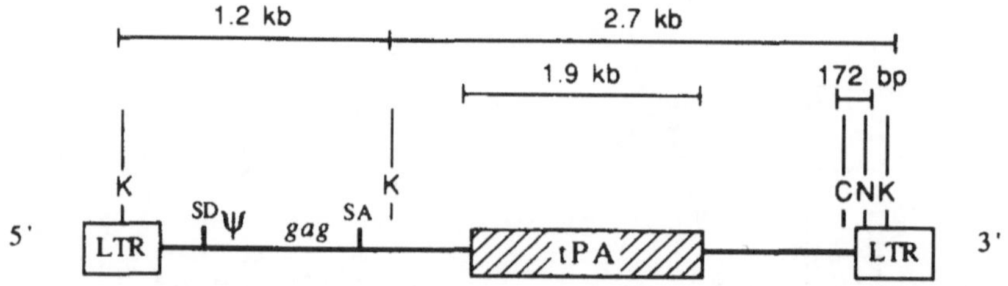

FIGURE 3. Schematic representative of the MFG-tPA vector. The backbone of the vector is the murine Moloney leukemia virus long-terminal repeat (LTR) and flanking sequences, with the human tPA cDNA. The Kpn I (K restriction site) is depicted, as well as the Cla I (C) to Nhe I (Frag). SD: splice donor sequence; ψ: retroviral packaging sequence; gag: sequences for viral core protein; SA: splice acceptor sequence; kb: kilobases; bp: base pair. (With permission from Reference 38.)

urokinase genes.[35] They examined extended periods of exposure to the vector, repeated exposures to the vector, maximization of the ratio of vector particles to endothelial cells, cocultivation of endothelial cells with the vector-producing cells, and finally variations in the type and concentration of polycation used with the retroviral vector. Only one experimental condition significantly improved the efficiency of gene transfer: use of a more concentrated, higher-titer vector-containing supernatant and use of the polycation DEAE-dextran. A 60-sec exposure to 1 mg/ml DEAE-dextran followed by a single 6-hr exposure to the supernatant with a titer of 10^5 to 10^6 CFU (colony-forming units)/ml resulted in transduction efficiencies in the range of 50 to 90%. Reducing the exposure from 6 hr to 15 min resulted in a reduction of transduction efficiency to the range of 15 to 20%.

A different approach to optimization of transduction by retroviral vectors was used by Podrazik and colleagues, who employed a retroviral vector containing the human tPA gene, as an MFG-tPA construct (Figure 3).[38] This MFG-tPA vector expressed tPA from a transcript initiated at the 5′ long-terminal repeat (LTR) sequence. A ψ CRIP producer cell line was used to generate the MFG-tPA vector at titers of 5×10^5 to 1.5×10^6 CFU/ml. In order to identify the duration and number of retroviral exposures yielding optimal recombinant tPA expression, adult canine jugular venous endothelial cells subjected to different exposure regimens had human tPA antigen secretion and functional activity determined at 2 and 14 days following transduction. Persistent secretion of human tPA was detected by transduced endothelial cells in all experimental groups, ranging from 6.8 to 53 ng/ml/10^6 cells/hr. The highest level of antigen secretion and functional activity followed two 12-hr exposures to the retroviral vector. However, quantitative Southern blot analysis did not show a higher number of gene copies per cell in the latter exposure regimen. Nonetheless, this study documented highly efficient gene transfer. The percentage of cells that were positive by immunohistochemical staining for cytoplasmic tPA was between 87 and 95% (Figure 4). These very high rates of transfection were accomplished without the need for selection in neomycin. The latter requires an additional 2 to 4 weeks and carries a risk of potential injury to the cells.

Concerns have been raised about the very high levels of plasminogen activators secreted by these transduced cells. The formation of plasmins catalyzed by plasminogen

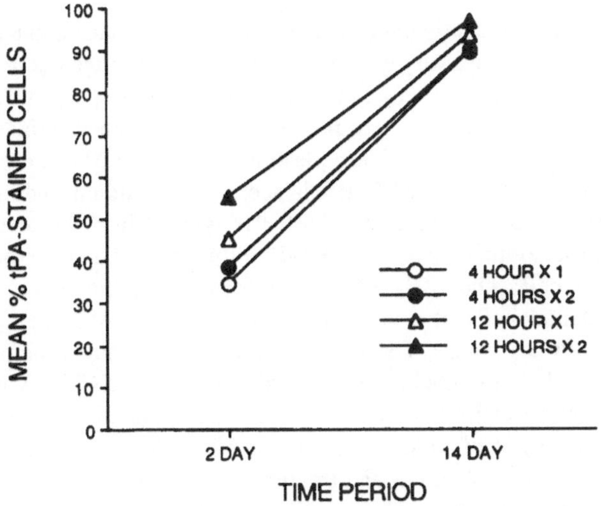

FIGURE 4. Difference in immunohistochemical staining of cellular tPA among exposure regimens at 2 and 14 days after transfection of endothelial cells. At both 2 days after and 14 days after exposure to the retroviral vector, the percentage of the exposed cells that stained positive for the tPA antigen increased as the duration of exposure increased. All exposure regimens showed a significantly higher rate of successful transduction at 14 days than at 2 days. (With permission from Reference 34.)

activators released from cancer cells, in combination with other proteolytic enzymes, appears responsible for the degradation of surrounding extracellular matrix that characterizes the invasive growth of malignant tumors. Plasminogen activators also induce changes in the cell phenotype that appear receptor-mediated.[36] Therefore, an important issue is whether a 30-fold increase in tPA secretion will alter endothelial cell phenotype.

To address this issue, Jaklitsch and colleagues examined the effect of increased tPA activity on endothelial cell morphology, adhesion, proliferation, migration, and invasion.[36] *In vitro* assays of these cellular functions were undertaken on endothelial cells transduced with the B2NSt vector, cells transduced by a control vector, and nontransduced cells. The morphology of the two transduced cell populations was unchanged. Only a small decrease in the horizontal migration rate of transduced cell populations was found. In all other assays, there were no significant differences between the nontransduced and transduced endothelial cells. Thus, high levels of tPA found in the B2NSt-transduced cells did not significantly affect endothelial cell phenotype *in vitro*. These results support the view held by many that the increases in plasminogen activator associated with malignant transformation of cells are relevant but alone not sufficient to alter cell phenotype.[36]

The aforementioned studies established that endothelial cells transduced with the human tPA gene resulted in increased fibrinolytic activity in conditioned medium *in vitro*. However, it remains unclear whether transduced endothelial cells would be able to exert a functional effect *in vivo*. Will the secreted recombinant tPA be diluted by the large circulating blood volume and cleared from the surface of the transduced endothelium? Will the secreted recombinant tPA be inactivated rapidly by its binding with PAI-1?

Lee and co-workers recently undertook a series of experiments that addressed some of these concerns.[37] A mutant single-chain urokinase plasminogen activator (SCU-PA) was constructed by addition of an apical membrane-targeting signal from the decay-accelerating factor to the carboxy terminus of a single-chain urokinase plasminogen activator. Normal single-chain urokinase plasminogen activator is a proenzyme known to resist binding by PAI-1. The SCU-PA is secreted by human endothelial cells and circulates in plasma at a concentration of 5 to 10 ng/ml. Thus, the recombinant SCU-PA represents a proenzyme resistance to PA-I inhibition anchored to the apical surface of the endothelial cell. The SCU-PA was expressed on the apical surface of the transduced endothelial cells at a concentration of 10^6 per cell, being converted by plasmin to two-chain urokinase. These results showed that apical membrane targeting can be accomplished in endothelial cells. Furthermore, the SCU-PA anchored to the endothelial cell surface caused significantly increased plasminogen activator activity. These novel modifications of endothelial cell plasminogen activator activity may eventually be of value in the prevention and treatment of intravascular thrombosis, as well as in reducing the thrombogenicity of small-caliber synthetic grafts.

An alternative method of reducing blood vessel surface thrombogenicity is to enhance prostacyclin (PGI_2) secretion by the luminal endothelium.[39] Prostacyclin synthesis is regulated at each of three enzymatic steps in its biosynthesis. Although phospholipase A_2 is the initial rate-limiting step in prostanoid synthesis, it is prostaglandin H synthase (PGHS) that is the most important in regulation of PGI_2 production.

Xu and colleagues recently used the BAG retroviral vector containing PGHS-1 cDNA to transduced human endothelial cells to augment the expression of human prostacyclin H synthase as a means of enhancing endothelial cell prostaglandin synthesis.[39] A number of interesting findings emerged from this study. Initial transduction with the BAG vector containing the PGHS-I cDNA in the sense orientation relative to the retroviral promoter resulted in a 30-fold increase in mRNA. However, no increase in PGHS protein or PGI_2 synthesis occurred due to a reading frame shift. When the PGHS-I cassette was placed in a reversed orientation relative to the viral promoter, a ten-fold increase in PGHS mRNA resulted, and increases occurred in the PGHS protein enzymatic activity resulting in an increase of PGI_2. This high level of PGI_2 synthesis by PGHS-transduced endothelial cells was increased even further by administration of ionophore, arachadonic acid, and thrombin.

The experimental studies discussed above show that gene transfer to endothelium should have considerable application to treatment of vascular diseases. Gene transfer to endothelial cells can also be used to treat noncardiovascular diseases, as well as to study vascular wall biology. An example of the latter is illustrated in the work of Ramos and colleagues, who recently transferred the human macrophage colony-stimulating factor (M-CSF) gene using a retroviral vector into rabbit endothelial and smooth muscle cells.[40] PCR confirmed the presence of the M-CSF gene in the cells, and Western blot of conditioned media confirmed secretion of the intact recombinant protein at a concentration of 2 to 8 ng/ml. Its functional activity was confirmed by stimulation of mouse bone marrow cell proliferation. One hypothesis of the development of atherosclerosis holds that the initial event is the migration of circulating monocytes into the arterial wall, where they differentiate into macrophages, accumulate lipids, and develop into foam cells. M-CSG directs the terminal differentiation and survival of these macrophages. Thus, this *in vitro* model employing gene transfer could provide substantial insight into the mechanisms controlling the development of atherosclerotic plaque.

Retroviral vector-mediated gene transfer to endothelial cells has been proposed as an approach to somatic gene therapy for hemophilia B.[41] This method is based on the strategic location of endothelial cells at the blood-surface interface as well as on the capacity of the endothelium (10^{13} cells per individual) to secrete therapeutic levels of the recombinant protein into the systemic circulation. Hemophilia B is an X chromosome-linked recessive factor IX-deficient bleeding disorder. Restoration of factor IX levels to just 15% of normal can convert a severe bleeding disorder to a mild bleeding disorder. In fact, restoration of factor IX levels to 30% of normal results in near complete amelioration of the clinical complications of this genetic disorder.

Yao and colleagues recently described the recombinant expression of active human factor IX produced in rat fat pad capillary endothelial cells.[42] A Moloney murine leukemia virus-derived retroviral vector that contained human factor IX cDNA linked to heterologous promoters and the neomycin-resistance gene was constructed for these studies. Transfected endothelial cells exhibited high levels of expression, up to 3.6 $\mu g/10^6$ cells/24 hr. The recombinant factor IX produced by these rat capillary endothelial cells showed full clotting activity. These results further reinforce the potential of endothelial cells in somatic gene therapy for noncardiovascular diseases.

B. Gene Transfer by Adenoviral Vectors

Although most initial efforts to achieve gene transfer into endothelial cells for gene therapy were undertaken using retroviral vectors, a number of other viruses are now available for this purpose. Replication-deficient adenoviruses are one of these alternative vectors. As previously noted, advantages of adenoviral vectors are that they can carry large quantities of cDNA, they do not require cell division for transfection, they are known to infect human cells efficiently but with low pathogenicity, and they are not associated with malignant transformation of the infected cells. In addition, adenoviruses are stable and can be obtained in relatively high titers, something of considerable importance for techniques of direct *in vivo* gene transfer to endothelial cells. The major disadvantage of adenoviral vectors is that the recombinant gene is not incorporated into chromosomal DNA.

Lemarchand and colleagues recently assessed the feasibility of using a replication-deficient adenovirus for gene transfer to human endothelial cells *in vitro*.[25] Human umbilical vein endothelial cells were infected *in vitro* with adenoviral vectors containing the *lac-Z* gene or the human α_1-antitrypsin (α_1-AT) cDNA. The α_1-AT cDNA is a prototype of a human gene that directs synthesis of a protein not normally secreted by endothelial cells that requires post-translational modification. The transduced endothelial cells synthesized and secreted fully functional α_1-AT within 6 hr of transfection. Secretion rates for this functionally active protein ranged from 0.3 to 0.6 $\mu g/10^6$ cells/24 hr, being constant for at least 14 days. The transduced endothelial cells also expressed β-galactosidase. These investigators also evaluated the feasibility of direct gene transfer of these genes to endothelial cells in human blood vessels. Adenoviral vectors at high titer were placed into the lumen of intact human umbilical veins *ex vivo*. At 24 hr, histologic evaluation showed transfer and expression of the lac-Z gene to the endothelial cells lining the umbilical vein. In addition, the α_1-AT protein could be quantitated in 24-hr perfusates of the vein at a level of 13 μg/ml. Thus, this study documented that endothelial cells can be transfected by adenoviral vectors to secrete proteins not normally secreted by endothelium and that the specific requirements for transfection using adenoviral vectors strongly support the feasibility of direct gene transfer *in vivo.*

C. Gene Transfer of Plasmid DNA

Gene transfer to endothelial cells can also be accomplished using plasmid DNA by a variety of means.[43-46] Powell and colleagues recently used DEAE-dextran and lipofectin to transduce endothelial cells so as to express the tissue plasminogen activator type-I inhibitor (PAI-1) and firefly luciferase genes under the control of several different promoters, including the Rous sarcoma virus. DEAE-dextran-mediated transductions resulted in inefficient and low-level transient expression of the RSV-luciferase gene. In contrast, lipofectin-mediated transductions resulted in a five-fold higher expression of RSV-luciferase that persisted for up to 14 days.

Etchberger and associates used calcium phosphate to augment plasmid DNA uptake by endothelial cells.[44] These investigators used pSV_2 plasmids with the neomycin gene and the chloramphenicol transferase gene inserted 3′ to SV40 fragment. These investigators transfected nearly confluent endothelial cell cultures. Both glycerol shock and passaging of cells substantially increased the number of transfected cells. Expression of the recombinant genes was confirmed by using fluorescently labeled antibodies to the recombinant proteins, confirming the usefulness of nonviral vector-mediated gene transfer.

III. GENE TRANSFER TO THE ARTERIAL WALL

A number of experimental models have been developed to determine the feasibility of gene transfer to endothelium as a means of gene therapy for both cardiovascular and noncardiovascular diseases. Two fundamentally different approaches have been used to achieve this aim. In the first approach, the endothelial cells undergo gene transfer *ex vivo* and are subsequently transplanted into the arterial wall *in vivo.* Under these circumstances, the endothelium first needs to be harvested from the individual and successfully cultured *in vitro* . Gene transfer is then accomplished *in vitro,* usually using a retroviral vector. The second approach is to transduce endothelial cells directly *in vivo.* Although this has been accomplished most successfully with lipofection and the use of adenoviral vectors, neither assure durable long-term expression.

A. Gene Transfer to Endothelium Ex Vivo and Subsequent Transplantation of Genetically-Modified Cells into the Arterial Wall In Vivo

In June of 1989, two articles were published in *Science* that addressed the question of whether endothelial cells that have undergone gene transfer *ex vivo* can be successfully transplanted into a vessel wall *in vivo.*

In the first study, Nabel and colleagues documented that genetically modified endothelial cells transplanted into the arterial wall survived and expressed recombinant proteins for up to 4 weeks after their transplantation.[47] These investigators first established an endothelial cell line from an inbred strain of Yucatan minipig *in vitro*. The endothelial cells were transfected with the BAG vector, a murine amphotrophic retroviral construct that carried the *lac-Z* gene coding for β-galactosidase protein and the Tn5 neomycin resistance gene. After transfection at relatively low efficiencies, the cells were selected in G418. After a sufficient number of transduced endothelial cells were grown in culture, minipigs were anesthetized and their iliofemoral arteries catheterized. To denude the pre-existing endothelium, a balloon catheter was inflated partially and withdrawn to remove the luminal cells mechanically. The denuded arterial segment was irrigated with heparinized saline, and any residual adherent endothelial cells were removed by installation of dispase (50 U/ml) for 10 min. After irrigating the artery, the BAG-transfected endothelial cells were instilled and allowed to incubate in the vessel for 30 min. At the end of this incubation, antegrade blood flow was restored. Artery segments were excised 2 to 4 weeks later, and X-gal staining of the fixed artery walls showed β-galactosidase primarily in the luminal endothelial cells. Importantly, these investigators addressed the potentially serious problem of production of a replication-competent retrovirus from the genetically modified endothelial cells. No helper-virus was detectable among these lines after 20 passages *in vitro*. Thus, this study established that endothelial cells could undergo gene transfer *ex vivo* and subsequently be transplanted into the arterial wall *in vivo*, where recombinant expression persisted for at least 4 weeks after transplantation.

In the second study, Wilson and colleagues transplanted genetically modified endothelial cells onto the surfaces of synthetic vascular grafts.[48] Endothelial cell seeding of synthetic vascular grafts with endothelial cells reduces platelet accumulation on the inner surface of the graft and improves graft patency. However, certain problems have prevented the application of this technique to the management of vascular disease in humans.[49] These problems include lower rates of endothelialization by human cells seeded on synthetic vascular grafts and persistently high thrombosis rates of small-caliber synthetic vascular grafts *in vivo*. These issues might be addressed by genetically modifying the endothelial cells used for seeding.

In the study by Wilson and colleagues, endothelial cells were derived from external jugular veins from adult mongrel dogs. These cells were cultured and transfected with a retrovirus containing the *lac*-*Z* gene. They achieved a variable transduction efficiency, ranging from 5 to 60%. These transduced endothelial cells were then seeded onto the luminal surfaces of 6 cm × 4 mm diameter Dacron synthetic vascular grafts that were subsequently placed as carotid interposition grafts into the mongrel dog from which the endothelial cells were derived. In addition, control grafts that were seeded with mock-infected autologous endothelial cells were placed in the contralateral carotid artery. Grafts were harvested 5 weeks after their implantation and stained with X-gal. *lac-Z*-transfected endothelial cells expressing β-galactosidase were found in substantial but variable numbers on each transduced graft, but not on control grafts. This study established that genetically modified endothelial cells could be seeded successfully onto synthetic vascular grafts and that recombinant gene expression occurred after graft implantation *in vivo*. A form of gene therapy, such as was established in this study, might eventually improve the clinical function of synthetic vascular grafts.

Podrazik and colleagues addressed a question of fundamental importance to the application of this technology to the clinical setting.[50] Does genetic modification of endothelial cells inhibit their capacity to proliferate and thereby reduce the extent of graft surface endothelialization? These investigators tested the hypothesis that *lac-Z*-transduced endothelial cells would achieve the same degree of surface endothelialization as nontransduced endothelial cells when seeded on expanded polytetrafluorethylene (ePTFE) grafts.

BAG-transduced endothelial cells following selection in G418, as well as nontransduced endothelial cells, were seeded on thoracoabdominal ePTFE grafts implanted for 6 weeks in dogs. Nonseeded ePTFE grafts served as controls. Cultures of endothelium derived from the ePTFE grafts seeded with endothelial cells, as well as en face graft staining, showed *lac-Z* expression from 5 of 7 grafts (Figure 5). Endothelial cells derived from grafts seeded with nontransduced cells and nonseeded grafts did not exhibit any X-gal staining. Total graft surface endothelialization was substantially less in grafts seeded with transfected endothelial cells than in those seeded with nontransduced endothelial cells. Transfected endothelial cells covered 5 to 12% of the graft surfaces, whereas the nontransfected endothelial cells covered more than 60% of the graft surface. This suggests that transfection with the BAG vector may impair the capacity of the BAG-transfected endothelial cell to adhere, migrate, or proliferate on the graft surface *in vivo*. This impairment of endothelial cell function could be due to (1) the process of transfection itself, including the potential toxicity of either the neomycin analogue G418 used to select the transduced cells or to the polybrene used to facilitate retroviral entry into cells, (2) the random integration of the retrovirus that alters cell function, or (3) recombinant proteins expressed by the *lac-Z* and Tn5 genes that affect cell function.

A prior study from the same laboratory by Brothers and colleagues documented that BAG-transfected endothelial cells showed reduced rates of cell proliferation and prostacyclin secretion during log phase growth.[51] These observations appear to have been extended to BAG-transfected endothelial cells growing on surfaces of ePTFE grafts *in vivo*. This underlines an important tenet, that the functional effects of gene transfer on the target cell are likely to vary for each gene and vector construct. The impact of genetic modification on a tissue must be evaluated for each *in vivo* application. Otherwise, incorrect conclusions concerning the potential application of gene therapy to human diseases may develop.

A recurring question concerning the application of retroviral vectors to human gene therapy has been their safety and toxicity. Complications for retroviral-mediated gene transfer that could potentially occur include overexpression of the recombinant gene. For instance, if a gene codes for a growth factor or angiogenic protein, uncontrolled cell proliferation might ensue. Alternatively, during incorporation of the retroviral vector into the host target cell genome, an insertional mutagenesis may occur. The latter may be particularly serious if a viral promoter inserts near a proto-oncogene.

Plautz and colleagues developed a novel approach to eliminate transfected cells containing recombinant genes whose unregulated expression would cause an undesired transformation of the target cell.[52] Their approach was to create a "suicide" gene that

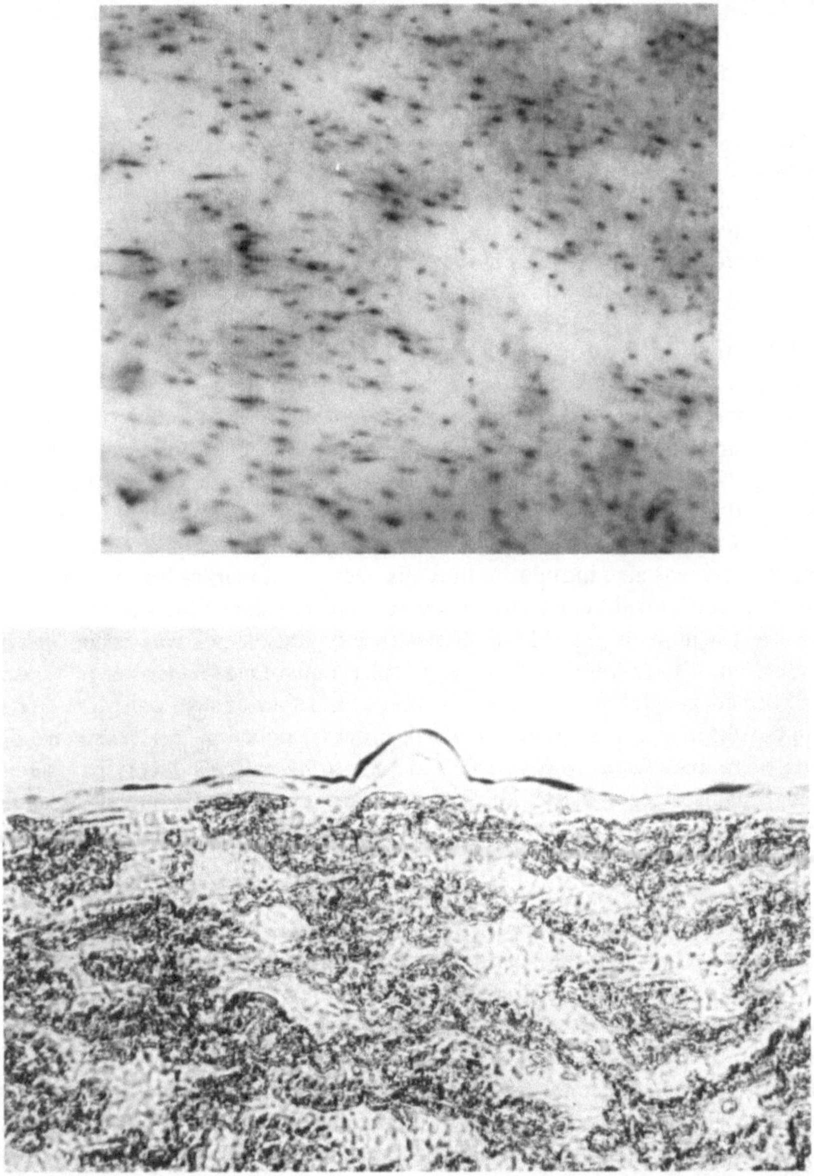

FIGURE 5. Photomicrograph of explant of a ePTFE thoracoabdominal graft that had been seeded with autologous BAG-transduced endothelium. (Top) *En face* preparation showing diffuse coverage of graft surface by *lac-Z* positive endothelial cells. (Bottom) Longitudinal section of graft showing continuous monolayer of *lac-Z* postive cells lining the graft.

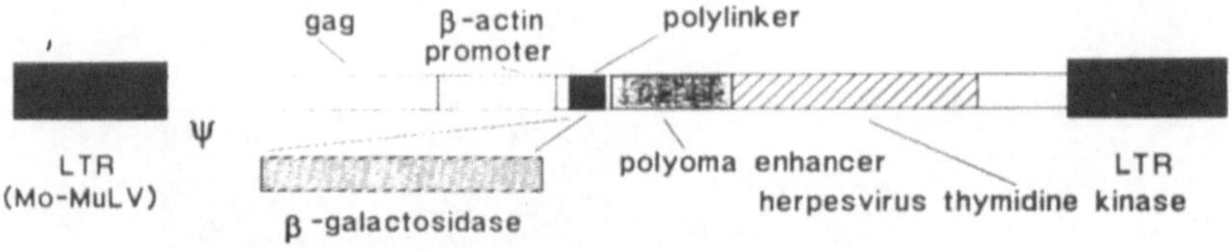

FIGURE 6. Suicide retroviral vector, derived from a murine Moloney leukemia virus plasmid that contained an internal chicken β-actin promoter. This plasmid was modified by the insertion of multiple cloning sites, into which was placed *HSV-tk* gene under the control of a modified polyoma enhancer. The *E. coli* β-galactosidase gene was inserted into this vector under the control of the β-actin promoter. LTR: long terminal repeat; ψ: retroviral packaging sequence; gag: sequences for viral core proteins. (With permission from Reference 52.)

rendered dividing cells sensitive to a specific drug but allowed the survival of normal nondividing cells. These investigators accomplished this using a mouse Moloney leukemia virus-derived retroviral vector containing a plasmid with cloning sites distal to the chicken β-actin promoter and proximal to the polyoma virus enhancer linked to the herpes simplex virus (HSV) thymidine kinase gene (Figure 6). Thymidine kinase confers sensitivity to the guanosine analogue ganciclovir that inhibits DNA synthesis and thereby eliminates proliferating cells. The *lac-Z* gene under the control of the chick β-actin promoter was also introduced into this vector as a marker for recombinant gene expression. β-Galactosidase activity in several cell-transduced lines *in vitro* was abolished by treatment with ganciclovir. Subsequently, the vector was tested *in vivo* by direct infection of the iliofemoral artery in four rabbits. One week after infection, two rabbits received ganciclovir orally for 4 weeks, and two control animals were maintained in the absence of this drug. Five weeks after transfection, the transfected arterial segments were incubated with X-gal. Arteries from animals receiving ganciclovir exhibited β-galactosidase activity similar to arteries in control animals that did not receive the ganciclovir, indicating that nondividing cells were unaffected. Thus, the transduced cells attached, expressed the recombinant gene, and were not proliferating at the time of ganciclovir administration. To test whether this vector would permit ganciclovir-mediated elimination of transformed cells, a mouse CT26 adenocarcinoma cell line was infected with the suicide vector. After implantation *in vivo*, the tumors regressed after administration of ganciclovir. Such a novel strategy may improve considerably the clinical safety and efficacy of gene transfer using retroviral vectors.

Bernstein and colleagues recently reported on pulmonary endothelial cells that underwent transfer of a human growth hormone fusion gene by calcium phosphate precipitation.[53] This study was the first to document detectable recombinant protein in the systemic circulation after intraperitoneal, subcutaneous, and intravenous transplantation of transfected endothelial cells *in vivo*. Furthermore, when the transduced endothelial cells were inserted beneath the renal capsule, large cysts formed that contained concentrations of human growth hormone several thousand-fold higher than that in serum. This experiment established that genetically modified endothelial cells could form a surface that resulted in polarized secretion of a new gene product *in vivo*.

Because capillaries constitute more than 80% of the surface area of the circulatory system, they are a logical recipient site to transplant genetically modified endothelial cells *in vivo*. However, transplantation of transduced endothelial cells into this tissue is problematic. Endothelial cells exist in tightly organized monolayers in which cell density is fixed, and mechanically strong interactions exist between endothelial cells and the basement membrane to regulate solute and protein transport. Although the endothelium of muscular arteries has been denuded and reseeded successfully, denudation of capillary endothelium as a means of facilitating cell transplantation might disrupt important normal microcirculatory hemodynamics and functions, as well as initiate a vessel wall injury response.

Messina and colleagues undertook a series of *in vitro* experiments to circumvent denudation of the capillary endothelium. They tested the hypothesis that activation of confluent endothelial cell monolayers by cytokines would promote adhesion and incorporation of the endothelial cells seeded onto these monolayers.[54] Cytokines increase the expression of a variety of endothelial cell adhesion molecules, cause loss of junctional integrity between endothelial cells, and increase the metabolic rate and proliferation of endothelial cells both *in vivo* and *in vitro*. In the course of their experiments in which endothelial cell monolayers were activated by tumor necrosis factor α, a previously unrecognized capacity of endothelial cells to adhere and incorporate spontaneously into *untreated* postconfluent endothelial cell monolayers *in vitro* was identified. The rate of endothelial cell adhesion onto postconfluent monolayers was rapid (Figure 7). Nearly 90% of seeded endothelial cells became adherent within 6 hr after seeding onto the surface of these monolayers, and nearly 100% were adherent by 24 hr. This rate of adhesion resulted in a 50% increase in the monolayer cell density by 24 hr. Although *lac-Z*-transduced endothelial cells became adherent at a significantly slower rate over the first 12 hr, adhesion by 24 hr was similar to that of nontransduced endothelial cells. Transmission electron microscopy of the monolayers documented unambiguously that the transduced endothelial cells became fully incorporated into the pre-existing monolayer.

The former *in vitro* property of spontaneous endothelial cell adhesion and incorporation also was documented by Messina and colleagues to occur *in vivo* and provides a strategy for somatic gene therapy.[54] Radiolabeled microvascular endothelial cells were tracked after being injected intra-arterially into the femoral artery of the hindlimb of the rat (Figure 8). One hr after the clamp was removed, 74% of the injected radioactivity was detected in the hindlimb. At 24 hr after injection, 27% of the injected radioactivity was present in the hindlimb, and at 7 days, 24% of the injected radioactivity was detected. At 28 days, 12% of injected radioactivity was still present in the hindlimb. Light microscopic analysis of the gastrocnemius and tibialis anterior muscle segments at each time point documented *lac-Z* gene expression within the skeletal muscle capillaries of the injected hindlimb. Discrete areas of blue appeared tubular, suggesting incorporation of the transfected cells into the capillary wall (Figure 9), and transmission electron microscopy revealed certain transduced endothelial cells to be incorporated into the capillary wall (Figure 10). However, the majority of transduced cells identified remain within the capillary lumen, where they formed multiple focal electron-dense contacts with the underlying host endothelium of the capillary wall. Although no major surface interaction with the capillary basal lamina could be demonstrated, these cells appeared healthy and viable.

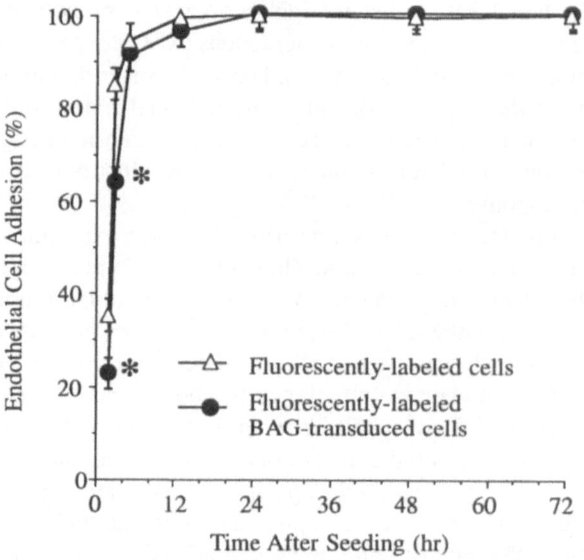

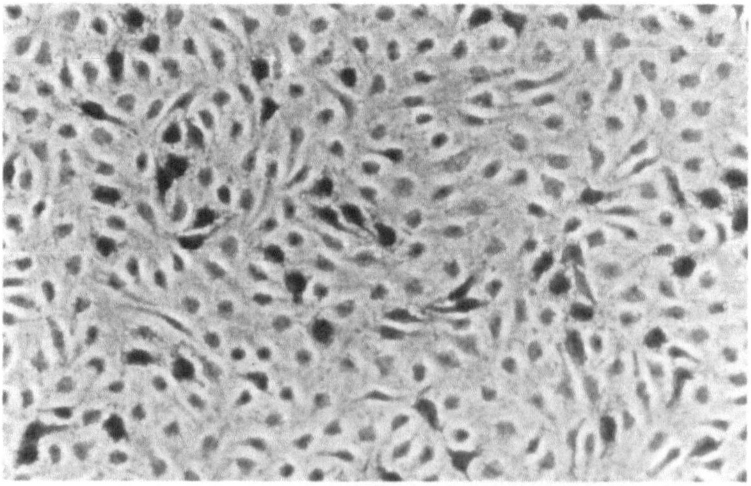

FIGURE 7. Rapid adhesion of endothelial cells to postconfluent monolayers *in vitro.* (Top) Nearly 90% of the seeded endothelial cellls became adherent within 6 hr after seeding onto the surface of the monolayers and nearly 100% of the seeded cells were adherent by 24 hr. This rate of adhesion resulted in a 50% increase in monolayer cell density by 24 hr. *lac-Z*-transduced endothelial cells became adherent at a significantly slower rate over the first 12 hr of the experiment, but by 24 hr their rate of adhesion was similar to that of nontransduced endothelial cells. (Bottom) Photomicrograph of *lac-Z*-transduced endothelium incorporated into monolayer 24 hr after seeding. (With permission from Reference 54.)

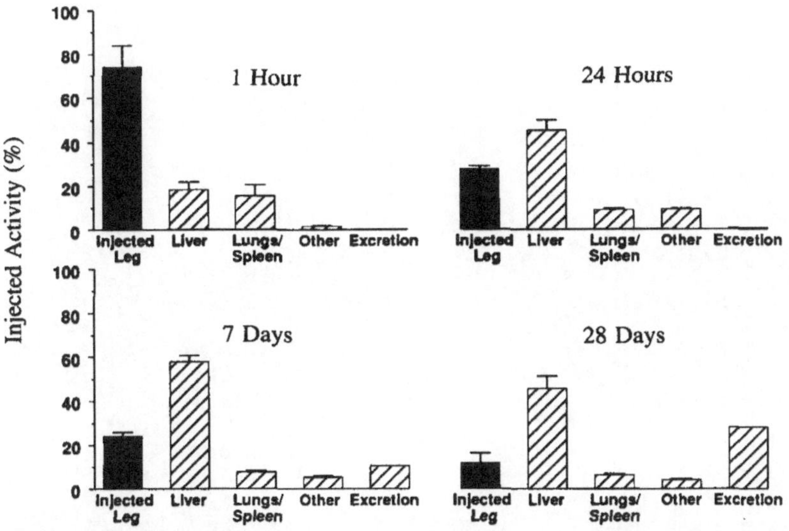

FIGURE 8. Distribution of ^{125}I-PKH-95-labeled endothelial cells in tissue and organs at 1 hr, 24 hr, 7 days, and 28 days after injection of transduced endothelial cells into rat femoral artery. At 1 hr after the clamp was removed, 74% of the injected radioactivity was detected in the hindlimb. At 24 hrs after injection, 27% of the radioactivity was present in the hindlimb, and at 7 days, 24% of the radioactivity was detected. At 28 days, 12% of the radioactivity was still present in the hindlimb. The remainder of the radioactivity resided in organs containing a portion of the reticuloendothelial system, the liver, lungs, and spleen. Selected histologic examination of the liver, lungs, and spleen did not establish definitively whether these cells were viable. (With permission from Reference 54.)

The long-term effectiveness of this approach to somatic gene therapy will depend upon the durability of recombinant gene expression in endothelium, the life span of the seeded endothelial cells, and the long-term consequences of this transplantation technique on capillary bed and skeletal muscle function.

B. Direct Gene Transfer to the Arterial Wall In Vivo

Considerable progress has been made in the development of experimental models in which direct gene transfer to the arterial wall *in vivo* has been accomplished. The impetus for this has been due in part to the realization of the complexity of *ex vivo* gene transfer and subsequent implantation of the transduced cells *in vivo*. It is also due to advances in techniques of successful gene transfer, specifically the development of adenoviral vectors, as well as efficient liposome-mediated gene transfer.

Direct gene transfer into a host target cell was first reported in 1989 by Palella and associates.[55] A viral vector was constructed from the herpes simplex virus-I, a neurotrophic DNA virus, that included the human gene for hypoxanthine phosphoribosyltransferase. After transfection of mice, the human hypoxanthine phosphoribosyltransferase was found to be expressed in the central nervous system. Others have since accomplished

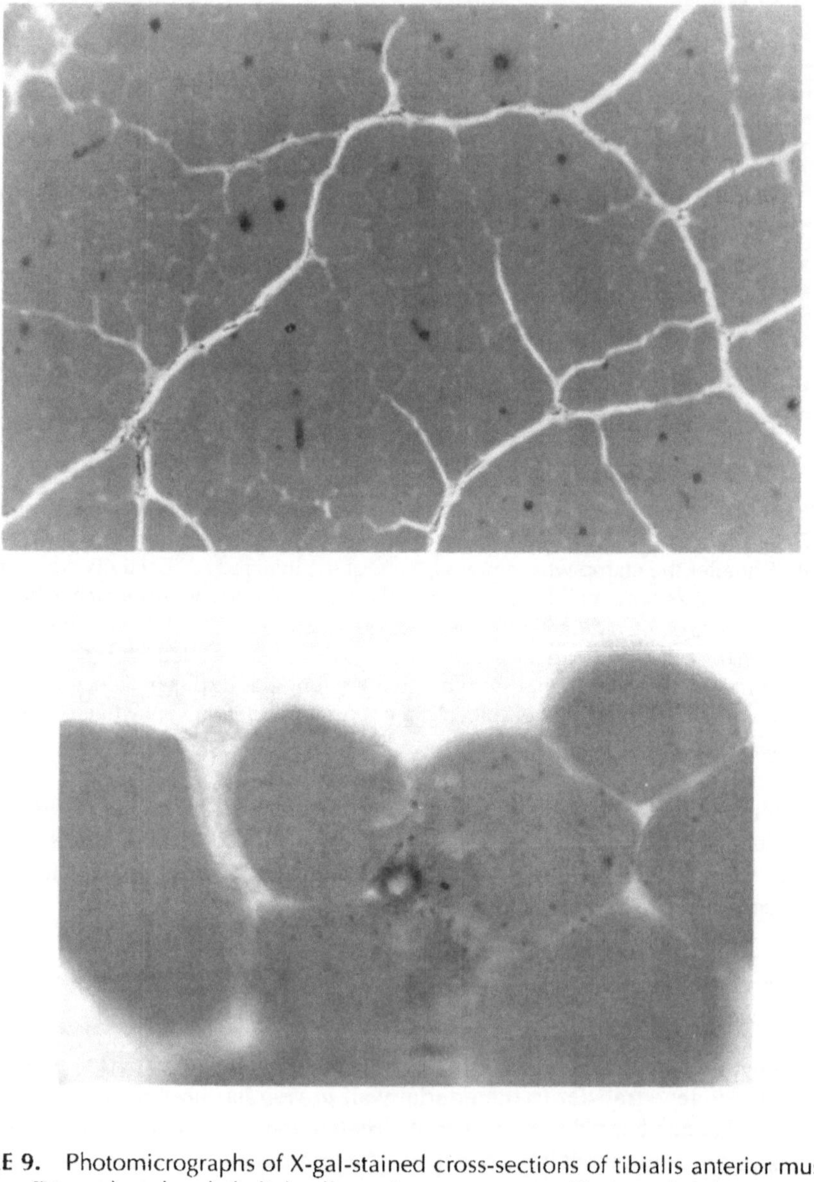

FIGURE 9. Photomicrographs of X-gal-stained cross-sections of tibialis anterior muscles. (Top) *lac-Z*-transduced endothelial cells are in numerous capillaries 1 hr after injection of transduced endothelial cells (original magnification 70×). Discrete areas of blue are noted only in the regions of capillaries between muscle fibers. (Bottom) *lac-Z*-transduced endothelial cell within skeletal muscle capillary 24 hr after injection. (With permission from Reference 54.)

direct gene transfer to other organs, such as with DNA injection into skeletal muscle[14] or receptor-mediated direct gene transfer to the liver.[16] Direct gene transfer avoids the problems of harvesting and culturing of target cells *in vitro.* The latter may alter cell function, cell surface receptors, as well as the life span of the cell, and may subject cells to the risk of contamination or infection.

Currently the major disadvantage of direct gene transfer is that it is more difficult to achieve both site and target cell specificity. In addition, the two most successful techniques for direct gene transfer to a target cell *in vivo*, adenoviral and liposome-mediated transfection, have not been shown to be capable of achieving long-term durable recombinant gene expression.

In September, 1990, Nabel and colleagues reported the first successful direct transfer of recombinant genes to the arterial wall *in vivo*.[56] Direct gene transfer was accomplished both with retroviral vectors as well as liposomes. The retroviral vector was derived from the Moloney murine leukemia virus and utilized the promoter from the chicken β-actin gene to express β-galactosidase mRNA. The viral particles from the supernatant of transfected ψ CRIP retrovirus packaging cells were concentrated by centrifugation to achieve titers of 10^4 to 10^6 particles/ml. The experimental model was the Yucatan or outbred pig iliofemoral artery, as described earlier by the same group.[47] Using a specially designed arterial balloon catheter, a portion of the iliofemoral artery was isolated into which the retroviral vector containing the *lac-Z* gene was introduced. Polybrene (10 μg/ml) was added to enhance transfection efficiency.

Arterial segments treated in the aforementioned manner revealed evidence of β-galactosidase activity from 10 days to 21 weeks after transfection. Optimal expression was observed 2 to 3 months after transfection. Microscopic analysis of the arterial segments shows that while the transfection was site-specific, that is, no other arterial segment was found to be transfected, it was not cell-specific, with blue staining cells found across the wall of the iliofemoral artery.

Site-specific gene transfer to the arterial wall was also achieved by transfecting iliofemoral arterial segments *in vivo*, using liposomes containing the *lac-Z* plasmid, prepared by combining 30 μg of DNA and 100 ml of lipofectin (Bethesda Research Laboratories) in serum-free medium. After irrigating the arterial segment with the serum-free medium, the liposomes were instilled into the artery through a catheter, and incubation was allowed for 30 min. Subsequently, flow was restored, and the catheter was removed. Arterial segments were removed between 4 and 42 days later. Microscopic examination of transfected arteries revealed β-galactosidase activity in the intima, medial, and adventia. No staining was observed in control arteries. No evidence of helper virus activity nor reverse transcriptase was noted in the serum, and no β-galactosidase activity was observed in gross or randomized microscopic sections from the liver, lung, or kidney. This study raises the possibility of achieving recombinant gene expression at any site within the arterial or venous circulation that is readily accessible by a catheter.

The technique of direct gene transfer to the arterial wall *in vivo* was extended to different species and different arteries by other investigators. Lim and colleagues evaluated the applicability of direct gene transfer to femoral and coronary arteries in dogs.[57] They noted variable but substantial endogenous β-galactosidase-like activity induced by

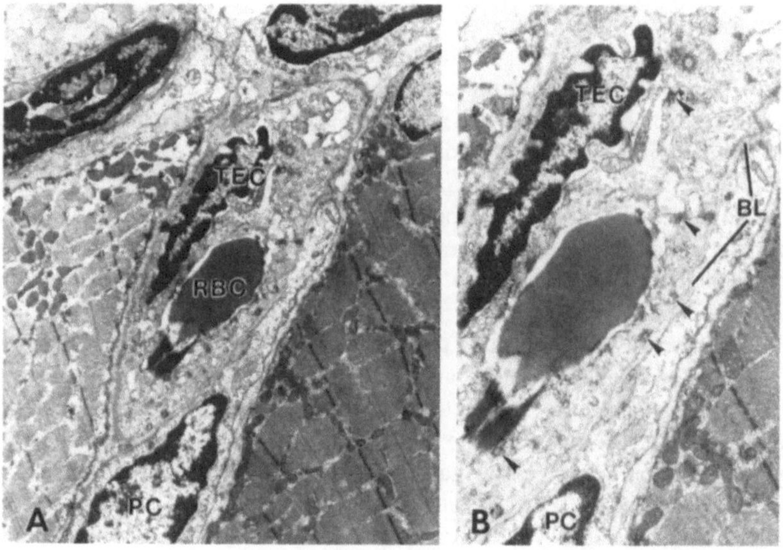

FIGURE 10. Electron micrograph of a transduced endothelial cell (TEC) within skeletal muscle capillaries of the rat hind limb 7 days after transplantation. Note the erythrocyte (RBC) within the capillary lumen and the pericyte (PC) adjacent to the capillary (original magnification 4350×).

simple manipulation of the canine peripheral arteries that precluded further use of *lac-Z* as a reporter or marker gene in evaluating the efficiency of gene transfer to the arterial wall in dogs. Subsequently, these investigators used a gene that encoded the firefly luciferase protein and found little or no background activity in the arterial wall of the dog even after arterial manipulation. These investigators demonstrated direct liposome-mediated gene transfer and expression up to 3 days in both the coronary and peripheral arteries of the intact dog. The liposome-DNA solution was introduced into these arteries by direct exposure and cannulation of a branch of the targeted arterial segment.

In a subsequent report from the same group, Chapman and associates undertook a second study with a liposome-plasmid complex containing the cDNA encoding for luciferase as in their prior study. The purpose of this second experiment was to develop a percutaneous method for direct gene transfer into the arterial wall of intact mongrel dogs.[58] Using percutaneous catheterization of the coronary arteries via a porous perfusion balloon system, recombinant gene expression was documented between 3 and 5 days in 8 of the 12 transduced arterial walls, averaging 4.3 pg luciferase. No activity was noted in 12 control coronary arteries. Using the same technique, these same investigators showed that the femoral artery could be transfected directly, both with naked DNA and DNA-liposome complexes. Unexpectedly, there was little difference in expression between these groups. In femoral arteries treated with naked DNA, 35 pg luciferase activity was found, whereas 42 pg luciferase activity was found in the arteries treated with the DNA complexed to liposomes. It is of note that their prior study on isolated femoral arteries was accomplished by an open surgical technique, and substantially higher rates of recombinant gene activity were shown. This raises the question of a lack

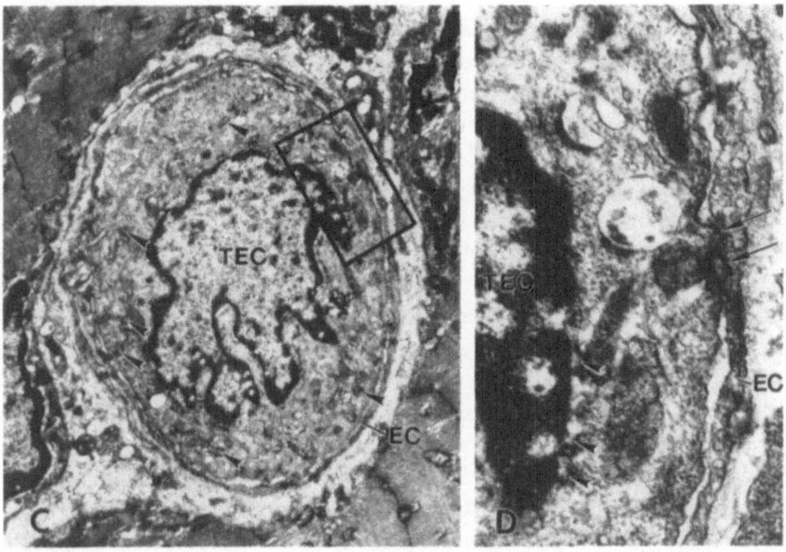

FIGURE 10 (continued). Enlargement of a portion of the capillary in Figure 10A. The transduced cell (TEC) is incorporated fully into the wall and against the capillary basal lamina (BL). Numerous X-gal-stained granules (arrowheads) are apparent (original magnification 7500×). (C) Electron micrograph of transduced endothelial cell within the lumen of a muscle capillary 7 days after transplantation. The TEC contains numerous X-gal-stained granules (arrowheads) and is closely opposed to the normal capillary endothelial cell (EC) (original magnification 5600×). (D) enlargement of the rectangular area of (C). In many focal regions, the cell (TEC) has membrane densities (double arrows) adjacent to the normal EC membrane. Note the X-gal-stained granules (arrowheads) (original magnification 25000×). (With permission from Reference 54.)

of complete isolation of the arterial segment by the catheter technique. Nonetheless, these results established the feasibility of direct gene transfer using a standard cardiovascular interventional technique.

Direct gene transfer to the arterial wall by an exclusively percutaneous strategy was supported further by the work of Leclerc and associates.[59] They undertook direct gene transfer in 10 normal and 12 atherosclerotic rabbit external iliac arteries by transfection with a solution of luciferase plasmid and liposomes, using a dual-balloon catheter system. Luciferase activity was detected in 10 of the 22 arteries in which direct gene transfer was attempted, including 4 of 10 normal arteries and 6 of 12 balloon-injured atherosclerotic arteries. No luciferase activity was detected in the contralateral control arterial segments. *In situ* hybridization of successfully transfected atherosclerotic sections documented luciferase gene mRNA in cells limited to the intima. The low efficiency of gene transfer in this study, less than 0.1%, would likely improve as more efficient techniques of direct gene transfer are developed. Thus, direct transfer can occur both to normal arteries as well as to diseased atherosclerotic arteries by an exclusively percutaneous approach.

Lemarchand and colleagues studied the feasibility of direct gene transfer using adenoviral vectors in the carotid arteries and jugular veins of sheep.[60] These vessels were isolated, occluded, and infused *in vivo* with a solution comprised of an adenoviral vector containing either the *lac-Z* gene or a human α_1-antitrypsin (α_1-AT) cDNA. After a 15-min incubation, the circulation was restored. Preliminary *in vitro* studies showed the adenoviral constructs to have a high rate of transduction efficiency, with 58% of endothelial cells transduced after only 15 min of exposure to the vector and 100% transduced after 24 hr of exposure. The vessels were harvested later and evaluated for the extent of gene transfer and expression. The contralateral artery and vein served as controls and were transfected by an adenoviral vector containing the human cystic fibrosis transmembrane conductance regulator (CFTR) cDNA. Northern blot analysis was used to determine the success of gene transfer and expression at the mRNA level. In order to evaluate the ability to achieve direct gene transfer and expression of a secreted protein, the biosynthesis and secretion of human α_1-AT in the vessels were evaluated *ex vivo*. Histochemical study of arterial segments transfected with the adenoviral vector containing the *lac-Z* gene showed both cell- and site-specific gene transfer. β-Galactosidase staining was observed only in endothelial cells. In other respects the arterial wall was reported to be normal. Importantly, no positive staining was seen in the contralateral artery exposed to the AdCFTR vector. In a subset of experimental animals, tissues from the cervical muscle, liver, lung, heart, spleen, brain, both kidneys, both gonads, and thyroid were excised and stained for β-galactosidase activity. None of these tissues showed any activity, except for the thyroid. The thyroid was included as a positive control, since light cytoplasmic granules of blue coloration are normally found in this organ. Northern blot analysis confirmed recombinant gene expression of the α_1-AT gene at 1, 7, and 14 days after transfection. Northern blots were negative in the 28-day animals. These results are consistent with the investigators' prior studies in which the same vectors were used for gene transfer to human umbilical vein endothelial cells *in vitro* and to intact umbilical veins in an *ex vivo* model.[25]

Adenoviral-mediated gene transfer was efficient in the above studies. Gene expression was easily detectable, being maximal at 7 days, after which it subsequently declined. The investigators hypothesized that the lack of expression at 28 days may have been caused by endothelial cell proliferation induced by minor trauma to the arterial wall during the procedure. Another possibility is that the heterologenous gene products, being tissue proteins, may have induced a cytocidal immune response. Finally, expression may have ceased because cellular enzymes destroyed the episomic transgene.

C. Direct Gene Transfer to Lung Endothelium after Intravenous Injection of Liposome-DNA Complexes

The lungs have a large endothelial cell surface area, and liposomes injected intravenously appear to home selectively to its microcirculation. Two recent studies pursued this technique in experimental models.

In the first study, Brigham and associates evaluated cultured pulmonary endothelial cells exposed to a plasmid containing the coding region for HGH (human growth hormone), driven by a metallothionein promoter.[61] After establishing regulation of recombinant gene expression *in vitro* by exposure to dexamethasone or cadmium, *in vivo* experiments with the same gene construct were undertaken. Twenty-four hr prior to the intravenous injection of a plasmid containing the metallothionein promoter complex to liposomes, mice were given 5,000 ppm $ZnSO_4$ in drinking water. Mice were sacrificed 1, 3, and 5 days after injection of the liposomes and HGH production was assayed in the lung, liver, and kidneys. Neither the kidneys nor the liver had detectable HGH. HGH was detected in the lungs beginning at 1 day, peaking on day 3, and declining subsequently. Lungs from control animals in which either DNA alone or liposome alone were injected did not produce HGH. In the experimental group, recombinant gene expression was confirmed by mRNA detection and PCR amplification of the cDNA followed by agarose gel electrophoresis. Thus, these investigators achieved direct gene transfer to the lung endothelium after intravenous infusion of a liposome-DNA complex. Furthermore, they showed that the recombinant gene expression could be controlled *in vivo* by induction of promoter regions. A shortcoming of this study was that no histologic analysis of the lung was undertaken to determine the cell specificity of the technique.

In the second study, Zhu and colleagues achieved widespread transfection of vascular endothelium and parenchymal tissues after the administration of a liposome-DNA complex in adult mice.[62] These investigators injected various mixtures of a cytomegalovirus chloramphenicol acetyltransferase (CAT) plasmid and DOTMA:dioleoylphosphatidylethanolamine (DOPE) liposomes. An optimal ratio of 1 μg of plasmid DNA to 8 nmol of liposomal lipid produced maximal CAT gene expression in lung, heart, and lymph node tissues. The ratio of DNA to cationic liposomes determines the net surface charge of the complex. The latter alters the interaction of the complex with potential opsonins in the circulation and alters the ability of the complex to bind and enter cells *in vivo*. A single intravenous injection of this cationic liposome-plasmid complex into adult mice efficiently transfected vascular endothelium and most of the extravascular parenchymal cells in many tissues, including the lung, spleen, lymph nodes, and bone marrow. The recombinant gene was expressed in a large number of cells in differing tissues for at least 9 weeks after a single injection. Nearly all the lung endothelium and parenchyma showed evidence of transfection. Recombinant gene expression was confined to the vascular endothelium in both the heart and kidney, with only a few extraparenchymal cells appearing transfected. In contrast, the bone marrow, liver, spleen, and lymph nodes showed extensive endothelial cell as well as parenchymal cell transfection. No evidence of tissue toxicity was identified.

Thus, this simple technique of *in vivo* gene transfer provided a rapid and reproducible transfer and expression of recombinant gene directly in adult animals. Such a technique could be modified to achieve endothelial cell-specific gene transfer. Although these results are preliminary, if the extent and durability of gene transfer can be achieved with similar ease in other experimental species, this would represent the most efficient degree of direct gene transfer to the endothelium identified at the time of this publication.

D. Direct Gene Transfer to the Arterial Wall to Study Vascular Biology

In a series of elegant experiments, Nabel and her colleagues have explored the use of gene transfer to the arterial wall to elucidate mechanisms that cause vasculitis and intimal hyperplasia.[63-65] Although these studies do not address the use of direct gene transfer to the arterial wall for gene therapy, they reveal the importance of direct gene transfer to the arterial wall as a means of increasing our understanding of vascular wall biology. In particular, gene transfer into the arterial wall provides an experimental model to define the role of specific gene products in vascular pathology.

In the first of their studies, the human class I major histocompatibility complex gene HLA-B7 was used to develop a large animal model of vasculitis, documenting for the first time that direct gene transfer into arteries can induce a biologic response *in vivo*.[63] A PLJ-retroviral vector containing the HLA-B7 cDNA was introduced into an amphotrophic Moloney murine leukemia virus packaging cell line by calcium phosphate transfection. After selection in the neomycin analogue G418, viral supernatants were concentrated by centrifugation. A similar vector containing the *lac-Z* gene was prepared in an identical manner. Direct gene transfer was accomplished by the use of DNA-liposome complexes. These complexes consisted of 30 μg of lipofectin (Bethesda Research Laboratories, MD) diluted in serum-free media just prior to instillation.

Direct gene transfer of HLA-B7 or *lac-Z* gene into porcine arteries was undertaken by exposure to supernatant containing the appropriate retroviral vector at titers of 10^5 to 10^6 particles/ml or plasmid DNA-liposome complexes of the same recombinant genes. This was accomplished by instilling these solutions into regions of normal femoral arteries of 15 pigs using a specially designed balloon catheter system. The pressure generated in the balloon and within the artery wall itself were controlled carefully.

These studies showed that porcine femoral arteries could express human HLA-B7 antigen on the cell surface. Autoradiography showed that the DNA-liposome complex reached all layers of the arterial wall, the highest expression being in the adventia. DNA-PCR confirmed the presence of the vector at 2 to 4 hr after instillation.

To determine whether the expressed human HLA-B7 protein induced a specific cellular immune response, lymphocytes were analyzed for cytolytic T-cell activity using a chromium release assay. After *in vitro* sensitization, cells from the experimental group in which the HLA-B7 gene was transferred to the pig arterial wall lysed radionuclide-labeled HLA-B7-positive but not HLA-B7-negative endothelial cells. Finally, in those vessels transduced with the HLA-B7 retroviral vector, an intense mononuclear infiltrate was observed starting at day 10 after gene transfer, with maximal changes noted at 2 to 4 weeks after gene transfer. By 10 weeks, this inflammation had subsided. A parallel reaction occurred in the arteries in which the HLA-B7 gene was delivered by a DNA-liposome complex. A prominent finding in the histologic sections of the arterial wall was an "onion skin pattern" of perivascular cuffing and granuloma formation surrounding small capillaries. This immune response in the adventia was noted with lesser degrees of inflammation in the media and intima.

An interesting and somewhat unexpected finding was that an inflammatory reaction of lesser intensity was observed in the HLA-B7 transduced animals that had undergone sham infection or balloon injury, but in which no vector was introduced in the control artery. No HLA-B7 DNA was detected by PCR in the sham-infected or balloon-injured arteries. The authors interpreted this to indicate that the immune response to the HLA-B7 antigen expressed in the experimental arteries enhanced the systemic sensitivity to antigens expressed during and after simple balloon injury in the arterial wall. The authors noted that this type of response in the control sham-infected arteries has not been seen with any other vector, including vectors for platelet-derived growth factor (PDGF), acidic fibroblast growth factor (FGF), or transforming growth factor (TGF) α and β. These studies show that direct gene transfer to the arterial wall *in vivo* can result in transfer of a gene that induces a potent, site-specific biologic response.

Heparin-binding FGF, acidic-FGF, and basic FGF are growth factors acting directly on endothelial cells to induce angiogenesis. The role of FGF growth factors in vessel wall pathology, specifically intimal hyperplasia, is not known. Nabel and colleagues have explored the effect of FGF-I in porcine arteries after direct gene transfer.[64] Because FGF polypeptide lacks a classic signal sequence for secretion, an expression vector was created by ligation of the signal sequence from the hst/KS3 (FGF-IV) gene to the 5′ end of the open reading frame of the FGF-I in a pmex neoeukaryotic expression vector. Porcine iliofemoral arteries were subjected to direct gene transfer with this vector, and control arteries were transfected with the *lac-Z* gene. The presence of the vector and mRNA of FGF-I was confirmed by PCR. Immunohistocytochemistry showed localization of recombinant gene expression primarily in the intima, including the endothelium, up to 21 days after transfection. Although intimal thickening was seen both in the arteries exposed to the *lac-Z* gene as well as to the FGF gene, a six-fold greater intimal-to-media ratio (a standard method for quantitating intimal hyperplasia) occurred in the arteries in which the FGF gene was expressed. In several animals, there appeared to be formation of capillaries in the neointima. Thus expression of secreted recombinant FGF-I induced intimal hyperplasia and angiogenesis *in vivo.*

The source of the endothelial cells for this neovascularization were most likely the luminal endothelial cells. The evidence for this was that the endothelial cells of the neocapillaries did not contain Von Willebrand's factor. Luminal endothelial cells of porcine iliofemoral arteries do not normally contain Von Willebrand's factor. However, endothelial cells of the capillaries of the adventia of the iliofemoral arteries are normally positive for Von Willebrand's factor. A similar type of neovascularization is seen in atherosclerotic plaques, and to date no mechanism accounting for such has been established. This study suggests that FGF may be important neovascularization of arterial wall tissue.

Platelet-derived growth factor β (PDGF-β) induces cell proliferation *in vitro* and has been implicated in a variety of experimental models to induce intimal hyperplasia *in vivo*. However, due to the complexity of cellular and protein interactions *in vivo* that occur when vessel wall injury is induced in an experimental model, it has been difficult to determine the role of specific gene products. In order to define the function PDGF-β within arteries, Nabel and colleagues introduced a eukaryotic expression vector plasmid encoding recombinant PDGF-β by direct means into porcine iliofemoral arteries using DNA-liposome complexes.[65] Expression of PDGF caused intimal hyperplasia 21 days after transfection with the recombinant gene. DNA PCR showed that approximately 0.1 to 1% of the cells in the artery segment contain the plasmid DNA.

At the time of direct gene transfer, the DNA-liposome complex was introduced in one iliofemoral artery at a pressure of 150 mmHg and in the other at a pressure of 350 mmHg for 30 min. This was done in order to determine an instillation pressure of the DNA-liposome complex that did not injure the vessel wall itself. This would allow a clearer understanding of the effect of PDGF independent of its effect on an already mechanically injured vessel. No intimal hyperplasia was observed in the iliofemoral arteries after instillation at 150 mmHg, but intimal thickening was seen during saline infusion at the 350 mmHg pressure.

Immunocytochemistry showed that at 150 mmHg pressure, no detectable PDGF-β was detected in the arteries transfected with the *lac-Z* gene. Thus, endogenous PDGF in the uninjured arteries transfected at the low pressure was undetectable. In contrast, arteries transfected with PDGF-β at low pressure revealed immunoreactive proteins in the intima, media, and adventia.

To determine whether macrophages could enter the transfected arteries to express recombinant PDGF protein, histochemical staining with a monoclonal antibody to porcine macrophages was undertaken. Macrophages in the intima were detected rarely in PDGF-transduced vessels and were not seen in *E. coli* β-galactosidase-transduced arteries. Thus, this study shows that expression of recombinant PDGF-β gene in normal arteries can induce intimal hyperplasia *in vivo*.

IV. POTENTIAL APPLICATIONS TO HUMAN DISEASE

The remarkable success of the techniques of gene transfer to endothelium as well as the experimental models of gene transfer to the arterial wall *in vivo* indicate that recombinant gene transfer to endothelial cells will play a role of fundamental importance in the development of somatic gene therapy for both cardiovascular and noncardiovascular human diseases. In addition, recombinant gene transfer to endothelium has become a powerful tool to study arterial wall pathobiology.

Use of vascular endothelium for human somatic gene therapy is enhanced because, unlike target cells in other organs, vascular endothelium can be accessed by a variety of percutaneous approaches that are already fully developed and of proven safety in humans. The applications of somatic gene therapy to vascular disease will likely be of considerable scope. Since the most common form of primary vascular disease is focal or segmental atherosclerosis, techniques of somatic gene therapy could be developed to reverse this process or that of intimal hyperplasia, a lesion that complicates many currently available interventional treatments. Gene transfer to the arterial wall directed at inhibition of the genes signaling or regulating cell proliferation will be important in treating many vascular diseases.

Somatic gene therapy may also be directed toward a reversal of the increased thrombogenicity of injured or diseased arterial wall as well as of small-caliber synthetic grafts and vascular stents. Evidence already exists that durable enhancement of the fibrinolytic activity of endothelial cells can be achieved without altering the phenotype and functional capabilities of the transduced cells.

Gene transfer to endothelium is likely to have application to a variety of nonvascular diseases. The ease of access and large number of target cells available make gene transfer to endothelium for the purpose of secreting a recombinant protein for a systemic effect an important possibility. This opportunity is best illustrated by efforts directed at gene therapy of hemophilia A and B, or α_1-antitrypsin deficiency.

The specific challenges for the clinical application of gene therapy biotechnology include (1) achieving a technique of gene transfer that is both efficient and durable, without long-term negative consequences on function of the transduced cell, (2) development of techniques of gene transfer that are site- and cell-specific, (3) achieving regulation of recombinant gene expression, including the availability of promoters directing cell-specific expression, of suicide vectors, and of inducible promoters, and (4) better characterization of the biologic fate of recombinant DNA that is not incorporated into host cell chromosomal DNA.

Recently, Ohno and associates[66] reported an important study concerning the application of somatic gene therapy to the treatment of vascular disease. These investigators were able to inhibit intimal hyperplasia after balloon injury to a porcine artery by transfecting the arterial wall with an adenoviral vector containing the enzyme thymidine kinase. Thymidine kinase expression in the transfected cells rendered them sensitive to the administration of the nucleoside analogue ganciclovir. Thus, transient expression of an enzyme that catalyzes the formation of a cytotoxic drug may limit intimal hyperplasia after balloon injury.

Another recent development that is of great significance to the field of somatic gene therapy is the approval by the Research Advisory Committee of the National Institutes of Health of the first gene therapy protocol in humans for cardiovscular disease. This trial will test the efficacy of vascular endothelial growth factor (vegF) delivered by an angioplasty catheter impregnated with the vegF genes to stimulate collateral artery development in patients with limb threatening ischemia.

Another recent event that is of great significance to the field of somatic gene therapy is the approval by the Research Advisory Committee of the National Institute of Health of the first gene therapy protocol in humans for cardiovascular disease on September 13, 1994. This trial will test the efficacy of vascular endothelial growth factor (vegF) delivered by an angioplasty catheter impregnated with the vegF genes to stimulate collateral artery development in patients with limb threatening ischemia.

REFERENCES

1. Vane, J. R., Anggard, E. E., and Botting, R. M., Regulatory functions of the vascular endothelium, *N. Engl. J. Med.*, 323, 27, 1990.
2. Ward, P. A., Mechanisms of endothelial cell injury, *J. Lab. Clin. Med.*, 118, 421, 1991.
3. Pober, J. S., Warner-Lambert/Parke-Davis award lecture. Cytokine-mediated activation of vascular endothelium. Physiology and pathology, *Am. J. Pathol.*, 133, 426, 1988.
4. Belloni, P. N. and Tressler, R. J., Microvascular endothelial cell heterogeneity: interactions with leukocytes and tumor cells, *Cancer Metast. Rev.*, 8, 353, 1990.
5. Center for Biologics Evaluation and Research, Food and Drug Administration, Points to consider in human somatic cell therapy and gene therapy, *Hum. Gene Ther.*, 2, 251, 1991.
6. Goldsmith, M. F., Tomorrow's gene therapy suggests plenteous, patent cardiac vessels, *JAMA*, 268, 3285, 1992.
7. Blaese, R. M., Progress toward gene therapy, *Clin. Immunol. Immunopathol.*, 61, S47, 1991.
8. Dichek, D. A., Retroviral vector-mediated gene transfer into endothelial cells, *Mol. Biol. Med.*, 8, 257, 1991.
9. Kelley, W. N., Gene therapy in humans: a new era begins, *Ann. Int. Med.*, 114, 697, 1991.

10. Nabel, E. G., Plautz, G., and Nabel, G. J., Gene transfer into vascular cells, *JACC*, 17, 189B, 1991.
11. Miller, A. D., Progress toward human gene therapy, *Blood*, 76, 271, 1990.
12. Anderson, W. F. and Kelley, W. N., Co-chairs, Summary of the Human Gene Therapy Conference, presented by the National Institutes of Health, 1991.
13. Mulligan, R. C., The basic science of gene therapy, *Science*, 260, 926, 1993.
14. Wolff, J. A., Malone, R. W., Williams, P., Chong, W., Acsadi, G., Jani, A., and Felgner, P. L., Direct gene transfer into mouse muscle in vivo, *Science*, 247, 1465, 1990.
15. Potter, H., Electroporation in biology: methods, applications, and instrumentation, *Anal. Biochem.*, 174, 361, 1988.
16. Wu, G. Y. and Wu, C. H., Delivery systems for gene therapy, *Biotherapy*, 3, 87, 1991.
17. Wu, C. H., Wilson, J. M., and Wu, G. Y., Targeting genes: delivery and persistent expression of a foreign gene driven by mammalian regulatory elements *in vivo*, *J. Biol. Chem.*, 264, 16985, 1989.
18. Felgner, P. L. and Ringold, G. M., Cationic liposome-mediated transfection, *Nature*, 337, 387, 1989.
19. Danos, O. and Mulligan, R. C., Safe and efficient generation of recombinant retroviruses with amphotropic and ecotropic host ranges, *Proc. Natl. Acad. Sci. U.S.A.*, 85, 6460, 1988.
20. Miller, A. D., Retrovirus packaging cells, *Hum. Gene Ther.*, 1, 5, 1990.
21. Miller, D. G., Adam, M. A., and Miller, A. D., Gene transfer by retrovirus vectors occurs only in cells that are actively replicating at the time of infection, *Mol. Cell Biol.*, 10, 4239, 1990.
22. Quantin, B., Perricaudet, L. D., Tajbakhsh, S., and Mandel, J.-L., Adenovirus as an expression vector in muscle cells in vivo, *Proc. Natl. Acad. Sci. U.S.A.*, 89, 2581, 1992.
23. Stratford-Perricaudet, L. D., Makeh, I., Perricaudet, M., and Briand, P., Widespread long-term gene transfer to mouse skeletal muscles and heart, *J. Clin. Invest.*, 90, 626, 1992.
24. Ragot, T., Vincent, N., Chafey, P., Vigne, E., Gilgenkrantz, H., Couton, D., Cartaud, J., Briand, P., Kaplan, J.-C., Perricaudet, M., and Kahn, A., Efficient adenovirus-mediated transfer of a human minidystrophin gene to skeletal muscles of *mdx* mice, *Nature*, 361, 647, 1993.
25. Lemarchand, P., Jaffe, H. A., Danel, C., Cid, M. C., Kleinman, H. K., Stratford-Perricaudet, L. D., Perricaudet, M., Pavirani, A., Lecocq, J.-P., and Crystal, R. G., Adenovirus-mediated transfer of a recombinant human α_1-antitrypsin cDNA to human endothelial cells, *Proc. Natl. Acad. Sci. U.S.A.*, 89, 6482, 1992.
26. Gerard, R. D. and Meidell, R. S., Adenovirus-mediated gene transfer, *Trends Cardiovasc. Med.*, 3, 171, 1993.
27. Faller, D. V., Kourembanas, S., Ginsberg, D., Hannan, R., Collins, T., Ewenstein, B. M., Pober, J. S., and Tantravahi, R., Immortalization of human endothelial cells by murine sarcoma viruses, without morphologic transformation, *J. Cell Physiol.*, 134, 47, 1988.
28. van Zonneveld, A.-J., Curriden, S. A., and Loskutoff, D. J., Type 1 plasminogen activator inhibitor gene: functional analysis and glucocorticoid regulation of its promoter, *Proc. Natl. Acad. Sci. U.S.A.*, 85, 5525, 1988.
29. Lee, M.-E., Bloch, K. D., Clifford, J. A., and Quertermous, T., Functional analysis of the endothelin-1 gene promoter, *J. Biol. Chem.*, 265, 10446, 1990.
30. Wilson, D. B., Dorfman, D. M., and Orkin, S. H., A nonerythroid GATA-binding protein is required for function of the human preproendothelin-1 promoter in endothelial cells, *Mol. Cell Biol.*, 10, 4854, 1990.
31. Zwiebel, J. A., Freeman, S. M., Kantoff, P. W., Cornetta, K., Ryan, U. S., and Anderson, W. F., High-level recombinant gene expression in rabbit endothelial cells transduced by retroviral vectors, *Science*, 243, 220, 1989.

32. Zwiebel, J. A., Freeman, S. M., Cornetta, K., Forough, R., Maciag, T., and Anderson, W. F., Recombinant gene expression in human umbilical vein endothelial cells transduced by retroviral vectors, *Biochem. Biophys. Res. Commun.*, 170, 209, 1990.
33. Wilson, J. M., Birinyi, L. K., Salamon, R. N., Libby, P., Callow, A. D., and Mulligan, R. C., Genetically modified endothelial cells in the treatment of human diseases, *Trans. Assoc. Am. Phys.*, 102, 139, 1989.
34. Dichek, D. A., Nussbaum, O, Degen, S. J. F., and Anderson, W. F., Enhancement of the fibrinolytic activity of sheep endothelial cells by retroviral vector-mediated gene transfer, *Blood*, 77, 533, 1991.
35. Kahn, M. L., Lee, S. W., and Dichek, D. A., Optimization of retroviral vector-mediated gene transfer into endothelial cells in vitro, *Circ. Res.*, 71, 1508, 1992.
36. Jaklitsch, M. T., Biro, S., Casscells, W., and Dichek, D. A., Transduced endothelial cells expressing high levels of tissue plasminogen activator have an unaltered phenotype in vitro, *J. Cell Physiol.*, 154, 207, 1993.
37. Lee, S. W., Kahn, M. L., and Dichek, D. A., Expression of an anchored urokinase in the apical endothelial cell membrane, *J. Biol. Chem.*, 267, 13020, 1992.
38. Podrazik, R. M., Whitehill, T. A., Ekhterae, D., Williams, W. D., Messina, L. M., and Stanley, J. C., High-level expression of recombinant human tPA in cultivated canine endothelial cells under varying conditions of retroviral gene transfer, *Ann. Surg.*, 216, 446, 1992.
39. Xu, X.-M., Ohashi, K., Sanduja, S. K., Ruan, K.-H., Wang, L.-H., and Wu, K. K., Enhanced prostacyclin synthesis in endothelial cells by retrovirus-mediated transfer of prostaglandin H synthase cDNA, *J. Clin. Invest.*, 91, 1843, 1993.
40. Ramos, T. K., Lauer, S. J., Goldstone, J., and Taylor, J. M., Human M-CSF production by rabbit vascular endothelial and smooth muscle cells, *Surg. Forum*, 43, 334, 1992.
41. Thompson, A. R., Palmer, T. D., Lynch, C. M., and Miller, A. D., Gene transfer as an approach to cure patients with hemophilia A or B, *Biotechnology of Plasma Proteins. Curr. Stud. Hematol. Blood Transf.*, No. 58, Albertini, A., Lenfant C. L., Mannucci, P. M., Sixma, J. J., Eds., Karger, Basel, 1991, 59.
42. Yao, S.-N., Wilson, J. M., Nabel, E. G., Kurachi, S., Hachiya, H. L., and Kurachi, K., Expression of human factor IX in rat capillary endothelial cells: toward somatic gene therapy for hemophilia B, *Proc. Natl. Acad. Sci. U.S.A.*, 88, 8101, 1991.
43. Powell, J. T., van Zonneveld, A. J., and van Mourik, J. A., Gene transfer into specific vascular cells, *Eur. J. Vasc. Surg.*, 6, 130, 1992.
44. Etchberger, K. J. and Taylor, M. W., Transfection with bacterial genes as a marker for cells seeded on vascular flow surfaces, *Ann. Vasc. Surg.*, 3, 123, 1989.
45. Brigham, K. L., Meyrick, B., Christman, B., Berry, L. C., and King, G., Expression of a prokaryotic gene in cultured lung endothelial cells after lipofection with a plasmid vector, *Am. J. Respir. Cell. Mol. Biol.*, 1, 95, 1989.
46. Flugelman, M. Y., Virmani, R., Leon, M. B., Bowman, R. L., and Dichek, D. A., Genetically engineered endothelial cells remain adherent and viable after stent deployment and exposure to flow in vitro, *Circ. Res.*, 70, 348, 1992.
47. Nabel, E. G., Plautz, G., Boyce, F. M., Stanley, J. C., and Nabel, G. J., Recombinant gene expression *in vivo* within endothelial cells of the arterial wall, *Science*, 244, 1342, 1989.
48. Wilson, J. M., Birinyi, L. K., Salomon, R. N., Libby P., Callow, A. D., and Mulligan, R. C., Implantation of vascular grafts lined with genetically modified endothelial cells, *Science*, 244, 1344, 1989.
49. Greisler, H. P., Endothelial cell transplantation onto synthetic vascular grafts: panacea, poison, or placebo, *New Biological and Synthetic Vascular Prosthesis.* Greisler, H. P., Ed., R. G. Landes Co., Austin, 1991, 47.

50. Podrazik, R. M., Whitehill, T. A., Komorowski, T. A., Karo, K. H., Messina, L. M., and Stanley, J. C., *In vivo* fate of *lacZ*-transduced endothelial cells seeded on ePTFE thoracoabdominal vascular prostheses in dogs, *Surg. Forum*, 44, 1993.
51. Brothers, T. E., Judge, L. M., Wilson, J. M., and Stanley, J. C., Effect of genetic transduction on *in vitro* canine endothelial cell prostanoid production and growth, *Surg. Forum*, 41, 337, 1989.
52. Plautz, G., Nabel, E. G., and Nabel, G. J., Selective elimination of recombinant genes *in vivo* with a suicide retroviral vector, *New Biol.*, 3, 709, 1991.
53. Bernstein, S. C., Skoskiewicz, M. J., Jones, R., Zapol, W. M., and Russell, P. S., Recombinant gene expression in pulmonary vascular endothelial cells: polarized secretion *in vivo*, *FASEB J.*, 4, 2665, 1990.
54. Messina, L. M., Podrazik, R. M., Whitehill, T. A., Ekhterae, D., Brothers, T. E., Wilson, J. M., Burkel, W. E., and Stanley, J. C., Adhesion and incorporation of *lacZ*-transduced endothelial cells into the intact capillary wall in the rat, *Proc. Natl. Acad. Sci. U.S.A.*, 89, 12018, 1992.
55. Palella, T. D., Hidaka, Y., Silverman, L. J., Levine, M., Glorioso, J., and Kelly, W. N., Expression of human HPRT mRNA in brains of mice infected with a recombinant herpes simplex virus-1 vector, *Gene*, 80, 137, 1989.
56. Nabel, E. G., Plautz, G., and Nabel, G. J., Site-specific gene expression *in vivo* by direct gene transfer into the arterial wall, *Science*, 249, 1285, 1990.
57. Lim, C. S., Chapman, G. D., Gammon, R. S., Muhlestein, J. B., Bauman, R. P., Stack, R. S., and Swain, J. L., Direct *in vivo* gene transfer into the coronary and peripheral vasculatures of the intact dog, *Circulation*, 83, 2007, 1991.
58. Chapman, G. D., Lim, C. S., Gammon, R. S., Culp, S. C., Desper, J. S., Bauman, R. P., Swain, J. L., and Stack, R. S., Gene transfer into coronary arteries of intact animals with a percutaneous balloon catheter, *Circ. Res.*, 71, 27, 1992.
59. Leclerc, G., Gal, D., Takeshita, S., Nikol, S., Weir, L., and Isner, J. M., Percutaneous arterial gene transfer in a rabbit model: efficiency in normal and balloon-dilated atherosclerotic arteries, *J. Clin. Invest.*, 90, 936, 1992.
60. Lemarchand, P., Jones, M., Yamada, I., and Crystal, R. G., *In vivo* gene transfer and expression in normal uninjured blood vessels using replication-deficient recombinant adenovirus vectors, *Circ. Res.*, 72, 1132, 1993.
61. Brigham, K. L., Meyrick, B., Christman, B., Conary, J. T., King, G., Berry, L. C., and Magnuson, M. A., Expression of human growth hormone fusion genes in cultured lung endothelial cells and in the lungs of mice, *Am. J. Respir. Cell Mol. Biol.*, 8, 209, 1993.
62. Zhu, N., Liggitt, D., Liu, Y., and Debs, R., Systemic gene expression after intravenous DNA delivery into adult mice, *Science*, 261, 209, 1993.
63. Nabel, E. G., Plautz, G., and Nabel, G. J., Transduction of a foreign histocompatibility gene into the arterial wall induces vasculitis, *Proc. Natl. Acad. Sci. U.S.A.*, 89, 5157, 1992.
64. Nabel, E. G., Yang, Z., Plautz, G., Forough, R., Zhan, X., Haudenschild, C. C., Maciag, T., and Nabel, G. J., Recombinant fibroblast growth factor-1 promotes intimal hyperplasia and angiogenesis in arteries *in vivo*, *Nature*, 362, 844, 1993.
65. Nabel, E. G., Yang, Z., Liptay, S., San, H., Gordon, D., Haudenschild, C. C., and Nabel, G. J., Recombinant platelet-derived growth factor B gene expression in porcine arteries induces intimal hyperplasia *in vivo*, *J. Clin. Invest.*, 91, 1822, 1993.
66. Ohno, T., Gordon, D., Hong, S., Pompilli, V. J., Imperiale, M. J., Nabel, G. J., and Nabel, E. G., Gene therapy for vascular smooth muscle cell proliferation, *Science*, 265, 781, 1994.

8 Gene Therapy by Myoblast-Mediated Gene Transfer

Kotoku Kurachi and Shou-Nan Yao

I. INTRODUCTION

In developing a gene therapy for a disease, the choice of the target tissue for gene transfer is crucial.[1,2] For some types of disease, such as hypercholesterolemia, cystic fibrosis, Duchenne-type muscular dystrophy, and others, gene transfer must be targeted to the specific tissues in which the gene products naturally function. For diseases that require the delivery of gene products through systemic circulation, however, target tissues for gene transfer may not have to be those that naturally express the genes.[3,4] If detrimental effects to the targeted tissue and the rest of the body are avoided and if proper processing required for the gene product, such as post-translational modifications, cleavages of signal peptide or propeptide, and transportation to the circulation, is efficiently carried out, tissues or cells other than the natural site of the gene expression may be targeted for ectopic expression of the gene.[3,4] By the ectopic expression of the gene, we may be able to develop a relatively simple (minimally invasive) and highly durable gene therapy method, which otherwise would be impossible or extremely difficult to achieve.

Hemophilia B, for instance, is caused by deficiency of factor IX in the circulation.[5,6] Factor IX is naturally synthesized in hepatocytes.[7] Therefore, the liver is the primary target tissue for transferring the factor IX gene in a gene therapy for hemophilia B. Gene transfer methods currently available for liver tissue, however, are not simple and require substantial surgical procedures.[8-10] In an ideal gene therapy, these procedures should be avoided, if possible. Supported with this rationale, a number of tissues other than liver, including skin fibroblasts,[4,11,12] endothelial cells and vascular sites,[13,14] keratinocytes,[15] and skeletal muscle cells,[3,16-19] have been tested for their feasibility to serve as an efficient and safe vehicle for *ex vivo* or direct *in vivo* gene delivery and ectopic production of factor IX transgenes *in vivo*. Among these tissues and cells, skeletal muscle cells have been extensively tested, and their promising potential as a durable drug delivery vehicle has been demonstrated.[3,16-19]

Skeletal muscles compose the largest tissue in the body, occupying about 40% of total body mass with almost 660 muscles.[20] Skeletal muscles can be accessed easily with

0-8493-2440-8/95/$0.00+$.50

minimal invasiveness for various manipulations, such as biopsy, cell implantation, or direct injection of expression vectors that may be involved in gene therapy. Because skeletal muscle is the tissue commonly used for delivery of various drugs, such as insulin and vaccines, its general safety for such a purpose has been well established. In addition, muscles have a number of interesting and advantageous properties as a drug delivery vehicle. Besides easy access for various manipulations, these include relatively easy growth of primary myoblasts in culture and efficient fusion of implanted myoblasts with host myofibers or with themselves, a natural mechanism for repairing muscle damage.[21,22]

There are two paradigms of gene transfer methods that target skeletal muscles. One is the direct *in vivo* gene transfer approach, which involves injecting naked vector DNAs or vectors in various forms, such as adenovirus, DNA-liposome complex, and others. Another is the indirect *ex vivo* approach, typically utilizing myoblasts isolated into culture for *in vitro* gene transfer, followed by their implantation into muscle tissue.

In this chapter, we review the recent development of gene transfer technology using muscles as a drug delivery vehicle. Specifically, the myoblast-mediated gene transfer method is discussed in detail for its advantages, potential problems, applications, and future prospects in human gene therapy.

II. GENE TRANSFER INTO MUSCLE TISSUE

A. Direct In Vivo Gene Transfer

Direct *in vivo* gene transfer, which is discussed in detail in Chapter 11, is only briefly reviewed here to contrast the differences of indirect *ex vivo* approaches.

Wolff et al.[23] originally described direct marker gene transfer into muscle cells. Later, Acsadi et al.[24] reported human dystrophin expression in *mdx* mice after intramuscular injection of the expression vector. The efficiency of DNA transfer, however, was very low, and only about 1% of cells at the DNA-injected site could receive the transgene. Lin et al.[25] also reported the direct gene transfer and the transgene expression in cardiac muscles. The gene transfer efficiency was about the same as that observed for skeletal muscles. More recently, production and systemic effects of cytokines (IL-2,4 and TGF-β1) by direct injection of plasmid expression vector DNA into muscles of BALB/C mice were reported.[26] In this experiment, significant immunologic effects were generated, and even an eight-fold increase of the basal level of systemic TGF-β1 was reported. Although this is an important observation, to see the effect required three injections of 300 μg DNA in total. This amount is equivalent to about 0.6 g of DNA for a 50-kg man. The transfected DNA may be present in the tissue up to at least 11 weeks after DNA injections. However, injected DNA is primarily extrachromosomal and eventually disappears. This direct DNA injection approach may prove to be useful for various diseases if the expression vectors and delivery conditions involved are significantly improved.[27,28]

For congenital disorders like hemophilia, this approach is not efficient enough currently to satisfy the requirement for persistent systemic production of transgene product at a therapeutically significant level. Such production should last for the entire life span of the patients. When 100 to 200 μg of pLIXSN (human factor IX expression vector under LTR promoter, which can express about 3 μg of factor IX per day per 1×10^6 cultured myoblast cells) was directly injected intramuscularly in our testing, no detectable systemic human factor IX was found in mice (unpublished data). Multiple injections, as in the case of cytokines,[26] may be required to see any detectable amount of factor IX produced in the circulation. The repeated injection of a vector DNA, however, may cause the potential problem of immune reactions. No detailed studies have yet been conducted on this important issue.

Several alternative approaches of direct gene transfer into muscles have also been described. Adenovirus vectors have been used to deliver a gene of interest into muscle cells.[29-31] A possible anti-adenovirus antibody development upon repeated injections of adenovirus may also limit the efficacy. Direct gene transfer in muscle tissues by injecting liposome-DNA complex or DNA-ligand-adenovirus complex[32] is yet to be tested to establish its feasibility for transfering genes into muscles. DNA transferred by these methods also remains extrachromosomal, and its expression is temporary. A possible use of papilloma virus episomal element for a prolonged maintenance of the transferred gene or similar approaches is attractive, but a substantial amount of testing is still required to assess its usefulness. Limitations of direct gene transfer that targets muscle cells, such as inefficient gene transfer and instability of the transferred gene *in vivo*, may be eliminated or minimized if an adenovirus-associated virus that does not require cell division for its efficient infection and a stable integration into the host genome is properly engineered.

Exciting applications of direct *in vivo* gene transfer approaches may be developed which do not need a prolonged expression of transgenes, such as tumor immunotherapies using cytokines. For much wider applications in gene therapy, however, the current direct gene transfer approaches need substantial improvements.

B. Myoblast-Mediated Ex Vivo Gene Transfer

As a more general method that assures a stable gene transfer and long-term production of transgenes, the myoblast-mediated gene transfer method has been tested extensively for its feasibility. As shown in Figure 1, this method involves genetic modification of myoblasts in culture, which are then implanted back into skeletal muscles. One of the advantages of using this method is that the basic safety of myoblast isolation and implantation has already been well tested with animals as well as with Duchenne muscular dystrophy patients.[33-39]

Viable myoblasts can be generated easily from satellite cells (dormant muscle stem cells) contained in small biopsy tissue sample in culture. Viable myoblasts in a range of 1×10^5 cells are usually obtained from 1 g of freshly isolated skeletal muscle tissue.[40] The yield, purity, and properties of myoblasts that can be obtained from different animals vary significantly. For instance, myoblasts can be isolated easily from mouse muscle but also are often contaminated with fibroblasts. These fibroblasts tend to overgrow myoblasts, and basic fibroblast growth factor produced by fibroblasts inhibits differentiation of myoblasts to myotubes.[41] To eliminate the difficulty of fibroblast

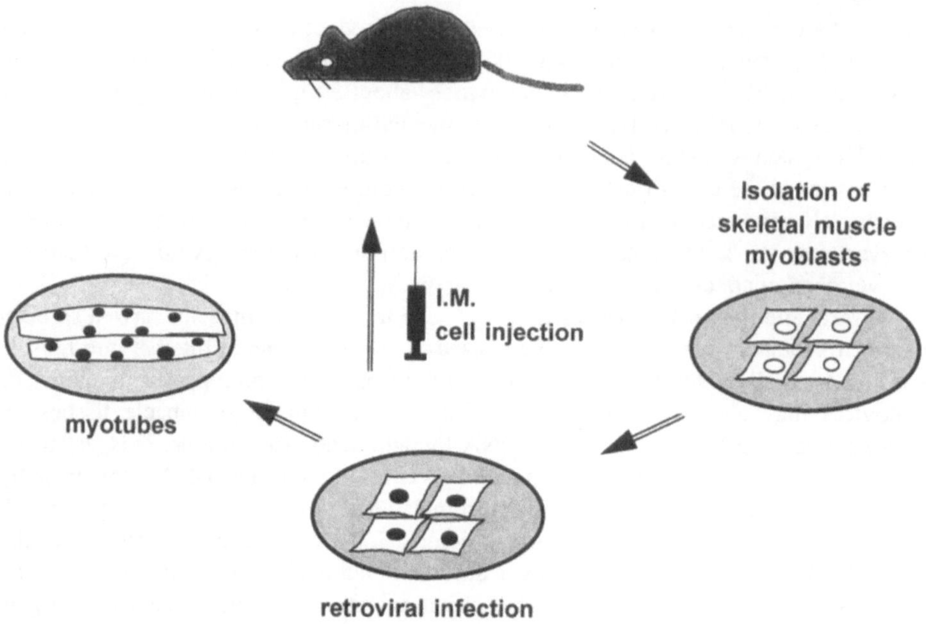

FIGURE 1. The basic scheme of myoblast-mediated gene transfer. Primary myoblasts are isolated from skeletal muscle biopsy into culture, genetically modified, and tested for the transgene expression before and after differentiation to myotubes. Myoblasts are then injected back into skeletal muscles to produce the recombinant transgene product *in vivo*.

overgrowth, myoblasts can be isolated from fibroblasts by various techniques, such as cell colony cloning and cell sorting.[40] A cell implantation protocol of gene therapy may require a large number of cells, in a range of 10^8 to 10^9; if fibroblast cell contamination is minimal, they may not be a practical problem for the procedure. A pure population of myoblasts is preferred, however, particularly if a muscle cell-specific promoter is used to obtain the optimal results for the expression of the gene transferred. In general, cleaner tissue biopsy obtained from large animals, such as dog and monkey, provide much cleaner myoblast preparations often requiring little further purification. Properties of primary myoblasts vary among different animals at different ages. Dog myoblasts appear to senesce sooner than mouse and monkey myoblasts, and cells isolated from older animals may also senesce sooner.[42] This indicates that myoblasts to be used for a gene transfer should be isolated from animals at a younger age.

In culture, myoblasts that are grown to confluence are easily induced to fuse with themselves, generating myotubes by lowering serum concentration in the medium to about 2%. This process is inhibited in the presence of high serum concentration (10% or higher) or alternatively in the presence of growth factor, such as basic FGF, which inhibits the differentiation process.[41] Such growth factors, therefore, may be utilized to enhance myoblast proliferation.

Myoblasts in culture can be modified genetically and reimplanted into muscle tissue, where they fuse with the host myofibers or with themselves, generating new recombinant myofiber cells, which can continue producing the recombinant product.

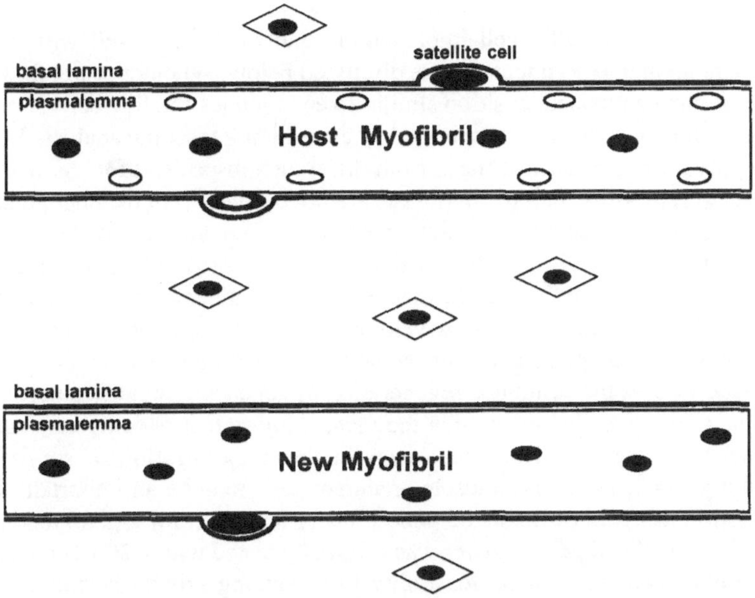

FIGURE 2. Fate of implanted myoblasts. Implanted myoblasts fuse with host myofibers donating recombinant myonuclei, depicted as dark nuclei near the center of myofibril. Implanted myofibers also fuse with themselves, generating new myofiber cells primarily with recombinant myonuclei. These myofibers continue producing recombinant products. Some implanted myoblasts may assume a physical location similar to the satellite cells and/or possibly an unknown unique physical position, which allows them to remain as quiescent myoblasts like satellite cells. A small fraction of implanted myoblasts become quiescent reserve muscle precursor cells, but the precise mechanism underlying this is not known.

The myoblast fusion is a natural muscle repair process and is highly efficient. Although there are some contradictory data, dependence of regeneration of specific muscle fiber cells on myoblast cell lineage has been reported.[43,44] Most skeletal muscles are composed of mixed cell types (fast- and slow-twitching muscles). Primary myoblasts obtained from a skeletal muscle should represent myoblasts of both muscle types. Upon implantation of the genetically modified cells, the mixed cell-type myoblasts should not give any significant problem in efficient generation of recombinant muscle fiber cells. After implantation into muscle tissue, most cells fuse with the host myofiber cells or with themselves. Some cells may die before having the chance to make fusions.

More recently, we also have demonstrated that a small population of the implanted retrovirus-marked myoblasts can survive in muscle tissue as quiescent reserve muscle precursor cells.[19] This situation is depicted in Figure 2. These reserve muscle precursor cells can be recovered in culture as viable myoblasts even 1 year after cell implantation. Once recovered in culture, these reserve cells continue expressing the transgenes contained in the retroviral vector (factor IX and β-galactosidase, as tested) at a level as high as that of the original retrovirus-marked primary myoblasts, with no experience of implantation. These results indicate that the promoter used (MoMLV LTR) is not inactivated over 1 year after cell implantation, and they agree well with the results obtained with animal experimentation as discussed below. Whether or not these reserve myoblasts assume a physical position similar to endogenous satellite cells is not known at the present time and must be determined by electron microscopic analysis.[22] As tested based on polymorphism of myoblasts from different subspecies of mice, it was found that recovered retrovirus-marked myoblasts are not derived from the endogenous satellite cells of the animal due to an unknown mechanism, but are actually derived from the originally implanted myoblast cells. These cells are also proven by extensive testing to be nontumorigenic. These findings further support the rationale employing a myoblast-mediated gene transfer approach for developing a gene therapy for a congenital disease like hemophilia B. The presence of genetically modified muscle precursor cells in the muscle tissue assures the continued regeneration of genetically modified myofiber cells (recombinant muscle cells), even upon the muscle injury that patients may suffer after the original gene therapy. These data suggest that maximization of survival of the implanted myoblasts as reserve muscle precursor cells may be an important factor for optimization in a myoblast-mediated gene transfer approach for a gene therapy.

Myoblast-mediated gene transfer was originally tested with C2C12 cells, an established myoblast cell line, for its feasibility in producing growth hormone genes,[45,46] factor IX gene,[16] and β-galactosidase.[16] It is reported that human growth hormones were produced at a peak level of 1 to 2 ng/ml plasma in mouse plasma after injection of retrovirally transduced C2C12.[45,46] We have shown that C2C12 cells in culture can be efficiently modified with retroviral vectors (BAG, β-galactosidase retroviral vector; and LIXSN, factor IX retroviral vector) and can express the transgenes at high levels.[16] Both vectors use LTR (long-terminal repeat) for transcriptional control of the transgenes and also contain neomycin-resistant gene under the transcriptional control of SV40 promoter. C2C12 cells can produce recombinant human factor IX at a very high level ($\geq$2.6 μg/10^6 cells/day). Furthermore, myotubes generated by differentiation of the recombinant myoblasts continued to express the factor IX transgene at a significant level (~0.7 μg/day/myotubes derived from 1×10^6 myoblasts), suggesting a great potential of producing recombinant factor IX *in vivo*. When retrovirus-transduced myoblasts were implanted into skeletal muscles (hind legs) of C3H mice, which were given with cyclosporine as an immunosuppressive agent, recombinant factor IX was produced at a substantial level *in vivo* (Figure 3). Up to about 1 week after cell implantation, recombinant human factor IX in the circulation was at a constant level (100 to 200 ng/ml plasma, depending on the cell numbers implanted), followed by a much elevated level on day 12 reaching $\geq$1 μg/ml plasma. This level is equivalent to about 25% of the normal plasma level in man and therapeutically very significant. However, the level gradually came down to the background level after about 1 month of cell implantation. This decrease was found to be due primarily to anti-human factor IX antibody produced in the mice. The antibody was first observed on day 12 and increased its titer as the recombinant factor IX level in the circulation decreased. These results indicated that cyclosporine administered to animals was not able to suppress completely the humoral immune reaction against human factor IX produced. A possibility of promoter inactivation, which may cause the decrease, was not substantiated based on several lines of other data, including that of implantation and recovery of retrovirus-marked myoblasts.[19]

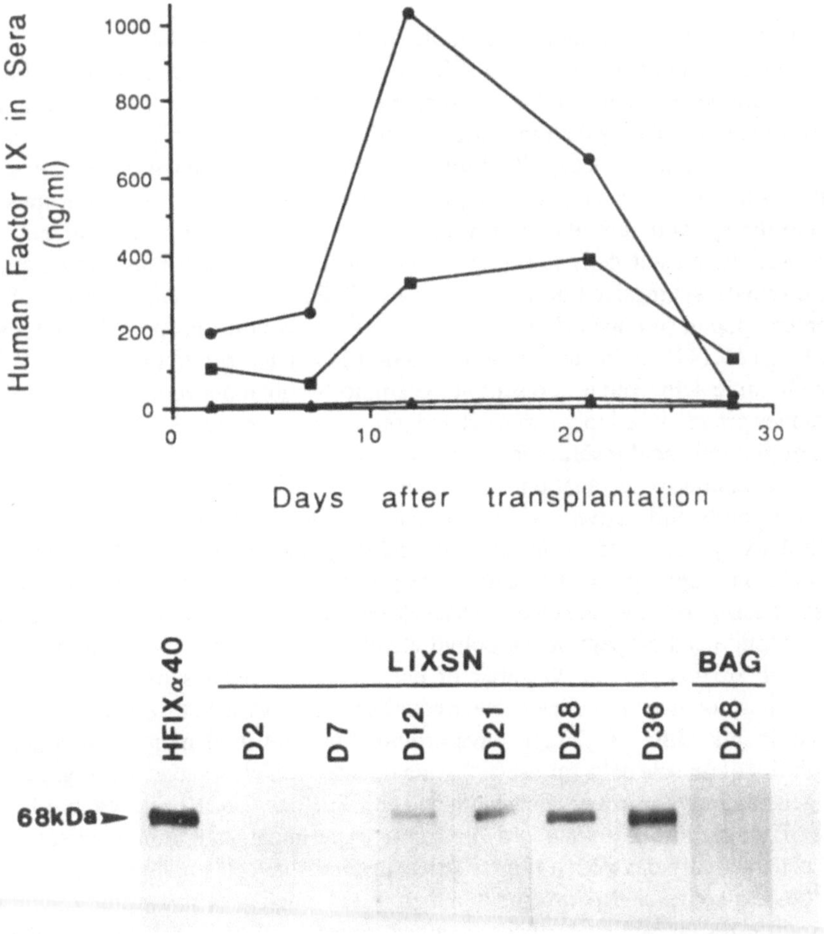

FIGURE 3. Systemic production of human factor IX in C3H mice by C2C12 cell-mediated gene transfer. (Top) Recombinant human factor IX in the systemic circulation Human factor IX was detected by ELISA. ▲: injected with 2×10^7 cells transduced with BAG at multiple sites of the hind leg skeletal muscles; ■: injected with 2×10^7 cells with LIXSN; ●: injected with 3×10^7 cells transduced with LIXSN. (Bottom) Detection of anti-human factor IX antibody in serum at various days (D) after cell implantation.

Another important finding in these experiments is that the efficiency of delivering the recombinant factor IX produced by muscle cells into the systemic circulation (systemic delivery efficiency) is very high.[16] This was a surprising finding for muscle tissue, which unlike liver, is not designed to secrete and transport a number of plasma proteins in large amounts. The systemic delivery efficiency observed, 29%, was calculated according to the two-compartment model kinetics for factor IX clearance from the vascular system, assuming the amount of factor IX produced in culture by the same number of myoblasts taken for implantation as 100%. The half-life of human factor IX in mice (17.8 hr) was determined to be similar to that in human and dog. The mechanism of recombinant protein transport into the blood circulation is not well established (lymphatic drainage system or more direct transportation into the bloodstream via a well-developed vascular system in muscles); however, the high efficiency of systemic delivery of recombinant factor IX observed is important and encourages the use of skeletal muscle tissue for an ectopic expression of factor IX and its efficient transportation into the systemic circulation, reaching a therapeutically significant level.

Importantly, muscle cells can carry out complex co- and post-translational modifications of newly synthesized polypeptide chains.[16] Factor IX is synthesized as single-chain prepro-peptide composed of modules for a signal peptide, propeptide, and mature plasma factor IX (415 amino acid residue in length), which is present in the circulation.[5] Twelve glutamic acid residues present in the amino-terminal portion of about 40 amino acid residues are modified to γ-carboxylated glutamic acid residues. This unique modification of glutamic acid residues is essential for generating the full biologic activity of factor IX. Quantitative evaluation of this modification has not been carried out yet; however, a nearly full activity of recombinant factor IX (81 to 90% normal plasma factor IX activity) produced by muscle cells and its good binding to $BaSO_4$ indicate that muscle cells have appropriate mechanisms responsible for this complex modification of newly synthesized polypeptide chains.[6] Muscle cells can also carry out proper cleavages of signal peptide and propeptide, carbohydrate chain attachment, and many others that are required for this protein. Whether or not these mechanisms of muscle cells are similar to those of liver in their efficiency and specificities is yet to be determined.

Experiments with C2C12 cells have provided us with fundamental data regarding the feasibility of muscle cells for use as a gene transfer vehicle in gene therapy. In actual human gene therapy, however, an established cell line like C2C12 cells cannot be used because of a possibility of tumor growth. In our experiments, a few mice injected with C2C12 cells actually developed tumors after 4 months of cell implantation.

In the next series of experiments, we then tested whether or not primary myoblasts can reproduce results similar to those obtained with C2C12 cells.[47] Primary myoblasts obtained from NIH III nude mice or SCID mice were easily modified with factor IX retrovirus (LIXSN). Myotubes derived from these primary myoblasts could continue expressing recombinant factor IX at a level (~750 ng/day/myotubes derived from 1×10^6 myoblasts) equivalent to or even slightly higher than that of C2C12 cells. When these myoblasts (1×10^7 cells) were implanted into muscle tissues, however, the level and temporal pattern of recombinant factor IX found in the circulation (Figure 4) were substantially different from those observed with C2C12 cells. On day 2 after cell implantation, the level of recombinant factor IX in the circulation was high (~50 ng/ml plasma) as observed for C2C12 cells. The factor IX level, however, drastically lowered after day 2, followed by a slight rebound, and then stabilized at a steady level. This pattern of systemic production of factor IX, which is significantly different from that observed for C2C12 cells (Figure 4), indicated that the lower expression level of recombinant factor IX is not due to the promoter inactivation, but to cell biologic properties, such as poorer survival and fusion efficiency of primary myoblasts compared to C2C12 cells.

This general pattern of recombinant factor IX expression by primary myoblast-mediated gene transfer was also observed by others who used a muscle-specific expres-

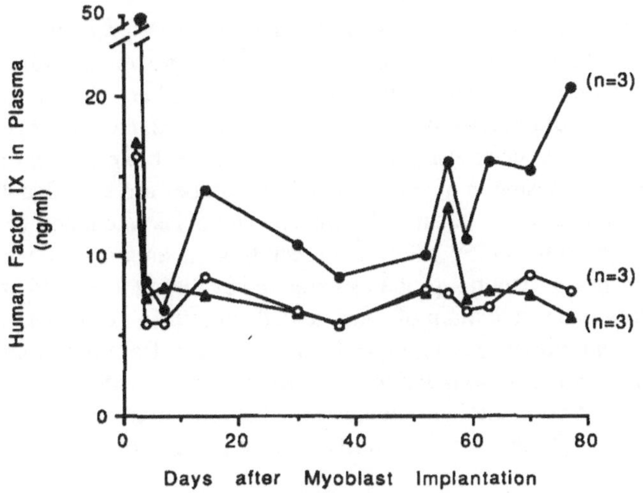

FIGURE 4. Systemic production of recombinant human factor IX in SCID mice by primary myoblast-mediated gene transfer. ▲: 1×10^7 cells transduced with LIXSN injected at a single site of the hind leg skeletal muscle; ○: 1×10^7 cells transduced with LIXSN injected at multiple sites; ●: 1×10^7 cells transduced with LIXSN injected at multiple sites with basic FGF treatment. Average values of human factor IX antigen detected by ELISA are plotted.

sion vector with muscle creatine kinase (MCK) enhancer.[17] In their experiment, the expression pattern over months was very similar to that we observed with LTR promoter-driven expression of factor IX. We also tested MMBAIX retrovirus, which contains 2 units of MCK enhancer linked to a β-actin promoter (a housekeeping gene promoter) as a transcriptional control sequence to express human factor IX. The overall pattern of expression with this was originally observed for LIXSN retroviral vector. These results also demonstrated that a retroviral vector promoter (LTR) is more active than a differentiated muscle cell-specific promoter with MCK enhancers and is not inactivated *in vivo* over a year.

The poor survival and fusion efficiency or primary myoblasts may be the major reasons for the different expression pattern with the lower expression level of factor IX gene compared to that with C2C12 cells; therefore, extensive optimization of the protocol of primary myoblast implantation combined with the treatment of the cells with various growth factors that can stimulate proliferation of these cells may improve the factor IX expression level and its long-term maintenance. Interestingly, multiple injections of 1×10^7 myoblasts into mouse muscle only slightly improved the expression level. This suggests that *in vivo* survival of the implanted myoblasts long enough to fuse with host myofiber cells or with themselves to generate new myotubes is an important factor for improving the expression level. When primary myoblasts are injected with basic FGF added to the cell suspension just before their implantation, a substantial improvement (up to a three- to four-fold increase) in the expression level has been observed. This further supports our hypothesis that a longer *in vivo* survival of the implanted myoblasts enhances fusion of the implanted myoblasts with the host myofibers as well as with themselves; this results in elevated factor IX production and therefore a higher level in the systemic circulation. This is in agreement with the results of experimentation with C2C12 cells, which as an established cell line can survive better and may even proliferate *in vivo* to enhance their fusion with the host myofiber cells as well as with themselves, generating a high-level, persistent expression of transgenes.

Marcaine (local anesthetic agent), to which myotubes are significantly more susceptible than myoblasts, may substantially improve myoblast-mediated gene transfer or direct *in vivo* gene transfer by retrovirus, resulting in an improved expression of transgenes.[48] This phenomenon is probably mediated by the cytotoxic effect of marcaine, which in turn stimulates activation of satellite cells and fusion of generated myoblasts with themselves and with surviving myofibers that suffers from only minor, reparable modifications. By testing various factors that can enhance myoblast proliferation and/or fusion singly or in combination, myoblast-mediated gene transfer may be improved substantially for its efficiency and stability in producing a recombinant protein at a therapeutic level.

An interesting paradigm of the myoblast-mediated gene transfer approach is to achieve ectopic expression of foreign genes by explanting genetically modified myoblasts in tissues other than muscles.[49] This was used to express tyrosine hydroxylase in brain.[50] Explanted myoblasts are differentiated to myotubes in the environment of brain tissue and continue expressing the transgene. Because adult brain cells are not dividing, fully differentiated muscle cells (myofiber cells) may provide a stable recombinant tissue for expressing a transgene required for treating brain diseases like Parkinson's disease for a longer period of time. Although little is currently known about its long-term safety and efficacy, this approach may find many interesting applications.

III. EPILOGUE

Advantages of using myoblasts as a vehicle for delivering recombinant gene product are summarized as follows:

1. Skeletal muscle is the largest tissue in a body and easily accessed for various manipulations, such as biopsy and cell implantation, minimizing invasive procedures involved.
2. Myoblasts in culture can be easily modified genetically by various methods, including retroviral vector and others.
3. Implanted myoblasts efficiently fuse with host myofibers or with themselves, generating recombinant muscle cells, which continue expressing transgenes, such as factor IX.
4. Muscle cells have efficient machineries to carry out complex co- and post-translational processing and modifications similar to those carried out by liver cells.
5. Muscle tissue can deliver the recombinant gene products into the systemic circulation efficiently.
6. A small fraction of the implanted myoblasts can survive as muscle precursor cells (muscle stem cells), assuring continuous renewal of recombinant myofibers (upon muscle injury), which can continue expressing transgenes.
7. Side-effects and potential problems, such as immune reaction, may be minimized by using autologous myoblasts in a human gene therapy.

Myoblast-mediated gene transfer is an *ex vivo* procedure and requires tissue biopsy and cell culture. This is a less ideal aspect of this approach compared to the direct *in vivo* gene transfer approach, which may potentially be developed as a much simpler method. The myoblast-mediated approach, however, assures an efficient and stable gene transfer, for instance by retroviral vector, which is currently the only vector guaranteeing a stable integration of transgenes.

The approach of myoblast-mediated gene transfer is still not fully optimized at present. However, it is highly likely that this approach can be fully optimized in the near future and may find a wide range of applications for congenital or acquired diseases that require a systemic delivery of a transgene product. Such diseases may include, in addition to hemophilia B, hemophilia A (factor VIII deficiency), cytokine therapy (for instance, erythropoietin for treating chronic anemia in patients with renal failure), diabetes, Duchenne muscular dystrophy, and many others.

ACKNOWLEDGMENT

This work was supported in part by research grants from National Institutes of Health (HL38644 and HL48313) and the Hemophilia Foundation of Michigan.

REFERENCES

1. Mulligan, R. C., The basic science of gene therapy, *Science*, 260, 926, 1993.
2. Morgan, R. A. and Anderson, W. F., Human gene therapy, *Annu. Rev. Biochem.*, 62, 191, 1993.
3. Kurachi, K. and Yao, S.-N., Gene therapy of hemophilia B, *Thromb. Haemostas.*, 70, 193, 1993.
4. St. Louis, D. and Verma, I. M., An alternative approach to somatic cell gene therapy, *Proc. Natl. Acad. Sci. U.S.A.*, 85, 3150, 1988.
5. Kurachi, K., Furukawa, M., Yao, S.-N., and Kurachi, S., Molecular biology of factor IX, *Blood Coagulation and Fibrinolysis*, 4, 953, 1993.
6. Roberts, H. R., Molecular biology of hemophilia B, *Thromb. Haemostas.*, 70, 1, 1993.
7. Salier, J.-P., Hirosawa, S., and Kurachi, K., Functional characterization of the 5′-regulatory region of human factor IX gene, *J. Biol. Chem.*, 265, 7062, 1990.
8. Chowdhury, J. R., Groossman, M., Gupta, S. J., Chowdhury, N. R., Baker, J. R., Jr., and Wilson, J. M., Long term improvement of hypercholesterolemia after *ex vivo* gene therapy in LDLR deficient rabbits, *Science*, 254, 1802, 1991.
9. Ponder, K. P., Gupta, S., Leland, F., Darlington, G., Finegold, M., DeMayo, J., Ledley, F. D., Chowdhury, J. R., and Woo, S. L. C., Mouse hepatocytes migrate to liver parenchyma and function indefinitely after intrasplenic transplantation, *Proc. Natl. Acad. Sci. U.S.A.*, 88, 1217, 1991.
10. Armentano, D., Thompson, A. R., Darlington, G., and Woo, S. L. C., Expression of human factor IX in rabbit hepatocytes by retrovirus-mediated gene transfer: potential for gene therapy of hemophilia B, *Proc. Natl. Acad. Sci. U.S.A.*, 87, 6141, 1990.
11. Hsueh, J. L., Lu, D. R., Zhou, J. M., Qui, X. F., Wang, J. M., Men, F. L., Han, F. L., Ming, B. H., Wang, X. P., Wang, J. P., Liang, J. Q., and Jiang, Z. S., Clinical trial (phase I) of gene therapy for hemophilia B using fibroblast, *Sci. China (Ser. B)*, 23, 53, 1993 (in Chinese).

12. Dai, Y.-F., Qui, X.-F., Hsueh, J.-L., and Liu, Z.-D., High efficient transfer and expression of human clotting factor IX cDNA in cultured human primary skin fibroblasts from hemophilia B patient by retroviral vectors, *Sci. China (Ser. B)*, 35, 183, 1992.
13. Yao, S.-N., Wilson, J. M., Nabel, E. G., Kurachi, S., Hachiya, H. L., and Kurachi, K., Expression of human factor IX in rat capillary endothelial cells: toward somatic gene therapy for hemophilia B, *Proc. Natl. Acad. Sci. U.S.A.*, 88, 8101, 1991.
14. Dichek, D. A., Gene therapy in the treatment of thrombosis, *Thromb. Haemostas.*, 70, 198, 1993.
15. Gerrard, A. J., Hudson, D. L., Brownlee, G. G., and Watt, F. M., Towards gene therapy for haemophilia B using primary human keratinocytes, *Nat. Genet.*, 3, 180, 1993.
16. Yao, S.-N. and Kurachi, K., Expression of human factor IX in mice after injection of genetically modified myoblasts, *Proc. Natl. Acad. Sci. U.S.A.*, 89, 3357, 1992.
17. Dai, Y., Roman, M., Naviaux, R. K., and Verma, I. M., Gene therapy via primary myoblasts: long-term expression of factor IX protein following transplantation *in vivo*, *Proc. Natl. Acad. Sci. U.S.A.*, 89, 10892, 1992.
18. Roman, M., Axelrod, J. H., Dai, Y., Naviaux, R. K., Friedmann, T., and Verma, I. M., Circulating human or canine factor IX from retrovirally transduced primary myoblasts and established myoblast cell lines grafted into murine skeletal muscle, *Somat. Cell Mol. Genet.*, 18, 247, 1992.
19. Yao, S.-N. and Kurachi, K., Implanted myoblasts not only fuse with myofibers but also survive as muscle precursor cells, *J. Cell Sci.*, 105, in press, 1994.
20. Spence, A. P., *Basic Human Anatomy*, 2nd ed., Benjamin/Cummings, Menlo Park, CA, 177, 192, 1986.
21. Partridge, T. A., Myoblast transfer: a possible therapy for inherited myopathies?, *Muscle Nerve*, 14, 197, 1991.
22. Mazanet, R. and Franzini-Armstrong, C., The satellite cell, in *Myology: Basic and Clinical*, Engel, A. G. and Banker, B. Q., Eds., McGraw-Hill, New York, 1986, chap. 9.
23. Wolff, J. A., Malone, R. W., Williams, P., Chong, W., Acsadi, G., Jani, A., and Felgner, P. L., Direct gene transfer into mouse muscle in vivo, *Science*, 247, 1465, 1990.
24. Acsadi, G., Dickson, G., Love, D. R., Jani, A., Walsh, F. S., Gurusinghe, A., Wolff, J. A., and Davies, K. E., Human dystrophin expression in mdx mice after intramuscular injection of DNA constructs, *Nature*, 352, 815, 1991.
25. Lin, H., Parmacek, M. S., Morle, G., Bolling, S., and Leiden, J. M., Expression of recombinant genes in myocardium in vivo after direct injection of DNA, *Circulation*, 82, 2217, 1990.
26. Raz, E., Watanabe, A., Baird, S. M., Eisenberg, R. A., Parr, T. B., Lotz, M., Kipps, T. J., and Carson, D. A., Systemic immunological effects of cytokine genes injected into skeletal muscle, *Proc. Natl. Acad. Sci. U.S.A.*, 90, 4523, 1993.
27. Wolff, J. A., Williams, P., Acsadi, G., Jiao, S., Jani, A., and Chong, W., Conditions affecting direct gene transfer into rodent muscle *in vivo*, *BioTechniques*, 11, 474, 1991.
28. Davis, H. L., Whalen, R. G., and Demeneix, B. A., Direct gene transfer into skeletal muscle *in vivo*: factors affecting efficiency of transfer and stability of expression, *Hum. Gene Ther.*, 4, 151, 1993.
29. Quantin, B., Perricaudet, L. D., Tajbakhsh, S., and Mandel, J.-L., Adenovirus as an expression vector in muscle cells *in vivo*, *Proc. Natl. Acad. Sci. U.S.A.*, 89, 2581, 1992.
30. Stratford-Perricaudet, L. D., Makeh, I., Perricaudet, M., and Briand, P., Widespread long-term gene transfer to mouse skeletal muscles and heart, *J. Clin. Invest.*, 90, 626, 1992.
31. Ragot, T., Vincent, N., Chafey, P., Vigne, E., Gilgenkrantz, H., Couton, D., Cartaud, J., Briand, P., Kaplan, J.-C., Perricaudet, M., and Kahn, A., Efficient adenovirus-mediated transfer of a human minidystrophin gene to skeletal muscle of *mdx* mice, *Nature*, 361, 647, 1993.

32. Wagner, E., Zatloukal, K., Cotten, M., Kirlappos, H., Mechtler, K., Curiel, D. T., and Birnstiel, M. L., Coupling of adenovirus to transferrin-polylysine/DNA complexes greatly enhances receptor-mediated gene delivery and expression of transfected genes. *Proc. Natl. Acad. Sci. U.S.A.*, 89, 6099, 1992.
33. Gussoni, E., Pavlath, G. K., Lanctot, A. M., Sharma, K. R., Miller, R. G., Steinman, L., and Blau, H. M., Normal dystrophin transcripts detected in Duchenne muscular dystrophy patients after myoblast transplantation, *Nature*, 356, 435, 1992.
34. Huard, J., Bouchard, J. P., Roy, R., Malouin, F., Dansereau, G., Labrecque, C., Albert, N., Richards, C. L., Lemieux, B., and Tremblay, J. P., Human myoblast transplantation: preliminary results of 4 cases, *Muscle Nerve*, 15, 550, 1992.
35. Law, P. K., Bertorini, T. E., Goodwin, T. G., Chen, M., Fang, Q., Li, H.-J., Kirby, D. S., Florendo, J. A., Herrod, H. G., and Golden, G. S., Dystrophin production induced by myoblast transfer therapy in Duchenne muscular dystrophy, *Lancet*, 336, 114, 1990.
36. Law, P. K., Goodwin, T. G., and Li, H.-J., Histoincompatible myoblast injection improves muscle structure and function of dystrophic mice, *Transplant. Proc., 20*, 1114, 1988.
37. Karpati, G., Pouliot, Y., Zubrzycka-Gaarn, E., Carpenter, S., Ray, P. N., Worton, R. G., and Holland, P., Dystrophin is expressed in mdx skeletal muscle fibers after normal myoblast implantation, *Am. J. Pathol.*, 135, 27, 1989.
38. Law, P. K., Goodwin, T. G., and Wang, M. G., Normal myoblast injections provide genetic treatment for murine dystrophy, *Muscle Nerve*, 11, 525, 1988.
39. Couillard, M., Deschenes, L., and Tremblay, J. P., Cytomegalovirus and myoblast transplantation, *Lancet*, 337, 1411, 1991.
40. Webster, C., Pavlath, G. K., Parks, D. R., Walsh, F. S., and Blau, H. M., Isolation of human myoblasts with the fluorescence-activated cell sorter, *Exp. Cell Res.*, 174, 252,1988.
41. Florini, J. R., Ewton, D. Z., and Magri, K. A., Hormones, growth factors, and myogenic differentiation, *Annu. Rev. Physiol.*, 53, 201, 1991.
42. Webster, C. and Blau, H. M., Accelerated age-related decline in replicative life-span of Duchenne muscular dystrophy myoblasts: implications for cell and gene therapy, *Somat. Cell Mol. Genet.*, 16, 557, 1990.
43. Hughes, S. M. and Blau, H. M., Muscle fiber pattern is independent of cell lineage in postnatal rodent development, *Cell*, 68, 659, 1992.
44. DiMario, J. X., Fernyak, S. E., and Stockdale, F. E., Myoblasts transferred to the limbs of embryos are committed to specific fibre fates, *Nature*, 362, 165, 1993.
45. Dhawan, J., Pan, L. C., Pavlath, G. K., Travis, M. A., Lanctot, A. M., and Blau, H. M., Systemic delivery of human growth hormone by injection of genetically engineered myoblasts, *Science*, 254, 1509, 1991.
46. Barr, E. and Leiden, J. M., Systemic delivery of recombinant proteins by genetically modified myoblasts, *Science*, 254, 1507, 1991.
47. Yao, S. -N. and Kurachi, K., Expression of human factor IX by primary myoblast-mediated gene transfer in mice, submitted.
48. Schultz, E. and Lipton, B. H., The effect of marcaine on muscle and non-muscle cells in vitro, *Anat. Rec.*, 191, 351, 1978.
49. Jiao, S., Schultz, E., and Wolff, J. A., Intracerebral transplants of primary muscle cells: a potential 'platform' for transgene expression in the brain, *Brain Res.*, 575, 173, 1992.
50. Jiao, S., Gurevich, V., and Wolff, J. A., Long-term correction of rat model of Parkinson's disease by gene therapy, *Nature*, 362, 450, 1993.

9 Gene Delivery to Neurons of the Adult Mammalian Nervous System Using Herpes and Adenovirus Vectors

Julie K. Andersen and Xandra O. Breakefield

I. ABSTRACT

Neurons are an extremely important cell target for somatic gene therapy in neurological disorders of the central nervous system, but are difficult to manipulate genetically in the adult, since they do not divide postnatally and cell division is a prerequisite for gene delivery in most experimental procedures. However, foreign genes have been introduced into this cell type using virus vectors derived from herpes simplex type I virus and adenovirus. This chapter will review the scientific basis for the use of these vectors for foreign gene delivery to the adult nervous system, the work done *in vivo* in animal models, and the types of human diseases that are potential targets for this form of somatic gene therapy.

II. GENE DELIVERY TO THE NERVOUS SYSTEM: POTENTIAL METHODS AND USES

Gene transfer into the adult mammalian nervous system is theoretically useful for several purposes, including the evaluation of the actions of neural proteins, the creation of animal models of human neurological disease, and as a means of gene therapy in humans. There are several methods currently in use for gene transfer into the nervous system, including the creation of transgenic mice, in which foreign gene expression is under the control of a neuron-specific promoter, grafting of either fetal transgenic tissues or other genetically modified cells into the brain, and the use of viral vectors for transfer of foreign DNA into cells of the adult nervous system.

0-8493-2440-8/95/$0.00+$.50

The creation of transgenic mice involves the injection of DNA into the pronuclei of single-cell mouse embryos, resulting in occasional integration into the mouse genome and expression of transgene activity.[101,102] This method of gene transfer has proven to be useful for several purposes in the nervous system,[105] including delineating transcriptional control elements that determine tissue/temporal expression of neuronal genes,[7,46,70,71] expressing oncogenes in various tissues and/or at various times during development to create specific neuronal cell lines,[92] expressing toxic molecules to ablate a particular subset of neurons to examine the effects of cell loss during development,[21,100,109] and creating models of human neurological disease states.[60,63,107,124] Recently, a method has been devised for introducing specific null mutations into animals.[84,85,135] To create this special type of "knockout" transgenic, a mutation is introduced into a genomic fragment of the cloned gene of interest, usually by insertion of another gene encoding a marker enzyme into an exonic region of the original gene, thereby disrupting it. The mutated gene fragment is then introduced into pluripotent embryonic stem (ES) cells in culture by either direct injection or electroporation of DNA. At a low frequency, homologous recombination takes place between the endogenous gene in the ES genome and the mutated gene fragment. Selected cells containing a mutated copy of the endogenous gene of interest are then introduced into blastocysts, which are, in turn, implanted into pseudopregnant females. Chimeric animals are produced, and these are bred in the hopes that the ES cells retain their pluripotency and are able to contribute to the germ-line precursor cells. Heterozygotic mice are produced, containing one mutant copy of the gene of interest, and these are subsequently bred to create homozygous "knockouts". Knockout of gene function by homologous recombination in ES cells has proven to be an extremely useful tool in the creation of animal models of human disease states where people previously had to rely on spontaneous mutants. Transgenic technology, however, is not a viable technique for somatic gene therapy, as this method does not allow gene transfer into the adult nervous system directly but only indirectly through, for example, grafting of transgenic cells.

Cell grafting into the CNS involves implanting either normal tissue or genetically-modified cells into the brain, both as a means of examining the effects of environment on cell differentiation, migration, plasticity, and function, and to release factors that are needed for development/survival of other cells in the brain as a means of gene therapy.[16,49-51] Cell types used for grafting into postnatal animals have included fibroblasts, muscle cells, chromaffin cells, astrocytes, neuroblasts, and oligodendrocytes, as well as continuous cell lines derived from such cell types. Grafted cells may act in the brain to "replace" missing cells or supplement cell functions, i.e., hormone or transmitter synthesis. Genetically-modified cells that have been surgically implanted in the CNS include, for example, fibroblasts releasing neurotrophic factors like nerve growth factor (NGF),[119] brain-derived growth factor (BDNF),[48] and β-fibroblast growth factor (bFGF)[48] or enzymes involved in neurotransmitter synthesis like tyrosine hydroxylase (TH),[62,132,144] choline acetyl transferase (CAT),[7] and glutamic acid decarboxylase (GAD).[7] Immune rejection of the grafted cells has been circumvented in some of these cases by use of the animal's own cells as donors for genetic manipulation. Genetically modified cells, however, also frequently down-regulate transgene expression over time, may interfere with endogenous neuronal circuitry, may bring other unwanted factors into the brain, and, if cell lines are used, may cause tumor formation. In terms of fetal tissue, grafting of fetal dopaminergic neurons[12,20] or adrenal chrommaffin cells[14] has been used as a gene replacement therapy for Parkinson's disease. Many factors are involved in determining survival of fetal tissue grafts, such as immune rejection, the age of the donor, and the age of the host.

DNA has also been introduced into the brain using DNA-liposomal complexes,[67,97] which basically consist of plasmids contained in liposomal membranes. Because plasmids are used, they can be easily manipulated genetically in cultured bacterial cells and, following delivery, appear to be quite stable. However, they have a low transfection efficiency, and the cationic lipids used to form the liposomes may be toxic. Polylysine-DNA complexes tagged with ligands for cell-specific receptors are also available for gene delivery (see Section II.B). Direct injection of RNA[13] or of antisense oligonucleotides[141] has also been explored as a means of manipulating gene expression in the brain and other tissues.[94] However, it may only be effective in the cells into which these elements are injected directly and then only for a brief period of time, and the efficiency of uptake is unknown.

Viral vectors have also been used in delivery of foreign genes to cells of the mammalian nervous system. One of these is derived from a mouse RNA leukemia retrovirus, which is able to infect most cells and integrate its DNA into the cellular genome of dividing cells. Retroviruses have been found to be a highly efficient means of foreign gene transfer.[24] They have been modified for gene delivery into the brain by removal of viral genes encoding reverse transcriptase, capsid proteins, and internal antigens, and insertion of transgenes of interest. Foreign transgenes are introduced into the modified virus, usually behind the long-terminal repeat element (LTR), which normally functions to initiate transcription of the viral DNA and can serve as a promoter for the transgene, although other promoter elements can be used. The virus can accommodate 6 to 8 kb of foreign sequence. Plasmids containing the gene of interest within the retrovirus sequences are then transfected into packaging cells in culture, which bear the viral genes lacking in the vector sequences but cannot themselves produce virus due to a mutation in the packaging signal. These cells allow the vector, but not wild-type virus, to be packaged. Virus released from these packaging cells or the packaging cells themselves can be injected stereotactically into the brain, where the virus can integrate its genome into dividing cells. Retroviruses have been used for many purposes in the brain including mapping cell lineage by introduction of a modified retrovirus containing a marker gene into the brain early in development,[6,106,120,138,139] genetic modification of cultured cells used for grafting,[121] delivery of toxic products to kill cancerous cells, i.e., brain tumors,[41,111] and gene replacement therapy.[30,73] They are not ideal vectors for delivery into the brain since they cannot integrate into post-mitotic neurons and it is hard to achieve high-titer stocks of the virus. (Typical titer is 10^6 CFU/ml, and only small fluid volumes can be injected into the brain.) It is their ability to integrate into dividing cells that makes them such good candidate vectors for delivery of killer genes to cancer cells in the brain.[26,93]

Because of the various problems inherent with these other methods for gene delivery into the adult brain, attention has turned to the use of other virus vectors for this purpose.

Table 1

Use of HSV-1 as a Vector for Gene Delivery into Neurons

- The HSV-1 genome can carry multiple inserts, up to a total of 30 kb.
- Viral mutants are available which are less pathogenic than wild-type virus.
- The virus can travel by retrograde and anteriograde transport, and across synapses.
- The virus can enter latency in some neurons (where it exists in a benign, episomal state).
- The viral LAT promoter is active in latency and can confer stable gene expression.

Modified from Reference 5.

A. Use of Herpes Virus Vectors for Delivery to the Nervous System

1. General Background for Use of Herpes as a Vector

Herpes simplex virus type I (HSV-1) is a double-stranded DNA virus that is well suited as a vector for delivery of foreign genes into the adult nervous system by several criteria (Table 1).[5,17,55] First, it is readily taken up at nerve terminals and can infect virtually all mammalian cell types. However, it is spread preferentially in the nervous system *in vivo*, where the virus is passed by both retrograde and anteriograde transport, as well as by selective transsynaptic passage between neighboring neurons.[75] Therefore, the virus can be passed *in vivo* from a peripheral infection site into the CNS. The virus can enter a latent state in some postmitotic neurons, where it exists as a benign, "episomal" unit within the nucleus of the cell and, in this state, retains transcriptional activity, allowing expression of foreign genes without compromising the function of the neuron (Figure 1).[91,113,132] During latency, only one virus transcription unit, the latency associated transcript (LAT), of unknown function, is normally expressed.[113,142] The entire virus genome contains 72 endogenous genes in a large, fully sequenced 150 kb genome (Figure 2),[90,104] which can be extensively manipulated genetically. Specifically, many of these endogenous genes can be disrupted and replaced with up to 30 kb of foreign genes and promoter elements without loss of the virus' ability to propagate, i.e., replicate and package, in permissive cells.[83] Certain mutations of the virus's endogenous genes result in a reduction or elimination of its ability to replicate within neurons, but this property can be used to the investigator's advantage (see Section II.A.2.b). For all these reasons, HSV-1 has been used extensively as a vector for foreign gene delivery in the nervous system to deliver toxic genes to cancer cells, for neuroanatomical tracing studies, and to alter neuronal physiology (see Section II.A.3).

2. Types of Herpes Vectors

a. *Amplicon vs. Recombinant Vectors*

Two basic types of HSV vectors have been used for gene transfer: plasmid-derived vectors, termed amplicons,[127] and recombinant virus vectors. Amplicons consist of a small

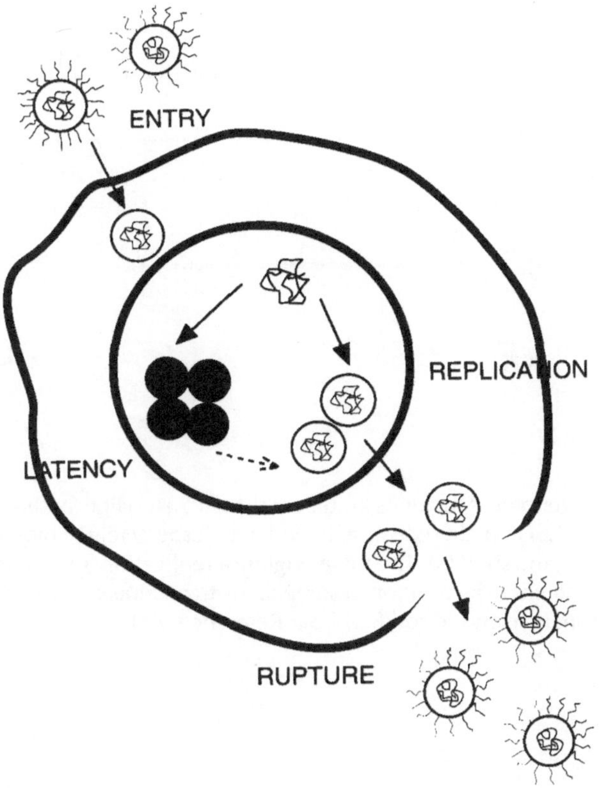

FIGURE 1. Herpes virus in neurons. Once herpes enters the neuron, it can either replicate or enter a benign, "episomal" latent state within the nucleus of the cell. The virus, however, can reactivate from latency and produce infectious particles. (Reproduced from Reference 5 with permission.)

(10 to 15 kb) plasmid, containing an HSV origin of replication and packaging signal (Figure 3).[47,53-56] This plasmid can be transfected into a herpes-permissive host cell, where it is packaged into HSV virus particles as concatamers (roughly 10 copies of the plasmid per capsid) with the aid of a HSV-1 helper virus.[127] Progeny virus from the permissive host cell consist of a mixture of amplicon vectors and helper virus. The advantage of amplicon vectors is their ease of construction, only requiring the manipulation of a plasmid element, and the fact that multiple copies of the gene of interest can be delivered to neurons within each viral capsid. However, the amount of foreign DNA that can be introduced into these vectors is limited (about 10 kb), and they cannot be produced at high titers. Also, amplicons cannot enter a latent state within the neuron and therefore are probably not useful for long-term gene therapy unless they integrate into the genome, which should occur at very low frequency. It should be noted, however, that expression up to 4 to 6 months has been reported in some cases using amplicon vectors *in vivo* (see Section II.A.3.c).

Recombinant vectors are created by direct alteration of the virus genome.[116] To generate recombinant vectors, the foreign DNA of interest is placed into a plasmid so as

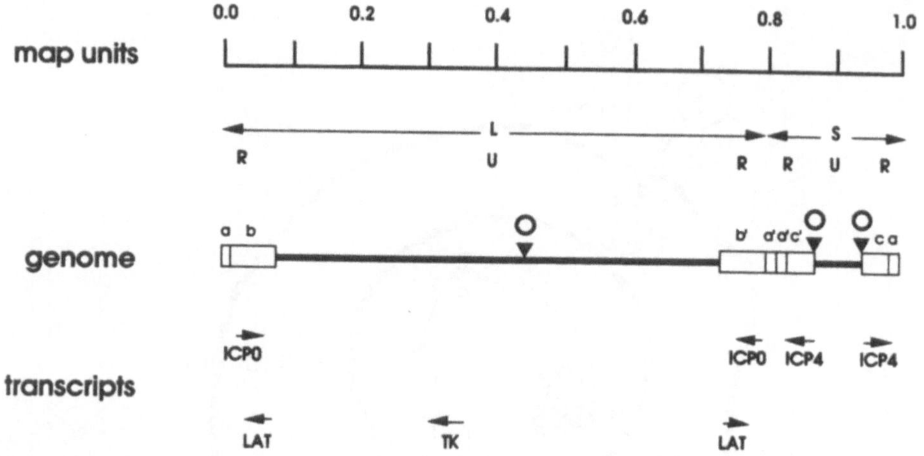

FIGURE 2. HSV-1 genome. Map units (m.u.) are shown on top line for this 150 kb genome. Repeat units (RL and RS) contain repeats a, b, and a, c, respectively. Unique regions (U) are designated as long (L) and short (S). The three origins of replication (O) are designated by the arrows on the genome line. The position of some of the transcripts described in the paper are indicated by horizontal arrows. (Modified from Reference 90.)

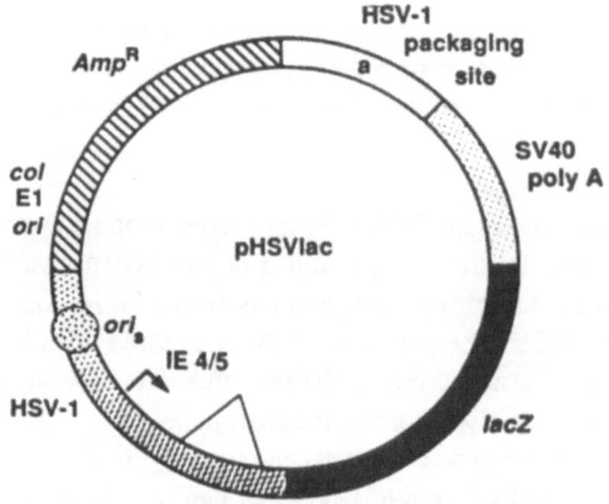

FIGURE 3. Structure of the amplicon plasmid used for generation of HSV vectors. The pHSVlac plasmid contains (1) an ampicillin-resistance gene (Amp^R) and origin of replication (col E1 ori) to allow propagation of the plasmid in bacteria, (2) an HSV-1 origin of replication (ori_s) and packaging signal (a) to allow propagation and packaging of plasmid DNA by HSV-1 helper-virus, and (3) a transcription unit containing the HSV E14/5 immediate-early promoter, the bacterial *lac-Z* gene, and SV40 polyadenylation signals. (Reproduced from Reference 54 with permission.)

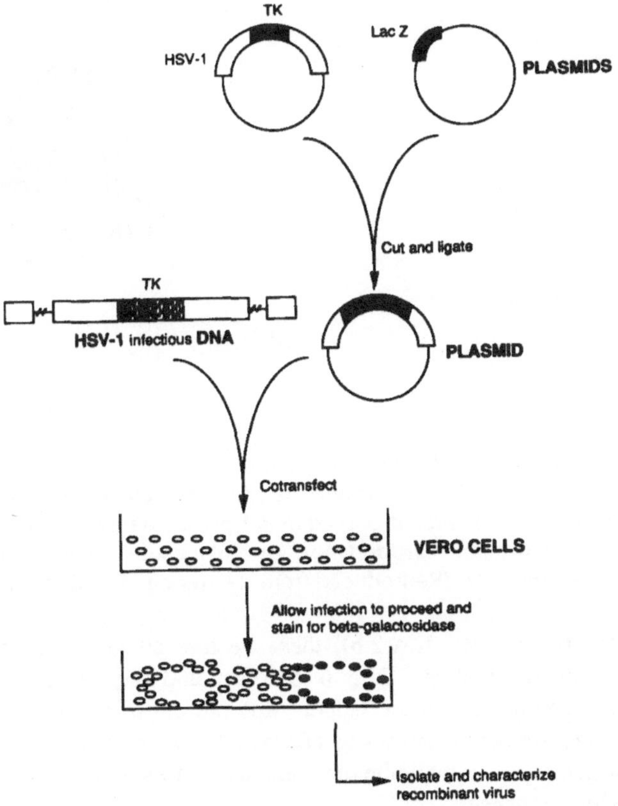

FIGURE 4. Generation of a TK⁻ recombinant HSV-1 vector expressing β-galactosidase. In this scheme, a DNA fragment containing the bacterial β-galactosidase gene (*lac-Z*) is cloned into a plasmid containing the HSV-TK gene (TK), at a restriction site within the TK gene thereby disrupting its function. This recombinant plasmid is then co-transfected into Vero cells along with infectious wild-type HSV-1 DNA, which contains functional TK. One to three days following infection, virus is harvested, and replated onto Vero cells in presence of acyclovir to select for TK^- virus. These cells are then stained for β-gal activity, to select for viral plaques containing recombinant virus expressing *lac-Z*.

to bear flanking portions of herpes genomic sequence (>500 bp) on either side of it. The foreign DNA is then introduced into the herpes viral genome by cotransfection of the plasmid and full-length infectious herpes viral DNA into cultured cells. Homologous recombination of plasmid sequences into the virus genome occurs, creating a full-length virus containing the desired mutation (Figure 4). Recombination frequencies vary from 10^{-1} to 10^{-6}. Recombinant vectors do not require the presence of a helper-virus and can carry large fragments of foreign DNA. They can also, unlike amplicon vectors, be grown to a high titer. Importantly, for the purposes of gene therapy, this type of HSV-1 virus vector theoretically should be capable of entering a true latent state in neurons *in vivo*. The disadvantages of using recombinant viral vectors are that manipulation of a large genome can be difficult from a technical standpoint and that, depending on the viral backbone used

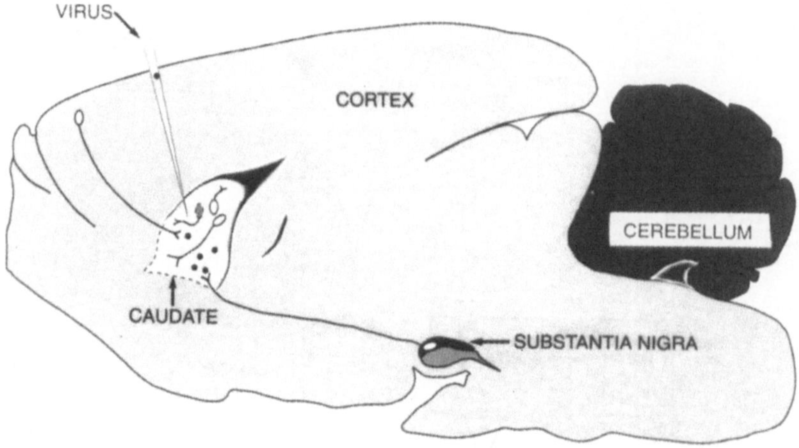

FIGURE 5. Gene delivery to adult rat brain by direct injection of virus particles. Following stereotactic injection of recombinant virus into the caudate nucleus of the striatum, particles infect cells at the injection site and can spread by retrograde transport into other regions of the brain including the substantia nigra and the cortex. SNC: substantia nigra pars compacta, SNR: substantia nigra reticulata. (Reproduced from Reference 19 with permission.)

for recombination (see Section II.A.2.b), these vectors contain viral genes, such as transcription factors, that can alter cellular metabolism and may cause cytotoxicity. This latter disadvantage also applies to the helper-virus present in the amplicon vectors. A major limitation of recombinant viruses as compared to amplicons is that the promoter used to drive transgene expression in the recombinant vectors may be strongly influenced by surrounding viral sequences.

Following construction, both types of herpes virus vectors are introduced directly into the brain by stereotactic injection (Figure 5). They can also potentially be delivered directly across the blood-brain barrier by osmotic shock.[95]

b. Uses of Various Recombinant Herpes Viruses as Backbone for Vectors

One major cause of toxicity of wild-type herpes virus is viral replication, which eventually causes cell lysis. Many genes are involved in this process.[115] Once the virus enters the cell and viral DNA is taken up into the cell nucleus, the *trans*-acting viral factor VP16, which is part of the viral capsid, initiates transcription of the HSV immediate early genes, ICP4, ICP0, ICP27, ICP22, and ICP47, in conjunction with cellular factors such as Oct-1.[129] These, in turn, act as transcription factors that regulate expression of early HSV genes whose products are primarily enzymes involved in replication of viral DNA, such as HSV thymidine kinase (TK), ribonucleotide reductase (RR), and dUTPase. Late gene products involved in capsid and viral assembly and including viral coat proteins are transcribed and translated later in infection. Replicative infection of a cell results in damage to the host cell chromosome, cellular degeneration, and cell death,[103] with the possible exception of replication associated with reactivation of the virus within a host neuron.

Less pathogenic viral mutants are available, created by deleting or disrupting genes involved in viral replication, which can be used as backbone vectors in the construction of recombinant herpes viral vectors and as helper-viruses for amplicon vectors. Vectors have been constructed containing disrupted immediate-early genes, rendering the resulting virus either replication-compromised or totally replication-defective.[22,27,59,68,122] Many other vectors have been constructed that are defective in enzymes involved in viral replication, such as TK,[3,61,99,123,125,134] DNA polymerase,[28] or UTPase.[9,136] Replication is not required for latency, as replication-defective viruses such as TK^- do enter latency and, in fact, are incapable of being reactivated from this state (Figure 6).[29,74,80] Replication-defective or compromised virus can be distinguished from wild-type on the basis of either the requirement for growth in special cell lines, such as in the case of $ICP4^-$ mutants, which will only grow in cells containing an integrated copy of the viral ICP4 gene,[122] or by growth in certain selective drugs, such as in the case of TK^- mutants, which are resistant to the drug, acyclovir. Another advantage of using TK^- mutants is that they can replicate in rapidly dividing cells in which the loss of the function of the HSV-TK enzyme is complemented by high levels of endogenous cellular enzyme, allowing ease of propagation in tissue culture; however, in nondividing cells *in vivo* in which the cellular enzyme is low, such as neurons and glia, the mutant virus cannot replicate. In fact, vectors in which the immediate-early gene, ICP4, or the early HSV-TK gene have been disrupted showed no viral spread and limited toxicity after inoculation into the frontal cortex of normal rodent brain and therefore appear to have little toxicity to neurons or normal glia[25,64] (see Section II.A.3.c). A double ICP4/ICP27 mutant also has been shown to have limited cytotoxicity after injection into the hippocampus.[57] A recombinant virus containing an insertion in the gene encoding the VP16 protein has been shown to have reduced expression of immediate-early genes, which interferes with viral replication but does not block entry into latency or reactivation from it.[1,130] A virus containing a mutation in the viral ribonucleotide reductase gene also shows decreased replication capacity.[58] Mutants in HSV-1 dUTPase have been found to be attenuated and cannot reactivate from latency.[108]

Other viral gene products are involved in disruption of host cell transcriptional activity and protein translation, which also contributes to viral toxicity. For example, the product of the herpes UL41gene is a cellular RNase that disrupts host cell translation.[43,76,112] Viral toxicity due to shut-down of host cell macromolecular synthesis can be eliminated by making vectors in which these viral genes are disrupted or eliminated.

Other viruses, termed nonneurovirulent or nonneuroinvasive, have been found to have reduced toxicity *in vivo* compared with wild-type virus.[27,66,112,137]

3. Uses of Herpes Viral Vectors

a. *Tumor Killing*

Herpes virus can be toxic to cells, both as a result of productive viral infection and through its ability to shut down host macromolecular synthesis and damage cellular DNA. It is possible to take advantage of HSV viral toxicity and the ability to modify the virus to make HSV vectors that can be used to selectively kill glioma brain tumor cells

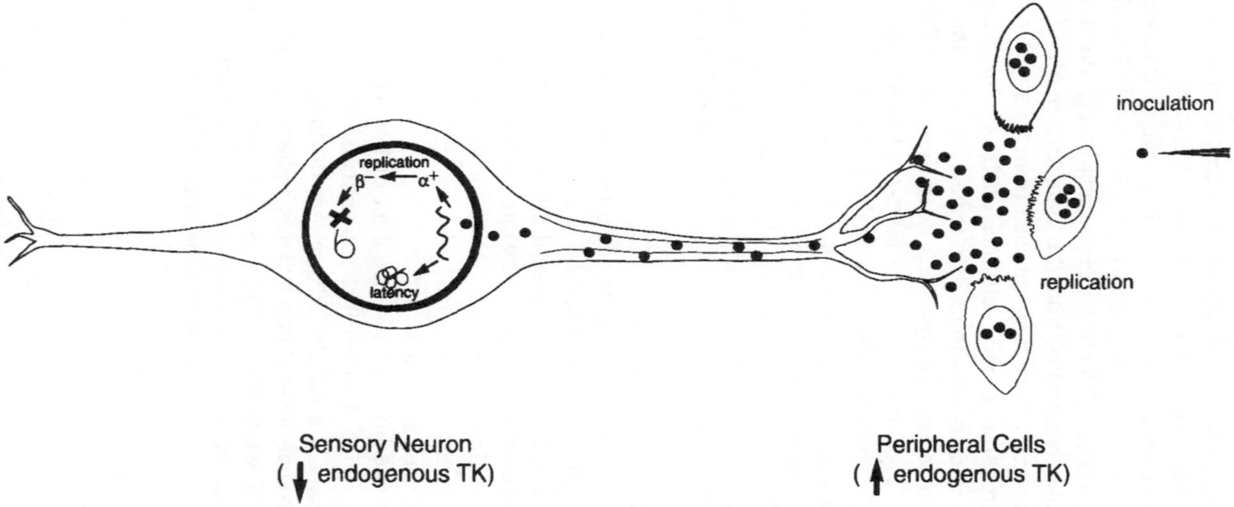

FIGURE 6. Life cycle of TK⁻ herpes virus. Following inoculation of TK⁻ virus onto nerve endings of, for example, sensory neurons, the virus can either replicate in peripheral cells, which have high endogenous TK activity, or it can be taken back to neuronal cell bodies by retrograde transport. Within sensory neurons, expression of immediate-early (α) genes are suppressed and the combined lack of HSV-TK encoded by an early (β) gene and low levels of endogenous TK cause the viral replication program to be aborted. The virus is, however, able to enter latency in these neurons, but does not reactivate from this state.

in vivo. Gliomas are the most common primary central nervous system neoplasm in adults, and are refractory to current available treatments, such as surgical excision, radiation therapy, and chemotherapy.[26] Therefore, there is great interest in exploring other forms of cancer therapy, including the use of herpes vectors.

The goal in making a recombinant HSV vector to destroy tumor cells selectively in the brain is to create a vector that will be toxic to rapidly dividing cells, like gliomas, but will spare nondividing cells, like neurons and glia.[26,32,72,88] It has been shown, for example, that a HSV-TK⁻ virus will selectively destroy glioma cells *in vitro* and *in vivo*.[88] The strategy takes advantage of the ability of rapidly dividing glioma cells to complement the loss of viral TK activity, by virtue of high cellular TK activity so that the virus can replicate in and kill these cells but not nondividing cells, like neurons and glia, in which cellular TK is low. In these experiments,[88] the survival rate of nude mice implanted intercranially with human U87 tumors was increased after injection of the tumors with HSV-TK deletion virus (dlsptk; one-third of these animals survived until sacrifice at 5 months, while all mock-injected control animals died by 6 to 8 weeks), apparently due to a decrease in the size of the tumors. In fact, histologic examination of herpes vector-injected animals revealed the absence of any surviving tumor cells. However, in animals infected with virus, some deaths did occur, perhaps from herpes-induced encephalitis due to replication of virus in the few dividing cells in the brain, such as reactive glia and endothelial cells. The dose of dlsptk virus that was lethal to 50% of the animals (LD_{50}), however, was 10^6 PFU, as compared to 10^3 PFU for wild-type herpes. Use of TK⁻ herpes vectors bearing an *E. coli*. β-galactosidase (*lac-Z*) transgene demonstrate that the virus spreads to and destroys tumor cells at some distance from the inoculation site[15] (see Section II.A.3.c). Several other mutants have also given promising results in this paradigm, including virus containing mutations in HSV DNA polymerase, ribonucleotide reductase, and a neurovirulence gene, γ34.5.[72,87]

A further refinement of the use of herpes vectors for tumor killing might include the use of cell-specific promoters to drive expression of genes involved in viral replication in order to target toxicity to a particular cell type, such as glioma cells.

b. Neuroanatomical Tracing

HSV-1 has been used as an effective method for visualizing trans-synaptic connections in the mammalian nervous system.[17,23,133,140] To be effective as a neuronal "tracer", a substance must transfer selectively across synapses and be present at high enough levels in the recipient cell to allow detection. Other compounds used as tracer molecules, such as horse radish peroxidase conjugated wheat germ agglutinin (WGA-HRP) and tetanus toxin C-fragment, while transferring preferentially at the synapse,[40,128] become diluted upon successive transfer between connecting neurons. HSV-1, on the other hand, is transfered to recipient neurons and then replicates, resulting in amplification of the tracer signal (Figure 7). The virus can easily be detectable by immunocytochemical techniques using antibodies directed against viral antigens,[77,140] autoradiography of ^{3}H-thymidine incorporated into replicating viral DNA,[86] or histochemical staining for marker enzymes such as *lac-Z* (see below) encoded within the HSV vector. *Trans*-synaptic transport of HSV-1 is rapid and can occur over long distances within the

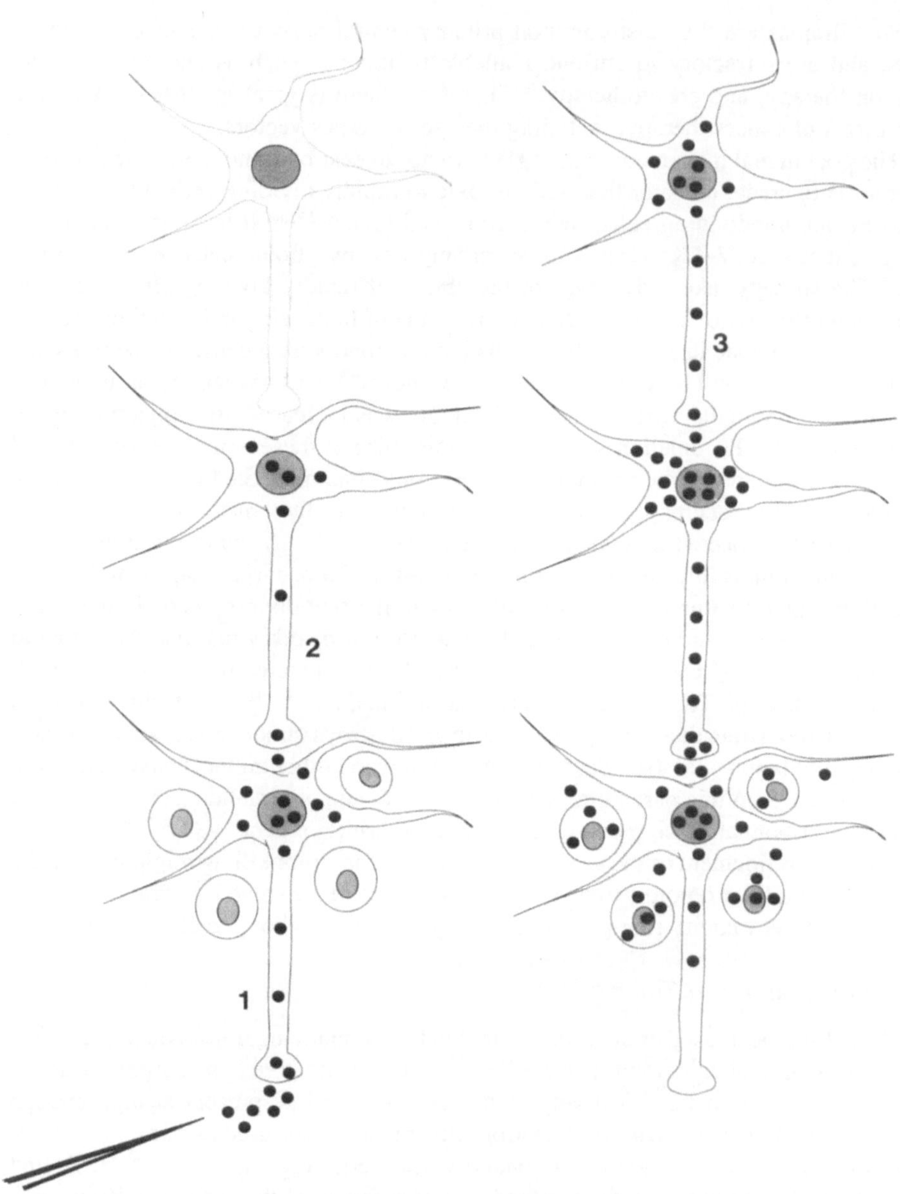

FIGURE 7. HSV as a transneuronal tracer. (1) Retrograde axonal transport of virus to the cell body of the primary neuron following viral inoculation at the nerve terminal. (2) Trans-synaptic transfer of the virus to a second order neuron, followed by retrograde transport to its cell body. (3) Replication of the virus in the secondary cell nucleus, leading to strong transneuronal labeling. Subsequent lysis of primary neuron and release of virus to surrounding cells. (Modified from Reference 75.)

nervous system over a period of a few days before the animal dies of encephalitis. HSV-1 has been shown to give effective labeling of trans-synaptic neuronal connections between the periphery and the CNS, for example between trigeminal ganglion sensory neurons and the contralateral thalamus and between hypoglossal motoneurons and the brainstem,[23,86,140] as well as between neurons within the brain itself, such as the hypothalamus and the retinal ganglion cell layer.[28,81]

c. *Gene Therapy*

The most exciting potential use of HSV-1 vectors is in focal delivery of biologically active foreign genes to neurons for the purposes of altering neuronal physiology so as to create and ameliorate animal models of human neurologic disease states, and as a gene replacement therapy for hereditary deficiency states affecting neurons. For this purpose, the objective is to develop vectors that allow both short- and long-term expression of foreign DNA, with little pathogenicity to animals.

Many viral and cellular promoters have been used to drive foreign gene expression in the context of both amplicon and recombinant herpes viral vectors following direct inoculation into the brain or uptake into peripheral ganglia after inoculation of the nerve endings. Marker genes, such as *lac-Z*, have also been included in both amplicon and recombinant virus constructs, as a means of marking those cells in which the foreign gene is expressed and as a means of identifying the vector during isolation of the recombinant virus.[61] *lac-Z*, which encodes the bacterial enzyme, β-galactosidase (β-gal), is a useful marker as it is readily detectable histochemically using the substrate 5-bromo-4-chloro-3-indolyl, β-D-galactosidase (X-gal). β-Gal can be distinguished from the mammalian enzyme on the basis of a higher optimal pH and by use of antibodies specific to the bacterial form. In addition, use of a *lac-Z* marker in the context of a herpes virus vector is theoretically a much more rapid method of evaluating neuronal-specific promoters than the creation of transgenic mice. Herpes vectors have been used to test both the tyrosine hydroxylase and brain type II sodium channel promoters in cultured cells and, in the case of the sodium channel promoter, in peripheral neurons.[79,96]

Using recombinant virus vectors, the extent of transgene delivery has been evaluated in several areas of the nervous system using different vector backbones and promoters. Transient expression of *lac-Z*, inserted into the coding region of either the ICP4, ICP0, or TK genes and under the control of the corresponding promoter in recombinant herpes vectors, was seen in neurons and other cells following inoculation of 200,000 PFU of virus into the adult rat caudate nucleus.[25] These viruses, moreover, were relatively nonpathogenic, i.e., no abnormal behavior, seizures, or untimely deaths were observed. Both the TK^- and $ICP4^-$ *lac-Z*-expressing vectors, which are replication-defective in neurons in the brain, demonstrated transient expression (3 days) only in those cells surrounding the site of injection. The $ICP0^-$ virus, which is replication-compromised, gave transient but widespread delivery to cells over an extended area spreading out some distance from the site of injection. Rats inoculated with either the $ICP4^-$ or TK^- virus showed no histologic or pathologic effects several months after injection.[64] However, rats inoculated with the $ICP0^-$ virus showed a histologic decrease in size of the caudate nucleus over 2 months, probably due to a "smoldering" infection.

Stereotactic injection of a *lac-Z*-containing recombinant herpes virus under the control of the HSV glycoprotein C (gC) late gene promoter into the brain also resulted in transient β-galactosidase expression.[44] Viral attenuation was achieved in this construct by insertion of the transgene into the ICP4 locus, thus rendering the virus replication-defective. Again, little damage to the normal brain morphology was noted. Ten days following injection, LATs were detected in the brain, indicating the ability of the mutant virus to establish latency. When the *lac-Z* gene was regulated by the defined LAT promoter[10,37,69,122] or a second putative TATA-less one that lies 5′ of the LAT transcription start site, both gave transient *lac-Z* expression in the brain.[45,57] Use of a non-HSV viral promoter, the long-terminal repeat of the Moloney murine leukemia virus (MoMLV-LTR), to drive *lac-Z* expression within the TK locus in a TK^- virus was also found to give transient but high levels of β-galactosidase activity in cells of the adult rat nervous system following injection into the caudate; many labeled cells were seen 3 days following inoculation, but by 14 days, no visible β-gal + cells were found.[18] The same vector inoculated on the snout and cornea was transported retrograde to the trigeminals ganglia, giving *lac-Z* expression in up to 2% of the neurons innervating these areas in decreasing numbers for 30 days after inoculation.[35] This expression also diminished with time. A similar vector had previously been found to give stable β-gal expression in sensory neurons of the peripheral nervous system (24 weeks) after inoculation of the sciatic nerve, and extended expression in motor neurons following injection into the tongue.[38] In this case, the recombinant virus was $ICP4^-$ with the transgene inserted in the LAT locus. In another set of studies, the use of a mammalian neuron-specific enolase (NSE) promoter to drive *lac-Z* expression in a TK^- HSV vector was found to give neuronal-specific expression in culture and *in vivo* following injection into the rat caudate.[3] A few neurons showed β-galactosidase expression for up to 30 days. Positive neurons were also noted in areas that project directly to the striatal injection site, in the frontal cortex, substantia nigra, and lateral thalamus, indicating uptake of the virus into these sites by retrograde transport.[4] The use of stronger promoters, such as pol III, is also currently being explored to see if they can give high-level, long-term expression in the brain during viral latency.[57]

In terms of delivery of biologically active molecules to the nervous system, promising results have been obtained using both amplicon and recombinant vectors. Specificity of expression using these viral vectors depends on many criteria, including the gene used in the construct, the promoter used to drive foreign gene expression, the inoculation site of the virus, and the subcellular localization of the resulting protein.[55] Introduction of a minigene encoding nerve-growth factor (NGF) driven by the NGF promoter prior to axotomy of neurons projecting from the superior cervical ganglion (SCG) in an amplicon vector was found to prevent subsequent decline of tyrosine hydroxylase (TH) levels.[42] It had been previously shown in culture that following infection of fibroblasts with this vector, NGF was secreted into the media at a level of 10 to 34 ng/ml over a period of 3 days, and the secreted NGF was proved to be biologically active by its ability to increase survival of sympathetic neurons. Sympathetic neurons in the SCG depend on target-derived NGF for maintenance of TH levels, and axotomy normally results in a 50% decrease in activity. Following inoculation of the NGF-containing amplicon into the SCG, transgenic NGF mRNA was readily detectable in the ganglia. Further, TH levels actually increased 18% in these neurons following axotomy 4 days after inoculation of the vector. In preliminary studies using a 6-hydroxydopamine rat model of Parkinson's disease, injection of an amplicon construct containing the TH gene, regulated by an HSV-1 immediate-early promoter, reduced apomorphine-induced asymmetric rotation by 80% over a 6-month period, supporting the idea that levels of dopamine in surviving neurons were augmented by TH activity confered by the transgene.[39] In a second related study utilizing the same model system, introduction of an amplicon vector containing the βII catalytic domain of protein kinase C (PKC) driven by the TH promoter was utilized to determine whether the PKC pathway controls dopamine release in the substantia nigra.[126] Direct inoculation of this vector encoding constituitively active PKC subunit was found to cause a dramatic increase of dopamine release for a period of 2 weeks, accompanied by an increased rotation rate in 10 out of 24 animals tested. In addition, these animals were also found to be immunopositive for this enzyme by immunocytochemical analysis 1 month after virus introduction. A recombinant TK^- vector bearing hypoxanthine phosphoribosyltransferase (HPRT) under control of the TK promoter injected at low dosages (2×10^5 PFU/ml) into mouse brains yielded transient expression of HPRT mRNA.[99] Unfortunately, however, injection of high doses (2×10^7 PFU/ml) resulted in encephalitis and death a few days following inoculation. This model might ultimately by used for treatment of the HPRT deficiency state in man, the Lesch Nyhan syndrome. A recombinant vector has also been used to introduce the gene for β-glucuronidase (GUSB) into a mouse model of mucopolysaccharidosis type VII (Sly disease).[143] Sly disease is a human genetic lysosomal storage disorder caused by a lack of this enzyme, which results in failure to degrade glycosaminoglycans (GAGs) and engorged lysosomes and leads to neuronal degeneration, mental retardation, and ultimately death. Corneal inoculation of a replication-competent virus containing the GUSB gene driven by the LAT promoter has resulted in expression of the gene in CNS neurons for up to 4 months postinoculation in latently infected animals and reduction of storage material in some cells expressing the enzyme.

In summary, herpes virus vectors are an efficient means to achieve short-term transgene expression on neurons, and they still appear to be the only means to confer long-term expression on these cells. Problems of cytotoxicity can be reduced dramatically by mutations in genes associated with virus replication, immediate-early gene expression, and neurovirulence. As yet, sustained (≥1 month) transgene expression has only been achieved in a few neurons in the peripheral and central nervous system. It is not clear if this is because only a subset of neurons will harbor the virus in latency or that the promoter elements used to date are being down-regulated over time. In its current state, these vectors are most effective *in vivo* for focal delivery to a subset of neurons following inoculation at peripheral nerve endings or direct injection into the brain. They are being tested in gene therapy protocols for animal models of brain tumors, Parkinson's disease, lysosomal storage diseases, Lesch-Nyhan syndrome, and pain syndromes. (This chapter does not cover the extensive use of herpes vectors in cell culture. In general, herpes vectors appear to be somewhat more cytotoxic to neurons in culture than *in vivo*. The reader is referred for this topic to References 11, 55, 68, 114.)

B. Use of Adenovirus as a Vector for Delivery to the Nervous System

1. General Background of Use of Adenovirus as a Vector

Adenovirus is known to be trophic for human epithelial cells, being the causative agent for such maladies as conjunctivitis and the common cold, and it is this propensity that has made it the vector of choice for gene delivery to the respiratory epithelium[117] for such genetic deficiency diseases as cystic fibrosis.[34,89] Despite this seeming preference for epithelial cells, the virus has a broad host-range, and because of this, its use as a vector is also being explored for gene delivery to such diverse cell types as hepatocytes,[110] liver cells,[65] and endothelial cells,[82] as well as brain cells (see below). The virus has been shown to be capable of robust, fairly stable gene expression over several months in postmitotic cells including neurons (for review see References 19, 94, and 98). Another advantage of the virus is that it can be grown to a high titer and is considered relatively safe, although it can alter host cell metabolism and produce malignant transformation in some animals. Adenoviruses replicate in an extrachromosomal state but are not known to enter latency like HSV-1, and mutants currently in use as vector backbones may continue low-level replication *in vivo*.[94] In rare cases, they may integrate into the host cell genome.

Recombinant adenovirus vectors are constructed in a similar manner to herpes virus, typically by homologous recombination using a replication-defective form of the virus that carries deletions in the E1A to E1B and E3 regions as a backbone for vector construction (Figure 8). These vectors will accommodate about 6 to 8 kb of insert under regulation of various viral and mammalian promoters.

Interestingly, adenovirus has also been used as conjugates for receptor-mediated gene delivery of polylysine DNA complexes[31,33,52,145] (Figure 9). This method involves the transfer of adenovirus-DNA-polylysine conjugates into cells via adenovirus receptor endocytosis, where the low pH of the endosomal compartment allows release of plasmid DNA into the cytoplasm via fusion of the adenovirus capsid proteins with vesicular membrane proteins. This technique has been used primarily for delivery of genes to cultured cells and epithelial cells *in vivo*, but it holds promise as a means of gene delivery to the brain as well, if the efficiency of delivery can be increased.

2. Uses of Adenovirus for Gene Delivery to the Nervous System

a. *Foreign Gene Delivery to Brain and CSF*

Injection of replication-defective adenoviruses under the control of strong viral promoters from the Rous sarcoma virus (RSV) and cytomegalovirus (CMV) into several regions of the rat brain resulted in extensive labeling of cells surrounding the site of injection, with the highest levels of expression being in the first week following inoculation with the virus, although some level of expression (10% of 1-week level) was

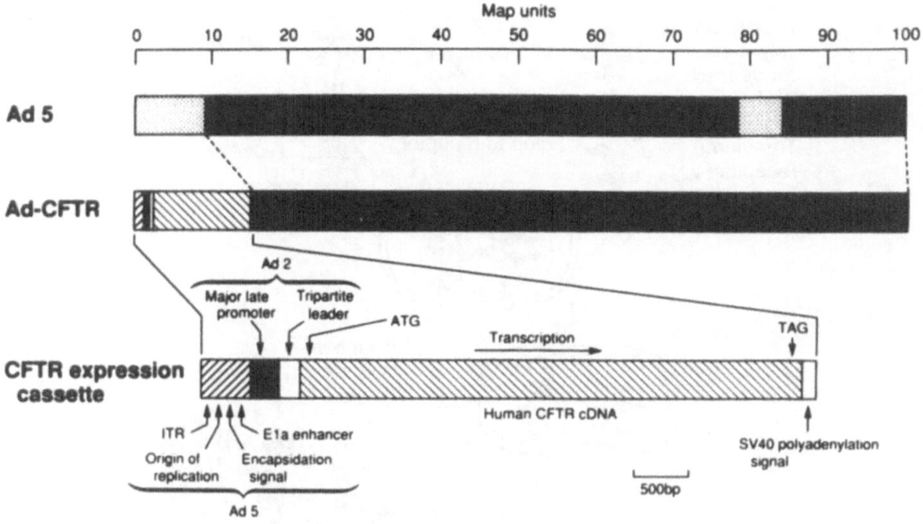

FIGURE 8. Schematic of a recombinant adenovirus vector containing the cystic fibrosis transmembrane conductance regulator. Shown is the adenovirus type 5 genome (Ad 5), along with the recombinant virus (Ad-CFTR), and an enlargement of the CTRF expression cassette. To construct this vector, the E1A to E1B and E3 regions (stippled segments) were removed from the adenovirus genome, and replaced with the CFTR expression cassette. The expression cassette contains adenovirus 5′ inverted terminal repeat (ITR), origin of replication, encapsidation signal, and E1A enhancer element from Ad5, the major late promoter and tripartite leader sequences from Ad2, and the entire 4.5 kb protein coding sequence of the human CFTR cDNA, including ATG translational start and TAG stop sequences, and the SV40 polyadenylation signal. (Reproduced from Reference 118 with permission.)

seen up to 2 months.[2,8,36,78] Loss of expression may be due to death of infected cells, down-regulation of heterologous promoter elements (as seen in retroviral vectors), or degradation of adenovirus vectors. Some retrograde transport of the virus was also noted through uptake of the vector at nerve terminals. Efficiency of expression was estimated at about 1 labeled cell for every 100 PFU injected; in which case, efficiency of delivery using replication-defective herpes vectors would be roughly 100-fold less.[4] Interestingly, at least for glioma cells in culture, gene delivery using herpes vectors is more efficient than using adenovirus vectors.[15] Cells labeled with adenovirus vectors in the brain included neurons, astrocytes, microglia, oligodendrocytes, and ependymal cells. Little neuropathology was observed in the brain following injection of adenovirus vectors at low dosages; however, inoculation with higher titers resulted in neuronal death, gliosis, and inflammation. A major potential drawback of use of this vector is the fact that it can elicit a strong immune response that might result in reaction against subsequent use of the vector. Another potential problem is that this virus can also recombine with other viruses such as SV40, and may cause reactivation of latent viruses such as herpes.

In terms of delivery of biologically active proteins, an adenovirus vector containing α-1-antitrypsin gene driven by the RSV promoter was delivered to ependymal cells by

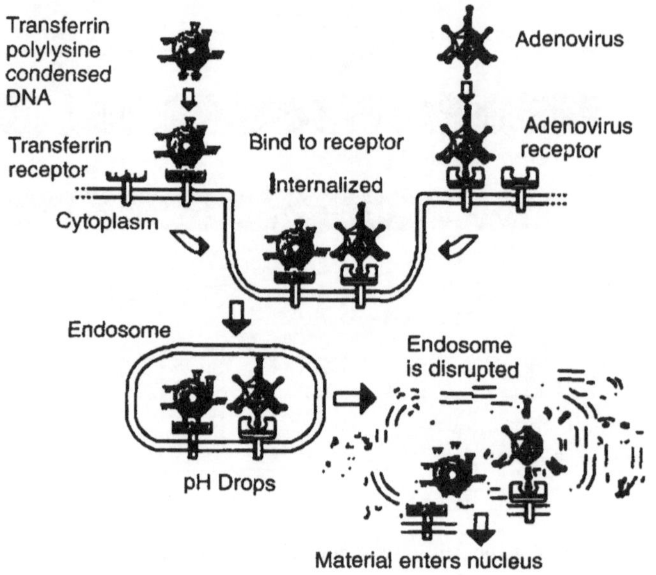

FIGURE 9. Adenovirus-enhanced receptor-mediated gene delivery. DNA condensed with transferrin-polylysine conjugate and inactivated adenovirus are introduced into cultured cells and taken up through receptor-mediated endocytosis, frequently into the same endosome. A low-pH-triggered change in the adenovirus capsid is believed to generate disruption of the endosome and release of the DNA complex. This prevents degradation of DNA in the lysosome and increases the efficiency of transfection. (Reproduced from Reference 31 with permission.)

inoculation of the lateral ventricles, and secreted protein was detected in cerebral spinal fluid for over a week following viral inoculation.[8] This is exciting, as it may be a new method for extended delivery of unstable secretable proteins into the brain. Although the use of these vectors in the nervous system is still very preliminary, they appear to be especially useful for high-level "bursts" of transgene expression. As such, they may have a role in gene therapy paradigms of brain tumors, stroke, and trauma.

III. CONCLUSIONS

Herpes and adenovirus are the only successful means to date by which foreign genes have been introduced into postmitotic neurons in the adult animal. The major goal of further development of vector technology using these viruses is to define optimal promoters and integration sites to achieve high-level and stable delivery of foreign genes into the adult nervous system while reducing cytotoxicity to the brain. Means of disseminated gene delivery will also be needed in some experimental paradigms. These types of vectors hold much promise as a means of modulating CNS nerve cell function, such as altering neurotransmission and promoting neuronal survival and regeneration, and as a means of gene replacement therapy for human hereditary deficiency states, as well as for treatment of brain tumors.

ACKNOWLEDGMENT

This effort was supported by NIH NRSA postdoctoral fellowship F32NS08810 (JKA) and NIH grant NS24279 (XOB).

REFERENCES

1. Ace, C. I., McKee, T. A., Ryan, J. M., Cameron, J. M., and Preston, C. M., Construction and characterization of a herpes simplex virus type 1 mutant unable to transinduce immediate-early gene expression, *Virology*, 63, 2260, 1989.
2. Akli, S., Caillaud, C., Vigne, E., Stratford-Perricaudet, L. D., Poenaru, L., Perricaudet, M., Kahn, A., and Peschanski, M. R., Transfer of a foreign gene into the brain using an adenovirus vector, *Nat. Genet.*, 3, 224, 1993.
3. Andersen, J. K., Garber, D. A., Meaney, C. A., and Breakefield, X. O., Gene transfer into mammalian central nervous system using herpes virus vectors: extended expression of bacterial lac Z in neurons using the neuron-specific enolase promoter, *Hum. Gene Ther.*, 3, 487, 1992.
4. Andersen, J. K., Frim, D. M., Isacson, O., and Breakefield, X. O., Herpes-mediated gene delivery into the rat brain: specificity and efficiency of the neuron-specific enolase promoter, *Cell. Mol. Neurobiol.*, 13, 503, 1993.
5. Andersen, J. K. and Breakefield, X. O., Herpes simplex virus and its use in neuroscience research, in *Supplement to Encyclopedia of Neuroscience*, Birkhauser, Boston, 1992, 79.
6. Austin, C. P. and Cepko, C. L., Cellular migration patterns in the developing mouse cerebral cortex, *Development*, 10, 713, 1990.
7. Baetge, E. E., Behringer, R. R., Messing, A., Brinster, R. L., and Palmiter, R. D., Transgenic mice express the human phenylethanolamine N-methyl transferase gene in adrenal medulla and retina, *Proc. Natl. Acad. Sci. U.S.A.*, 85, 3648, 1988.
8. Bajocchi, G., Feldman, S. H., Crystal, R. G., and Mastrangeli, A., Direct in vivo gene transfer to ependymal cells in the central nervous system using recombinant adenovirus vectors, *Nat. Genet.*, 3, 229, 1993.
9. Barker, D. E. and Roizman, B., Identification of three genes nonessential for growth in culture near the right terminus of the unique sequences of long component of herpes simplex virus 1, *Virology*, 177, 684, 1990.
10. Batchelor, A. H. and O'Hara, P., Regulation and cell-type specific activity of a promoter located upstream of the latency-associated transcript of herpes simplex virus type 1, *J. Virol.*, 64, 3269, 1990.
11. Battleman, D. S., Geller, A. I., and Chao, M. V., HSV-1 vector-mediated gene transfer of the human nerve growth factor receptor p75hNGFR defines high-affinity NGF binding, *J. Neurosci.*, 13, 941, 1993.
12. Bjorklund, A., Intracerebral transplantation: prospects for neuronal replacement in neurodegenerative disease, *Res. Publ. Assoc. Res. Nerv. Ment. Dis.*, 71, 361, 1993.
13. Bloom, F. E., Sanna, P., Macjiewski-Lenoir, D., Trembleau, A., Morales, M., and Melia, K., Direct RNA delivery into the brain, in *Gene Transfer Strategies in the Study of Brain Damage and Repair*, Wenner-Gren Center International Symposium, 1993.
14. Bohn, M. C., Cupit, L., Marciano, F., and Gash, D. M., Adrenal medulla grafts enhance recovery of striatal dopaminergic fibers, *Science*, 237, 913, 1987.

15. Boviatsis, E. J., Chase, M., Wei, M. X., Tamiya, T., Hurford, R. K., Jr., Kowall, N. W., Tepper, R. I., Breakefield, X. O., and Chiocca, E. A., Gene transfer into experimental brain tumors mediated by adenovirus, herpes simplex virus (HSV), and retrovirus vectors, *Hum. Gene Ther.*, submitted.
16. Breakefield, X. O. and Geller, A. I., Gene transfer into the nervous system, *Mol. Neurobiol. Rev.*, 1, 339, 1988.
17. Breakefield, X. O. and DeLuca, N. A., Herpes simplex virus for gene delivery to neurons, *New Biol.*, 3, 203, 1991.
18. Breakefield, X. O., Huang, Q., Andersen, J. K., Kramer, M. F., Bebrin, W. R., Davar, G., Vos, B., Garber, D. A., DiFiglia, M., and Coen, D., Gene transfer into the nervous system using recombinant herpes virus vectors, in *Gene Transfer and Therapy in the Nervous System*, Vol. 16, Gage, C., Ed., Springer-Verlag, Heidelberg, 1992, 45.
19. Breakefield, X. O., Gene delivery into the brain using virus vectors, *Nat. Genet.*, 3, 187, 1993.
20. Brundin, P., Bjorklund, A., and Lindvall, O., Practical aspects of the use of human fetal brain tissue for intracerebral grafting, *Prog. Brain Res.*, 82, 707, 1990.
21. Burton, F. H., Hasel, K. W., Bloom, F. W., and Sutcliffe, J. G., Pituitary hyperplasia and giganticism in mice caused by a cholera toxin transgene, *Nature*, 350, 74, 1991.
22. Cai, W. and Schaffer, P. A., Herpes simplex type 1 ICPO plays a critical role in the *de novo* synthesis of infectious virus following transfection of viral DNA, *J. Virol.*, 63, 4579, 1989.
23. Card, J. P., Rinaman, L., Schwaber, J. S., Miselis, R. R., Whealey, M. E., Robbins, A. K., and Enquist, L. W., Neurotropic properties of pseudorabies virus: uptake and transneuronal passage in the rat central nervous system, *J. Neurosci.*, 10, 1974, 1990.
24. Cepko, C., Retroviral vectors and their application in neurobiology, *Neuron*, 1, 345, 1988.
25. Chiocca, E. A., Choi, B. B., Weizhong, C., DeLuca, N. A., Schaffer, P. A., DiFiglia, M., and Breakefield, X. O., Transfer and expression of the lacZ gene in rat brain neurons mediated by herpes simplex virus mutants, *New Biol.*, 2, 739, 1990.
26. Chiocca, E. A., Andersen, J. K., Takamiya, Y., Martuza, R. L., and Breakefield, X. O., Virus-mediated genetic treatment of rodent gliomas, in *Gene Therapeutics*, Wolfe, J. H., Ed., Birkhauser, Boston, in press, 1994.
27. Chou, J., Kern, E. R., Whitley, R. J., and Roizman, B., Mapping of herpes simplex virus-1 neurovirulence to gamma 34.5, a gene nonessential for growth in culture, *Science*, 250, 1262, 1990.
28. Chrisp, C. E., Suntrum, J. C., Averill, D. R., Levine, M., and Glorioso, J. C., Characterization of encephalitis in adult mice induced by intracerebral inoculation of herpes simplex virus type 1 (KOS) and comparison with mutants showing decreased virulence, *Lab Invest.*, 60, 822, 1989.
29. Coen, D. M., Kosz-Vnenchak, M., Jacobson, J. G., Leib, D. A., Bogard, C. L., Schaffer, P. A., Tyler, K. L., and Knipe, D. M., Thymidine-kinase negative herpes simplex virus mutants establish latency in mouse trigeminal ganglia but do not reactivate, *Proc. Natl. Acad. Sci. U.S.A.*, 86, 4736, 1989.
30. Cooper, D. N., Murine retroviral vectors and human gene therapy, *Science*, 228, 650, 1985.
31. Cotten, M., Wagner, E., Zatloukal, K., Phillips, S., Curiel, D. T., and Birnstiel, M. L., High-efficiency receptor-mediated delivery of small and large (48 kilobase) gene constructs using the endosome-disruption activity of defective or chemically inactivated adenovirus particles, *Proc. Natl. Acad. Sci. U.S.A.*, 89, 6094, 1992.
32. Culver, K. W., Ram, X., Waeberg, S., Ishii, H., Oldfield, W. H., and Blaese, R. M., *In vivo* gene transfer with retroviral vector-producer cells for treatment of experimental brain tumors, *Science*, 256, 1550, 1992.
33. Cristiano, R. J., Smith, L. C., and Woo, S. L., Hepatic gene therapy: adenovirus enhancement of receptor-mediated gene delivery and expression in primary hepatocytes, *Proc. Natl. Acad. Sci. U.S.A.*, 2618, 1993.

34. Crystal, R. G., Gene therapy strategies for pulmonary disease, *Am. J. Med.*, 92, 44S-52S, 1992.
35. Davar, G., Kramer, M. F., Garber, D., Roca, A. L., Andersen, J. K., Bebrin, W., Coen, D. M., Kosz-Vnenchak, M., Knipe, D. M., Breakefield, X. O., and Isacson, O., Comparative efficiency of gene delivery to mouse sensory neurons using herpes virus vectors, *J. Comp. Neurol.*, in press.
36. Davidson, B. L., Allen, E. D., Kozarsky, K. F., Wilson, J. M., and Roessler, B. J., A model system for in vivo gene transfer into the central nervous system using an adenovirus vector, *Nat. Genet.*, 3, 219, 1993.
37. Dobson, A. T., Sederati, F., Devi-Rao, G., Flanagan, W. M., Farrell, M. J., Stevens, J. G., Wagner, E. K., and Feldman, L. T., Identification of the latency-associated transcript promoter by expression of rabbit beta-globin mRNA in mouse sensory nerve ganglia latently infected with a recombinant herpes virus, *J. Virol.*, 63, 3844, 1989.
38. Dobson, A. T., Margolis, T. P., Sedarati, F., Stevens, J. G., and Feldman, L. T., A latent, nonpathogenic HSV-1-derived vector stably expresses beta-galactosidase in mouse neurons, *Neuron*, 5, 353, 1990.
39. During, M. J., Geller, A. I., Deutch, A., and O'Malley, K. L., Recovery in the rat 6-hydroxydopamine model of Parkinson's disease by direct intrastriatal injection of HSV-1 vectors which express the human tyrosine hydroxylase gene, *Soc. Neurosci. Abstr.*, 331.8, 782, 1992.
40. Evinger, C. and Erichsen, J. G., Transsynaptic retrograde transport of fragment C of tetanus toxin demonstrated by immunohistochemical localization, *Brain Res.*, 380, 383, 1986.
41. Ezzeddine, Z. D., Martuza, R. L., Platika, D., Short, M. P., Malick, A., Choi, B., and Breakefield, X. O., Selective killing of glioma cells in culture and in vivo by retrovirus transfer of the herpes simplex virus thymidine kinase gene, *New Biol.*, 3, 608, 1991.
42. Federoff, H. J., Geschwind, M. D., Geller, A. I., and Kessler, J. A., Expression of nerve growth factor in vivo from a defective herpes simplex virus 1 vector prevents effects of axotomy on sympathetic ganglion, *Proc. Natl. Acad. Sci. U.S.A.*, 89, 1636, 1992.
43. Fenwick, M. L. and Everett, R. D., Transfer of UL41, the gene controlling virion-associated host cell shut-off, between different strains of herpes simplex virus, *J. Gen. Virol.*, 71, 411, 1990.
44. Fink, D. J., Sternberg, L. R., Weber, P. C., Mata, M., Goins, W. F., and Glorioso, J. C., In vivo expression of beta-galactosidase in hippocampal neurons by HSV-1 mediated transfer, *Hum. Gene Ther.*, 3, 11, 1992.
45. Fink, D. J., Mata, M., Sternberg, L. R., Goins, W., and Glorioso, J. C., Gene transfer into the brain using a herpes simplex virus vector, *Soc. Neurosci. Abstr.*, 399.3, 16, 1990.
46. Forss-Petter, S., Danielson, P. E., Catsica, S., Battenberg, E., Price, J., Nerenberg, M., and Sutcliffe, J. G., Transgenic mice expressing beta-gal in mature neurons under neuron-specific enolase promoter control, *Neuron*, 5, 187, 1990.
47. Freese, A., Geller, A. I., and Neve, R., HSV-1 vector mediated neuronal gene delivery, *Biochem. Pharmacol.*, 40, 2189, 1990.
48. Frim, D. M., Uhler, T. A., Short, M. P., Ezzedine, D., Klagsburn, M., Breakefield, X. O., and Isacson, O., Effects of biologically delivered NGF, BDNF, and bFGF on striatal excitotoxic lesions, *NeuroRep.*, 4, 367, 1993
49. Gage, F. H., Wolff, J. A., Rosenberg, M. B., Xu, L., Yee, J. K., Sults, C., and Friedmann, T., Grafting genetically modified cells to the brain: possibilities for the future, *Neuroscience*, 23, 795, 1987.
50. Gage, F. H and Fisher, L. J., Intracerebral grafting: a tool for the neurobiologist, *Neuron*, 6, 1, 1991.
51. Gage, F. H., Kawaja, M. D., and Fisher, L. J., Genetically modified cells: applications for intracerebral grafting, *Trends Neurosci.*, 14, 328, 1991.

52. Gao, L., Wagner, E., Cotten, M., Agarwal, S., Harris, C., Romer, M., Miller, L., Hu, P. C., and Curiel, D., Direct in vivo gene transfer to airway epithelium employing adenovirus-polylysine-DNA complexes, *Hum. Gene Ther.*, 4, 17, 1993.
53. Geller, A. I., Keyomarsi, K., Bryan, J., and Pardee, A. B., An efficient deletion mutant packaging system for defective herpes simplex virus vectors: potential applications to human gene therapy and neuronal physiology, *Proc. Natl. Acad. Sci. U.S.A.*, 87, 8950, 1990.
54. Geller, A. I. and Breakefield, X. O., A defective HSV-1 vector expresses E.coli beta-galactosidase in cultured rat peripheral neurons, *Science*, 241, 1667, 1988.
55. Geller, A. I., During, M. J., and Neve, R. L., Molecular analysis of neuronal physiology by gene transfer into neurons with herpes simplex virus vectors, *Trends Neurosci.*, 14, 428, 1991.
56. Geller, A. I., Neve, R. L., During, M. J., and O'Malley, K. L., Molecular analyses of neuronal physiology and potential for gene therapy using herpes simplex virus vectors, *Soc. Neurosci. Abstr.*, 453, 1085, 1992.
57. Glorioso, J. C., Goins, W. F., DeLuca, N., and Fink, D. J., Development of herpes simplex virus as a gene transfer vector for the nervous system, in *Gene Transfer Strategies in the Study of Brain Damage and Repair*, Weiner-Gren Center International Symposium, 1993.
58. Goldstein, D. J. and Weller, S. K., Herpes simplex type 1-induced ribonucleotide reductase activity is dispensible for virus growth and DNA synthesis: isolation and characterization of an ICP6 lacZ insertion mutant, *J. Virol.*, 62, 196, 1988.
59. Herz, C. and Roizman, B., The alpha promoter regulator-ovalbumin chimeric gene resident in human cells is regulated like the authentic alpha 4 gene after infection with herpes simplex virus mutants in alpha 4 gene, *Cell*, 33, 145, 1983.
60. Hinrichs, S. H., Nerenberg, M., Reynolds, R. K., Khoury, G., and Jay, G., A transgenic mouse model for human neurofibromatosis, *Science*, 237, 1340, 1987.
61. Ho, D. and Mocarski, E. S., Beta-galactosidase as a marker in the peripheral and neural tissues of the herpes-simplex virus-infected mouse, *Virology*, 167, 279, 1988.
62. Horellou, P., Lundberg, C., Le Bourdelles, B., Wictorin, K., Brundin, P., Kalen, P., Bjorklund, A., and Mallet, J., Behavioural effects of genetically engineered cells releasing dopa and dopamine after intracerebral grafting in a rat model of Parkinson's disease, *J. Physiol.*, 85, 158, 1991.
63. Hsiao, K. K., Scott, M., Foster, D., Groth, D. F., DeArmound, S. J., and Prusiner, S. B., Spontaneous neurodegeneration in transgenic mice with mutant prion protein, *Science*, 250, 1587, 1990.
64. Huang, Q., Vonsattel, J.-P., Schaffer, P. A., Martuza, R. L., Breakefield, X. O., and DiFiglia, M., Introduction of a foreign gene (Escherichia coli lacZ) into rat neurostriatal neurons using herpes simplex virus mutants: a light and electron microscopic study, *Exp. Neurol.*, 115, 303, 1992.
65. Jaffe, H. A., Danel, C., Longnecker, G., Metzger, M., Setoguchi, Y., Rosenfeld, M. A., Gant, T. W., Thorgeirsson, S. S., Stratford-Perricaudet, L. D., Perricaudet, M., Pavirani, A., Lecocq, J.-P., and Crystal, R. G., Adenovirus-mediated *in vivo* gene transfer and expression in normal rat liver, *Nat. Genet.*, 1, 372, 1992.
66. Javier, R. T., Izumi, K. M., and Stevens, J. G., Localization of a herpes simplex neurovirulence gene dissociated from high-titer virus replication in the brain, *J. Virol.*, 62, 1381, 1988.
67. Jiao, S., Acsadia, G., Jani, A., Felgner, P. L., and Wolffe, J. A., Persistence of plasmid DNA and expression in rat brain cells *in vivo*, *Exp. Neurol.*, 115, 400, 1992.
68. Johnson, P. A., Miyanohara, A., Levine, F., Cahill, T., and Friedmann, T., Cytotoxicity of a replication-defective mutant of herpes simplex virus type 1, *J. Virol.*, 66, 2925, 1992.
69. Jones, C., Delhon, G., Bratanich, A., Kutish, G., and Rock, D., Analysis of the transcriptional promoter which regulates the latency-associated transcript of bovine herpesvirus, *J. Virol.*, 64, 1164, 1990.

70. Julien, J.-P., Tretjakoff, I., Beaudet, L., and Peterson, A., Expression and assembly of a human neurofilament protein in transgenic mice provide a novel neuronal marking system, *Genes Dev.*, 1, 1085, 1987.
71. Kanada, N., Sasaoka, T., Kobayashi, K., Kiuchi, K., Nagatsu, I., Kurosawa, Y., Fujita, K., Yokoyama, M., Nomura, T., Katsuki, M., and Nagatsu, T., Tissue-specific and high-level expression of the human tyrosine hydroxylase gene in transgenic mice, *Neuron*, 6, 583, 1991.
72. Kaplitt, M., Tjuvajev, J., Berk, J., Rabkin, S. D., Posner, J. B., Pfaff, D. W., and Blasberg, R. G., Treatment of W256 tumors in immunocompetent rats using herpes simplex virus mutants, in *Gene Therapy*, Anderson, W. F., Friedmann, T., and Mulligan, R., Eds., Cold Spring Harbor Laboratory Press, New York, 1992, 81.
73. Kohn, D. B. and Kantoff, P. W., Potential applications of gene therapy, *Transfusion*, 29, 812, 1989.
74. Kosz-Vnenchak, M., Coen, D. M., and Knipe, D. M., Restricted expression of herpes simplex virus lytic genes during establishment of latent infection by thymidine kinase negative mutant viruses, *J. Virol.*, 64, 5396, 1990.
75. Kuypers, H. G. J. M. and Ugolini, G., Viruses as neurotracers, *Trends Neurosci.*, 13, 71, 1990.
76. Kwong, A. D., Kruper, J. A., and Frenkel, N., HSV virion host shutoff function, *J. Virol.*, 62, 912, 1989.
77. LaVail, J. H., Zhan, J., and Margolis, T. P., HSV (Type 1) infection of the trigeminal complex, *Brain Res.*, 514, 181, 1990.
78. Le Gal La Salle, G., Robert, J. J., Berrard, S., Ridoux, V., Stratford-Perricaudet, L. D., Perricaudet, M., and Mallet, J., An adenovirus vector for gene transfer into neurons and glia in the brain, *Science*, 259, 988, 1993.
79. Leib, D. A., Blatt, A. N., Pepose, J. S., and Mandel, G., Sodium channel promoter sequences direct neural-specific expression in vivo: creation of a herpes simplex virus-sodium channel promoter vector, 1994, submitted.
80. Leist, T. P., Sandri-Goldin, R. M., and Stevens, J. G., Latent infections in spinal ganglia with thymidine kinase-deficient herpes simplex virus, *J. Virol.*, 63, 4976, 1989.
81. Leistma, J. E., Viral infections of the nervous system, in *Textbook of Neuropathology*, Davis, R. L. and Robertson, D. M., Eds., Williams and Wilkins, Baltimore, 1985, 704.
82. Lemarchand, P., Jaffe, H. A., Danel, C., Cid, M. C., Kleinman, H. K., Stratford-Perricaudet, L. D., Perricaudet, M., Pavirani, A., Lecocq, J.-P., and Crystal, R. G., Adenovirus-mediated transfer of a recombinant human alpha 1-antitrypsin cDNA to human endothelial cells, *Proc. Natl. Acad. Sci. U.S.A.*, 15, 6482, 1992.
83. Longnecker, R., Roizman, B., and Meignier, B., Herpes simplex viruses as vectors: properties of a prototype vaccine strain suitable for use as a vector, in *Viral Vectors*, Gluzman, Y. and Hughs, S.H., Eds., Cold Spring Harbor Laboratory Press, New York, 1988, 68.
84. Mansour, S. L., Thomas, K. R., and Capecchi, M. R., Disruption of the proto-oncogene int-2 in mouse embryo-derived stem cells: a general strategy for targeting mutations to non-selectable genes, *Nature*, 336, 348, 1988.
85. Mansour, S. L., Gene targeting in murine embryonic stem cells: introduction of specific alterations into the mammalian genome, *Genet. Anal. Tech. Appl.*, 7, 219, 1990.
86. Margolis, T. P., LaVail, J. H., Setzer, P. Y., and Dawson, C. R., Selective spread of herpes simplex virus in the central nervous system after ocular inoculation, *J. Virol.*, 63, 4756, 1989.
87. Markert, J. M., Malick, A., Coen, D. M., and Martuza, R. L., Reduction and elimination of encephalitis in an experimental glioma therapy model with attenuated herpes simplex mutants that retain susceptibility to ganciclovir, *Neurosurgery*, 32, 1, 1993.
88. Martuza, R. L., Malick, A., Markert, J. M., Ruggner, K. L., and Coen, D. M., Experimental therapy of human glioma by means of a genetically engineered virus mutant, *Science*, 252, 854, 1991.

89. Mastrangeli, A., Danel, C., Rosenfeld, M. A., Stratford-Perricaudet, L., Perricaudet, M., Pavirani, A., Lecocq, J. P., and Crystal, R. G., Diversity of airway epithelial cell targets for in vivo recombinant adenovirus-mediated gene transfer, *J. Clin. Invest.*, 91, 225, 1993.
90. McGeough, D. J., Dalrymple, M. A., Davison, A. J., Dolan, A., Frame, A. C., McNab, D., Perry, L. J., Scott, J. E., and Taylor, P., The complete DNA sequence of the long unique region in the genome of herpes simplex virus type 1, *J. Gen. Virol.*, 69, 1531, 1988.
91. Mellerick, D. M. and Fraser, N. W., Physical state of the latent herpes simplex virus genome in mouse model system: Evidence suggesting an episomal state, *Virology*, 158, 265, 1987.
92. Mellon, P. L., Windle, J. J., Goldsmith, P. C., Padula, C. A., Roberts, J. L., and Weiner, R. I., Immortalization of hypothalmic GnRH neurons by genetically targeted tumorigenesis, *Neuron*, 5, 1, 1990.
93. Moolten, F. L., Tumor chemosensitivity conferred by inserted thymidine kinase genes: paradigm for a prospective cancer control strategy, *Cancer Res.*, 46, 5276, 1986.
94. Mulligan, R. C., The basic science of gene therapy, *Science*, 260, 926, 1993.
95. Neuwelt, E. A., Pagel, M. A., and Dix, R. D., Delivery of ultraviolet-inactivated ^{35}S-herpesvirus across on osmotically modified blood-brain barrier, *J. Neurosurg.*, 74, 475, 1991.
96. Oh, Y. J., Wong, S. C., Moffat, M., Ullrey, D., Geller, A. I., and O'Malley, K. L., Delineation of CNS and PNS DNA response elements responsible for cell-specific expression of tyrosine hydroxylase, *Soc. Neurosci. Abstr.*, 578.14, 1379, 1992.
97. Ono, T., Fujino, Y., Tsuchiya, T., and Tsuda, M., Plasmid DNAs directly injected into mouse brain with lipofectin can be incorporated and expressed by brain cells, *Neurosci. Lett.*, 117, 259, 1990.
98. Neve, R. L., Adenovirus vectors enter the brain, *Trends Neurosci.*, 16, 251, 1993.
99. Pallela, T. D., Hidaki, Y., Silverman, L. J., Schroll, C. T., Homa, F. L., Levine, M., and Kelley, W. N., Expression of human HPRT mRNA in brains of mice with a recombinant vector, *Gene*, 80, 137, 1989.
100. Palmiter, R. D., Behringer, R. R., Quaife, C. J., Maxwell, I. H., Maxwell, F., and Brinster, R. L., Cell lineage abation in transgenic mice by cell-specific expression of a toxin gene, *Cell*, 50, 435, 1987.
101. Palmiter, R. D. and Brinster, R. L., Transgenic mice, *Cell*, 41, 343, 1985.
102. Palmiter, R. D. and Brinster, R. L., Germ-line transformation of mice, *Annu. Rev. Genet.*, 20, 465, 1986.
103. Peat, D. S. and Stanley, M. A., Chromosome damage induced by herpes simplex virus type 1 in early infection, *J. Gen. Virol.*, 67, 2273, 1986.
104. Perry, L. J. and McGeogh, D. J., The DNA sequences of the long repeat region and adjoining parts of the long unique region in the genome of herpes simplex virus type 1, *J. Gen. Virol.*, 69, 2831, 1988.
105. Popko, B., Germ-line manipulation of the mouse in neuroscience, in *Molecular Genetic Approaches to Neuropsychiatric Diseases*, Academic Press, New York, 1990, 429.
106. Price, J., Turner, D., and Cepko, C., Lineage analysis in the vertebrate nervous system by retrovirus-mediated gene transfer, *Proc. Natl. Acad. Sci. U.S.A.*, 84, 156, 1987.
107. Propst, F., Rosenberg, M. P., Cork, L. C., Kovatch, R. M., Rauch, S., Westphal, H., Khillan, J., Schulz, N. T., and Neumann, P. E., Neuropathological changes in transgenic mice carrying copies of a transcriptionally activated mos proto-oncogene, *Proc. Natl. Acad. Sci. U.S.A.*, 87, 9703, 1990.
108. Pyles, R. B., Sawtelle, N. M., and Thompson, R. L., Herpes simplex virus type 1 dUTPase mutants are attenuated for neurovirulence, neuroinvasiveness, and reactivation from latency, *J. Virol.*, 66, 6706, 1992.
109. Radovick, S., Wray, S., Lee, E., Nicols, D. K., Nakayama, Y., Weintraub, B. D., Westphal, H., Cutler, G. B., and Wondiford, F. E., Migratory arrest of gonadotrophin-releasing hormone neurons in transgenic mice, *Proc. Natl. Acad. Sci. U.S.A.*, 88, 3402, 1991.

110. Ragout, T., Vincent, N., Chafey, P., Vigne, E., Gilgenkrantz, H., Couton, D., Cartaud, J., Briand, P., Kaplan, J.-C., Perricaudet, M., and Kahn, A., Efficient adenovirus-mediated transfer of a human minidystrophin gene to skeletal muscle of mdx mice, *Nature*, 361, 647, 1993.
111. Ram, Z., Culver, K. W., Walbridge, S., Blaese, R. M., and Oldfield, E. H., *In situ* retroviral-mediated gene transfer for the treatment of brain tumors in rats, *Cancer Res.*, 52, 83, 1993.
112. Read, G. S. and Frenkel, N., Herpes simplex virus mutants defective in the virion-associated shut-off of host polypeptide synthesis and exhibiting abnormal synthesis of alpha (immediate-early) viral polypeptides, *J. Virol.*, 46, 498, 1983.
113. Rock, D. L. and Fraser, N. W., Detection of HSV genome in central nervous system, *Nature*, 302, 523, 1987.
114. Roemer, K., Johnson, P. A., and Friedmann, T., Recombination between a herpes simplex virus type 1 vector deleted for immediate early gene 3 and the infected cell genome, *J. Gen. Virol.*, 73, 1553, 1992.
115. Roizman, B. and Batterson, W., Herpes viruses and their replication, in *Virology*, Fields, B. N., Ed., Raven Press, New York, 1985, 497.
116. Roizman, B. and Jenkins, F. J., Genetic engineering of novel genomes of large DNA viruses, *Science*, 229, 1208, 1985.
117. Rosenfeld, M. A., Siegfried, W., Yoshimura, K., Yoneyama, K., Fukayama, M., Stier, L. E., Pakko, P. K., Gilardi, P., Stratford-Perricaudet, L. D., Perricaudet, M., Jallat, S., Pavirani, A., Lecocq, J.-P., and Crystal, R. G., Adenovirus-mediated transfer of a recombinant α1-antitrypsin gene to the lung epithelium in vivo, *Science*, 252, 431, 1991.
118. Rosenfeld, M. A., Yoshimura, K., Trapnell, B. C., Yoneyama, K., Rosenthal, E. R., Dalemans, W., Fukayama, M., Bargon, J., Stier, L. E., Stratford-Perricaudet, L., Perricaudet, M., Guggino, W. B., Pavirani, A., Lecocq, J.-P., and Crystal, R. G., In vivo transfer of the human cystic fibrosis transmembrane conductance regulator gene to the airway epithelium, *Cell*, 68, 143, 1992.
119. Rosenberg, M. B., Freidmann, T., Robertson, R. C., Tuszynski, M., Wolff, J. A., Breakefield, X. O., and Gage, F. H., Grafting genetically modified cells to the damaged brain: restorative effects of NGF expression, *Science*, 242, 1675, 1988.
120. Sanes, J. R., Rubenstein, J. L. R., and Nicolas, J. F., Use of recombinant retrovirus to study post-implantation cell lineage in mouse embryos, *EMBO J.*, 5, 3133, 1986.
121. Schumaker, J. M., Short, M. P., Hyman, B. T., Breakefield, X. O., and Isacson, O., Intracerbral implantation of nerve growth factor-producing fibroblasts protects striatum against neurotoxic levels of excitatory amino acids, *Neuroscience*, 45, 561, 1991.
122. Shepard, A. A., Imbalzano, A. N., and DeLuca, N. A., Separation of primary structural components conferring autoregulation, transactivation, and DNA-binding properties to the herpes simplex virus transcription regulation protein ICP4, *J. Virol.*, 63, 3714, 1989.
123. Shih, M.-F., Arsenak, P., Tiollais, P., and Roizman, B., Expression of hepatitis B virus S gene by herpes simplex type 1 vectors carrying alpha and beta-regulated gene chimeras, *Proc. Natl. Acad. Sci. U.S.A.*, 81, 5867, 1984.
124. Small, J. A., Scangos, G. A., Cork, L., Jay, G., and Khoury, G., The early region of human papovavirus JC induces dysmyelination in transgenic mice, *Cell*, 46, 13, 1986.
125. Smiley, J. R., Construction in vitro and rescue of a thymidine kinase-deficient deletion mutation of herpes simplex virus, *Nature*, 285, 333, 1980.
126. Song, S., Hartley, D., Bryan, J., Ullrey, D., Ashe, O., O'Malley, K., Neve, R., Geller, A., and During, M., A HSV-1 vector expressing an unregulated protein kinase C from the tyrosine hydroxylase promoter causes rotational behavior following stereotaxic injection into the substantia nigra pars compacta of unlesion rats, *Soc. Neurosci. Abstr.*, 363.17, 872, 1992.
127. Spaet, R. R. and Frankel, N., The herpes simplex virus amplicon: Analyses of cis-acting replication functions, *Proc. Natl. Acad. Sci. U.S.A.*, 82, 694, 1985.

128. Spatz, W. B., Differences in transneuronal transport of horseradish peroxidase conjugated wheat germ agglutinin in the visual system: marmoset and monkey and guinea pig compared, *J. Hirnforsch*, 30, 375, 1989.
129. Stern, S., Masafumi, T., and Herr, W., The Oct-1 homeodomain directs formation of a multiprotein-DNA complex with the HSV transactivator VP16, *Nature*, 341, 624, 1989.
130. Steiner, I., Spivak, J. G., Lirette, R. P., Brown, S. M., Maclean, A. R., Subak-Sharpe, J. H., and Fraser, N. W., Herpes simplex type 1 latency associated transcripts are evidently not essential for latent infection, *EMBO J.*, 8, 505, 1989.
131. Steiner, I., Spivak, J. G., Deshmane, S. L., Ace, C. L., Preston, C. M., and Fraser, N. W., A herpes simplex virus type 1 mutant containing a nontransducing Vmw65 protein establishes latent infection in vivo in the absence of viral replication and reactivates efficiently from explanted trigeminal ganglia, *J. Virol.*, 64, 1990.
132. Stevens, J. G., Human herpesviruses: a consideration of the latent state, *Microbiol. Rev.*, 53, 318, 1989.
133. Strack, A. M. and Loewy, A. D., Pseudorabies virus: a highly specific transneuronal cell body marker in the sympathetic nervous system, *J. Neurosci.*, 10, 2139, 1990.
134. Tackney, C., Cachianes, G., and Silverstein, S., Transduction of the Chinese hamster ovary aprt gene by herpes simplex virus, *J. Virol.*, 52, 606, 1984.
135. Thomas, K. R. and Capecchi, M. R., Site-directed mutagenesis by gene targeting in mouse embryo-derived stem cells, *Cell*, 51, 503, 1987.
136. Thompson, R. L. and Wagner, E. K., Partial rescue of herpes simplex virus neurovirulence with a 3.2 kb cloned DNA fragment, *Virus Genes*, 1, 261, 1988.
137. Thompson, R. L., Rogers, S. K., and Zerhusen, M. A., Herpes simplex virus neurovirulence and productive infection of neural cells is associated with a function which maps between 0.820 and 0.832 map units on the HSV genome,*Virology*, 172, 435, 1989.
138. Turner, D. L. and Cepko, C., A common progenitor for neurons and glia persists in rat retina late in development, *Neuron*, 328, 131, 1987.
139. Turner, D. L., Snyder, E. Y., and Cepko, C., Lineage-independent determination of cell type in the embryonic mouse retina, *Neuron*, 4, 833, 1990.
140. Ugolini, G., Kuypers, H. G., and Strick, P. L., Transneuronal transfer of herpes virus from peripheral neurons to cortex and brainstem, *Science*, 40, 359, 1980.
141. Whalsted, C., Pich, E. M., Koob, G. F., Yee, F., and Helig, M., Modulation of anxiety and neuropeptide Y-Y1 receptors by antisense oligodeoxynucleotides, *J. Biol. Chem.*, 268, 2300, 1993.
142. Wagner, E. K., Devi-Rao, G., Feldman, L. T., Dobson, A. T., Zhang, Y.-F., Flanagan, W. M., and Stevens, J. G., Physical characterization of the herpes simplex virus latency-associated transcript in neurons, *J. Virol.*, 62, 1194, 1988.
143. Wolfe, J. H., Deshmane, S. L., and Fraser, N. M., Herpesvirus vector gene transfer and expression of beta-glucuronidase in the central nervous system of MPS VIII mice, *Nat. Genet.*, 1, 379, 1992.
144. Wolff, J. A., Fisher, L. J., Xu, L., Jinnah, H. A., Langlais, P. J., Iuvone, P. M., O'Malley, K. L., Rosenberg, M. B., Shimohama, S., Friedmann,T., and Gage, F. H., Grafting fibroblasts genetically modified to produce L-dopa in a rat model of Parkinson's disease, *Proc. Natl. Acad. Sci. U.S.A.*, 86, 9011, 1991.
145. Yoshimuri, K., Rosenfeld, M. A., Seth, P., and Crystal, R. G., Adenovirus-mediated augmentation of cell transfection with unmodified plasmid vectors, *J. Biol. Chem.*, 268, 2300, 1993.
146. Zwaastra, J. C., Ghiasi, H., Slanina, S. M., Nesburn, S. B., Wheatley, S. C., Lillycrop, K., Wood, J., Latchman, D. S., Patel, K., and Wechsler, S. L., Activity of herpes simplex virus type 1 latency-associated transcript (LAT) promoter in neuron-derived cells: evidence for neuron specificity and for a large LAT transcript, *J. Virol.*, 64, 5019, 1990.

10 Implantation of Genetically Modified Cells in the Brain

Jasodhara Ray, Lisa J. Fisher, and Fred H. Gage

I. SOMATIC GENE THERAPY: PRESENT STATUS

Somatic gene therapy, or the genetic manipulation of nongerm line cells for therapeutic applications, has emerged as one of the most promising strategies for treating human disease. There are two methods for performing somatic gene therapy: either genes may be transferred into cells in culture and then transplanted into an organism (*ex vivo* approach) or genes may be directly delivered into an organism for *in situ* gene transfer into cells (*in vivo* approach). Since techniques for directly delivering genes to cells *in vivo* are not yet well established, most studies that have explored somatic gene therapy have focused on *ex vivo* approaches. Currently, there are several clinical trials that are pursuing such a strategy, with therapy predominantly directed toward diseases resulting from either inherited genetic defects or cancer. Of particular note is work on severe combined immune deficiency (SCID) disease (see Chapter 3) and familial hypercholesterolemia (see Chapter 5). Results from initial gene therapy trials on patients suffering from these disorders have indicated that engineered cells can effectively replace deficient enzymes (adenosine deaminase) or reduce toxic compounds (cholesterol) for several months after the peripheral infusion of the modified cells.[1,2]

In contrast to the numerable studies that have focused on the application of gene therapy to peripheral targets, there has been much less work done on this strategy in the central nervous system (CNS). At least some of the reasons for the slower development of gene therapy for the CNS lie in the complexity of most human neurological diseases and the relatively difficult access to dysfunctional areas of the brain. However, despite the complicated phenotype of most CNS disorders, work with intracerebral grafts of fetal neurons offers compelling support for the usefulness of an *ex vivo* somatic gene therapy approach for CNS disease. Research over two decades has indicated that cells of diverse origin survive well within the brain and that graft-derived products can effectively replace and/or supplement deficient compounds within the brain and reverse behavioral abnormalities in animal models of CNS damage[3] and human neurological disease.[4]

Although fetal tissue grafts offer a promising therapeutic strategy for CNS disorders, far greater benefits may be realized by using genetically modified cells for intracerebral

0-8493-2440-8/95/$0.00+$.50

transplantation. Foremost, cells engineered for gene therapy may be derived from the patients themselves to minimize problems with cellular rejection after grafting. Second, cells may be modified to produce a purer and broader range of factors than can be obtained with nonengineered tissues. Finally, genetically modified cells may be constructed in such a manner that the safety of the host organism is ensured (e.g., through the insertion of an easily activated "suicide" gene along with the therapeutic gene).

In general, somatic gene therapy for the CNS can be directed at several points in the disease process. Damage to neuronal systems that may be initiated and/or exacerbated by toxic compounds (i.e., free radicals, excess excitatory amino acids) may be limited by grafting engineered cells that function by reducing or eliminating the toxic element from the brain (i.e., cells that produce free radical scavengers or express amino acid receptors). Alternatively, cells may suffer injury due to the loss of a vital source of trophic support. In such cases, cells engineered to produce one or more neurotrophic factors that are specific for the compromised system (e.g., nerve growth factor for the cholinergic system) can be grafted adjacent to the injured somata. Later in the disease process, when neuronal degeneration has progressed to the point where intercellular communication becomes dysfunctional, cells can be genetically modified to produce critical neurotransmitters and/or modulators. For several reasons, this particular point of intervention is more complicated than the two earlier strategies, primarily because engineered cells grafted into the brain will generally be unable to mimic the dynamic functioning and recapitulate the precise point-to-point contacts between cells that occurs during neurotransmission. However, the striking success with which neurotransmitter-rich fetal tissue grafts have been found to restore neural functioning in the damaged brain[3] provides a strong precedent for proceeding with such a strategy. In addition to these direct applications to neural systems, somatic gene therapy for the CNS may also be useful for the treatment of intractable cancers (see Chapters 14 and 15). Such an approach is accomplished by engineering cells to carry and/or deliver a drug-sensitive gene product (e.g., cells producing tyrosine kinase would be vulnerable to treatment with ganciclovir). This genetic information would be targeted preferentially to rapidly dividing cells within the brain (adult neurons do not divide and would be immune) and, once incorporated, result in the destruction of tumor cells following systemic administration of the appropriate drug.[5-8]

In this chapter, we review the development and current applications of somatic gene therapy to the CNS, with particular emphasis on the *ex vivo* strategy. Rationales for selecting a gene delivery method will be described, and the advantages and disadvantages of different target cells for genetic manipulation will be discussed. Then, examples of work to date that have applied the somatic gene therapy strategy to animal models of neurological disease will be presented. Based on results from *in vivo* animal models, human disorders of the CNS that may be amenable to somatic gene therapy are discussed. Finally, we conclude with a note on some of the future directions of somatic gene therapy for the CNS.

II. GENE TRANSFER IN VITRO

Successful gene therapy depends on a number of factors, including the type of cell used for gene delivery, the methods of transduction of target cells, and most importantly, the elements that control the long-term expression of new genetic material (transgenes) both *ex vivo* and *in vivo*. Issues to consider in selecting these factors are discussed below.

A. Choice of Target Cells

There are several considerations when selecting a target cell for *ex vivo* genetic manipulation and intracerebral grafting. Foremost, the cells must be able to survive through the gene transfer process and synthesize transgene products at levels that are biologically relevant. Second, engineered cells must be able to survive noninvasively in the CNS environment and continue to express the inserted transgene. In early work that explored the feasibility of *ex vivo* gene delivery to the CNS, studies focused on the use of immortal cell lines for genetic manipulation, since the robust proliferation of these cells *in vitro* made them easy to grow and to transduce efficiently. However, many of these cell lines continued to proliferate after grafting, which limited their usefulness for long-term therapeutic applications. Thus, more recent studies have focused on the use of primary cells as targets for gene transfer, since such cells show good but not pathologic survival within the CNS.

1. Immortalized or Established Cell Lines

The stable incorporation of transgenes into target cells often requires that the cellular population is actively proliferating. The highly mitotic nature of immortalized or established cell lines thus made them ideal candidates for initial studies of gene transfer to the CNS. There are a variety of cells lines of both neural and nonneural origin that have been manipulated genetically and grafted into the brain. These include C6 glioma cells, neuroblastoma NS20 Y cells, neuroendocrine AtT 20 cells, several fibroblast cell lines (208F, NIH 3T3, Rat-1) and pancreatic RIN cells.[9-16] These early studies revealed that cells could be transduced efficiently to produce and release neurotrophic factors and neurotransmitters both *in vitro* and after implantation into discrete regions of the CNS. However, the growth characteristics of the cells *in vivo* was not ideal. Specifically, several of the cell lines (C6, NIH 3T3, Rat1, RIN) continued to show mitotic activity within the brain that was frequently lethal to the experimental animals.[12-15,17] Thus, although this work confirmed that genetically modified cells provided a useful method for delivering specific compounds to the CNS, the use of cell lines in this strategy is limited unless the tumorigenic properties of the cells can be altered or arrested.

2. Primary Nonneuronal Cells

The cells that subsequently became a focal target population for gene transfer were primary fibroblasts and muscle cells. In addition to the limited proliferative capability of these cells, an advantage of primary cells in general is the potential for using them for autologous grafting. Specifically, cultures can be generated from an individual, genetically modified *in vitro,* and then implanted into the same individual. Alternatively, inbred animal strains may be used in experimental work as both donors and recipients of cells (isologous graft). Both methods will limit the immune rejection of the modified cells postimplantation. One possible disadvantage of fibroblasts or muscle cells for gene transfer to the CNS is the inability of these cells to form functional connections with the host brain. Therefore, the phenotypic or biologic effects of the grafts *in vivo* will rely on cell-specific uptake mechanisms of the gene products or metabolites that are passively secreted from the engineered cells within the CNS region of interest.[18]

The highly secretory nature of fibroblasts has made them an excellent vehicle for delivering gene products to the CNS. Primary fibroblasts can be obtained from a skin biopsy and can be maintained in culture under standard tissue culture conditions. Although the growth rate of primary fibroblasts is slower than that of their immortalized counterparts, the gene transfer efficiency and long-term gene expression in the two cell types are comparable.[19,20] However, primary cells should be maintained for a limited time in culture to avoid possible antigen shifts.[21] Primary fibroblasts have been reported to immortalize spontaneously when maintained in culture for a prolonged period of time or following alterations in culture conditions.[22,23] Rapid and efficient methods of gene transfer have made it possible to keep cells in culture for only a short period of time before implantation. Primary fibroblasts from mouse, rat, and human have been modified genetically and used for gene therapy in both peripheral systems and in the CNS.[19,20,24-31] In all cases, the fibroblasts have been found to survive well in a nonmitotic state for periods ranging up to 2 years. Further, there is clear evidence that modified fibroblasts continue to produce and release engineered products *in vivo* that affect the anatomic and/or functional properties of the host brain.[28-30]

Primary muscle cells are another population that has been explored for *ex vivo* somatic gene therapy, since they can be cultured easily from muscle biopsies and expanded *in vitro*.[32,33] *In vitro* and *in vivo* studies have shown that while muscle cells are typically nonsecretory, they release growth hormones following genetic modification and intramuscular implantation.[34,35] Muscle cells engineered to express the catecholamine enzyme tyrosine hydroxylase (TH) have also been found to produce and release the neurotransmitter dopamine after intracerebral grafting.[33] As seen for primary fibroblasts, muscle cells are quiescent after grafting and survive well for extended periods both in the periphery and within the CNS.[33-35] Thus, studies using primary cells as target populations for somatic gene therapy have demonstrated that fibroblasts and muscle cells can be manipulated easily *in vitro* to express foreign genes and that both provide an effective cellular vehicle for delivering gene products to the brain for extended periods.

3. Neural Cells

Cells derived from the CNS may be a more appropriate target cell for gene delivery to the CNS, since their survival may be enhanced within the familiar surroundings of the brain. Further, such cells may secrete factors that are beneficial to the host brain and functionally incorporate into the CNS more effectively than nonneural cells. However, there are also disadvantages to the use of these cells. Neural tissues used for genetic manipulation are typically not autologously derived and may evoke adverse immune responses after grafting. Further, unlike nonneural cells, both astrocytes and neurons can migrate after intracerebral implantation and may deliver factors to inappropriate regions of the brain. Although grafted neurons can form synaptic contacts with host cells, there is also the possibility that the implanted cells will form aberrant connections. Perhaps most critical, neural tissues are difficult to grow and manipulate in culture, especially once the cells have differentiated and become postmitotic. To overcome these difficulties, a number of cell lines of neural origin have been explored for their possible use in gene therapy in the CNS.[17] As discussed above, work with such cells has indicated that neural-derived tissues can be transduced effectively but that the immortal nature of the lines has often resulted in destructive growth after grafting.[12,14,15] An alternative to the use of established CNS-derived cell lines for gene transfer is the more recent approach of using primary neural tissues that have been immortalized with an oncogene to induce continued proliferation *in vitro*. Although a series of immortalizing oncogenes have been used,[36] most studies have focused on the use of avian *v-myc*[37,38] or a temperature sensitive allele of SV40 large T-antigen.[39,40] Cells immortalized by these oncogenes retain many of their *in vivo* characteristics, but terminal differentiation of the cells is arrested.[41] A number of immortalized cell lines have been generated from different regions of the brain.[37,41,42] These cells can be cultured easily and transduced with genes of interest. When implanted into the brains of neonates, such cells have been shown to survive and integrate with the host tissues to a limited extent.[38,39] Further, cells immortalized with the SV40 oncogene have shown transgene expression for up to 6 weeks postgrafting,[40] whereas those immortalized with *v-myc* appear to express a transgene for at least 22 months after implantation.[38] Although oncogene-immortalized CNS cells do not appear to show inappropriate mitotic activity after grafting, they have been reported to have chromosomal damage and different morphologic and/or phenotypic properties than their primary counterparts.[36,41,42] Thus, while oncogene-immortalized cells of CNS origin may be a better alternative than the established cell lines for *ex vivo* gene therapy to the CNS, it would be most optimal to use primary cells directly for gene transfer. Recent advances in the culturing and manipulation of primary neural cells *in vitro*[43] will contribute to the development of primary CNS-derived cells for gene therapy.

B. Choice of Expression Vectors and Promoters

The expression of transgenes within target cells is regulated by a number of factors including the *cis*-acting (or flanking) elements, such as promoters, enhancers, polyadenylation signals, splice signals and signals that control the half-life of messenger RNA (mRNA). These elements exert their effects in a cell-type-specific manner and determine when and where a gene will be expressed. To construct an expression vector, one should mix and match these elements with each other to achieve optimum levels of transgene expression. In studies using retroviral vectors, it has been shown that the presence of splice sequences controls both the efficiency of transfection of the packaging cell lines by the retroviral vectors and also the infection efficiency of the recombinant virus.[44] The factors that govern the stability and the half-life of mRNA appear to be structural in nature, such as polyadenylation signals and the AU-rich instability regions.[45-48] However, among all of these factors, promoter/enhancer systems have been the most studied. A number of viral and cellular promoter/enhancer systems have been used to express genes of interest in a variety of cell lines for gene therapy purposes.[13,25,26,32,49-51] All of these studies used retroviral vectors in which the transgene was expressed from either the promoter within the Moloney murine leukemia virus long-terminal repeat (MLV-LTR) region or from an internal viral or cellular promoter. The strength of the promoter has been found to be dependent on the transgene assessed and the cell type used to express the transgene. For example, in work conducted by Hock and colleagues in which adenosine deaminase (ADA) was expressed from a variety of promoters, including MLV-LTR, cytomegalovirus immediate-early (CMV-IE), simian virus (SV) 40 early region, lymphotropic papovavirus (LPV), and human β-globin, the MLV-LTR was found to be the strongest promoter, CMV was of intermediate strength, and the SV 40 early region promoter was the weakest promoter in all of the cell lines assessed.[51] However, when purine nucleoside phosphorylase (PNP) was expressed from some of the same promoters and cell lines, the SV 40 and CMV promoters were found to be equally strong in inducing PNP activity.[44] In comparing promoter strength in different cell populations, Palmer and his colleagues[20] explored the expression of factor IX within rat and human fibroblasts when the transgene was driven from either the MLV-LTR or the CMV promoter. They found that factor IX expression from the MLV-LTR promoter was higher in rat compared to human fibroblasts, whereas the expression of factor IX within human fibroblasts was highest when driven from the CMV promoter. The species specificity of the promoters reflected the origin of the viral elements, since the Moloney murine leukemia virus infects rodents and the cytomegalovirus infects human. Thus, promoter activity within different cell types may, at least partially, reflect species-specific efficiency of the virus in infecting different cell types.

In addition to vector elements and target cell types, the growth state of genetically modified cells and extrinsic (environment) factors also play a role in transgene expression. This was seen in a study by Schinstine and colleagues, who reported an 80% decrease in the activity of choline acetyltransferase expressed from the MLV-LTR promoter in Rat-1 fibroblasts under confluence-induced quiescence or serum starvation conditions.[52] In addition, a number of cytokines have been shown to regulate adversely the expression of genes expressed from the MLV-LTR promoter.[53] Many if not most of these environmental influences are likely to exist *in vivo*. Transduced primary cells assume a quiescent state after grafting and are often found to be surrounded by infiltrating lymphocytes and macrophages that are known to express and release cytokines. Indeed, in model systems in which genetically modified fibroblasts have been grafted on the skin, transgenes driven from MLV-LTR or CMV promoters were expressed for only a short period of time.[19,20,26] The changes observed in gene expression did not reflect a deletion of the gene, since the vector sequences were evident within the cells for up to 8.5 months.[20] Similarly, engineered cells implanted into the brain have also shown a gradual reduction in transgene expression from the MLV-LTR promoter.[54] Decreased gene expression from retroviral promoters *in vivo* did not appear to be due to immune responses to the graft or cell death but rather reflected permanently altered functioning of the promoters, since the suppression could not be reversed by reculturing the cells.[20] The work from a number of groups suggests that at least some changes in gene expression are related to the type of target cell used for gene transfer.[26,32] These combined results indicate that transgene expression is susceptible to numerable influences and that achieving long-term stabilility of transgene expression *in vivo* will require continued optimization of the many components of the gene transfer strategy (i.e., cell type, promoter/enhancer system, gene transfer method). Recently, the advantages and disadvantages of using retroviral vectors have been discussed elegantly.[55] One area of particular interest is the search for promoters that can function effectively in adverse environmental conditions. In this regard, the housekeeping gene promoters are currently being scrutinized, since they are generally considered to be very stable, even with changes in the surrounding environment or during different cellular growth conditions. In support of the notion that such promoters may be stable *in vivo*, Scharfmann and colleagues have reported that the dihydrofolate reductase promoter maintains stable transgene activity within fibroblasts grafted on the skin.[26] In addition to identifying promoters that are immune to extrinsic influences, the development of promoters that can be externally modulated to alter the level of transgene production after grafting is also underway.

C. Choice of Gene Transfer Methods: In Vitro

1. Retroviral Vectors

Foreign genes can be transduced into cells by a number of physical and chemical methods.[17,56] However, the use of these methods is quite limited, due to their low transfection efficiency. In some cases, particularly when using primary cells that grow slowly and can be kept in culture for a limited time, it is necessary to use gene transfer methods with high efficiency. One of the gene transfer methods that can reach almost 100% efficiency is the retroviral gene transfer technique. For most of the work on gene therapy, retrovirus-mediated gene transfer to target cells has been widely utilized[18,57-60] and is reviewed in detail elsewhere.[59-61] Retroviral vectors have a broad host- and cell-type range, at least partly dictated by the envelope (*env*) protein of the retrovirus and specific receptors on the recipient cells. Retroviruses that can infect only rodent cells are termed "ecotropic", whereas "amphotropic" viruses can infect their natural hosts as well as cells of a number of other mammalian species, including human. Unlike the chemical transduction methods, such as calcium phosphate precipitation or lipofection, which may result in multiple copies of the gene within the target cells, retroviral-mediated gene transfer can be modulated in such a way that a single copy of the transgene is introduced and integrates into cells.[61]

Retroviral vectors are made by using retroviral packaging cell lines in the absence of replication-competent helper virus.[61] Although there is a possibility that the production of replication-competent virus may occur, improved design of the packaging lines and vectors has effectively eliminated this problem.[44,61] A disadvantage of retrovirus-mediated gene transfer is that it requires at least one round of cell division for effective integration of the transduced genes. Thus, this method is not suitable for gene transfer into postmitotic cells, such as neurons. Another potential disadvantage of the retroviral vector is its limited DNA-carrying capacity, since it can only accommodate up to ~7 kb of genetic material.

2. Gene Transfer by Herpes Simplex Virus Vector

To deliver transgenes into postmitotic cells that are refractory to common methods of gene transfer, a replication-defective herpes simplex virus type 1 (HSV-1) vector has been developed.[62-66] Neurons infected with the HSV-1 vector expressing β-galactosidase (HSV-β-gal) from the HSV-1 immediate-early (IE) 4/5 promoter have been reported to show transgene activity for 2 weeks *in vitro*.[62] The activity of the promoter appears to be dependent on the cell type used, as evidenced by the finding that although high levels of β-gal expression are observed in fibroblasts, neuroblastomas, pheochromocytomas, and pituitary cells, there was a five-fold variation in β-gal levels among the cell lines.[65] A disadvantage of the herpes gene transfer method is that the HSV-1 vector has been reported to be toxic to cells. This was demonstrated in a study by Johnson and colleagues who used the HSV-1 vector in which β-gal was expressed from the CMV-IE promoter to infect primary cortical neurons in culture.[67] Glial cells present in these cultures were also infected with this vector. The infection was cytopathic to both cell types and killed the cultures within 3 days. Further, expression of β-gal was detected only transiently, either due to cell death or shut-off of transgene expression. These results suggested that a replication-defective virus can cause cytopathic effects in the absence of a lytic infection, perhaps due to the expression of a number of viral gene products that are toxic to cells.[68] However, reports that herpes-infected cells can survive successfully and express a transgene after CNS grafting[69,70] have encouraged the continued development of this method for delivering genes to neural target cells.

D. Choice of Gene Transfer Methods: In Vivo

In designing gene therapy for neurologic diseases in the brain, it may be important to deliver genes focally in the adult brain. A number of vector systems have been developed to accomplish such an *in vivo* somatic gene therapy approach. Vectors derived from herpes simplex virus type 1 and adenovirus have both been used for direct gene delivery into the CNS and will be described briefly. Additional details about these strategies may be found elsewhere[64,71] (see Chapter 9).

1. Herpes Simplex Virus

Recombinant herpes virus vectors that express transgenes from different HSV promoters or from other viral or cellular promoters have also been used for gene delivery *in vivo*.[66,72-74] Studies with HSV-1-derived recombinant vectors have shown that stable (>1 month) transgene expression can be achieved in peripheral and central neurons when transgenes are expressed from the herpes virus latency-associated transcript (LAT), ICP4, or thymidine kinase (TK) promoters. Nonherpes promoters may also be inserted into the herpes vector and used to express transgenes of interest. For example, the *lac*-Z gene (encoding for β-gal) placed under the control of the MLV-LTR or the neuron-specific enolase promoter has been found to express for 2 to 3 weeks in the adult rat brain.[66] Initially, stereotaxic injection of the virus into the caudate-putamen of adult rats led to extensive expression of β-gal in neurons. However, 17 days postinoculation, the expression was restricted to nonneuronal cells. This shift in expression patterns may indicate that the virus is replicating and distributing *in vivo*. Indeed, consistent with the results obtained with this virus *in vitro*,[68] some HSV-infected animals displayed extensive cytotoxicity. Thus, although there have been some promising results with HSV vectors, widespread use of this gene transfer approach will be restricted until the safety of the virus can be ensured.

2. Adenovirus

Adenovirus vectors, like herpes simplex virus vectors, can efficiently infect postmitotic cells. In contrast to the herpes virus, adenovirus appears to be less cytotoxic and thus safer for *in vivo* gene transfer. Specifically, cytotoxicity is generally not evident when less than 10^9 PFU/ml are injected into the brain.[75] Adeno vectors have been constructed with the *lac-Z* gene expressed from either the Rous sarcoma virus (RSV) LTR or the CMV promoter/enhancer.[75-78] Regardless of the promoter employed, there is evidence that transgene activity is not stable within the adeno vectors. When virus was injected into different regions of the brain, a large number of neurons, astrocytes, and microglia were found to express β-gal as early as 24 hr postinoculation.[75,78] However, activity was substantially reduced 1 to 2 months after injection. Similarly, when adenovirus was injected into the striatum of 7-week-old mice, *lac-*Z expression from the CMV promoter declined at 8 weeks postinoculation to 10 to 20% of the levels observed 1 week after the virus was injected.[77] These results indicate that although adenovirus vectors can be used for *in vivo* gene transfer, transgene expression within these vectors is susceptible to the same adverse regulatory influences that are experienced by the retroviral and herpes vectors.

3. Nonviral Method

Most methods developed for *in vivo* gene transfer have been based on the use of viruses. However, there are a few reports that have described successful gene transfer into the CNS using nonviral methods. For example, the firefly luciferase gene expressed from the RSV-LTR promoter was mixed with a cationic lipid (Lipofectin) and then injected into *Xenopus* brain.[79] Luciferase activity was then detected for up to 28 days postinoculation. A Lipofectin-DNA mixture has also been injected into the mouse brain and has been found to result in transgene expression in both neurons and glia.[80] While these methods for direct gene delivery offer an alternative to the viral technique, they are typically characterized by significantly lower transduction efficiencies.

4. Advantages and Disadvantages of In Vitro and In Vivo Gene Transfer Methods

The *ex vivo* gene transfer approach has been the main strategy for somatic gene therapy. There are several advantages of this method: large numbers of genetically modified cells may be easily generated; the expression of the transduced gene(s) can be examined and assayed *in vitro* prior to therapeutic use; the expression of the transgene under different experimental conditions can be readily determined; clonal cultures can be generated to express varied amounts of the transgenes; and once the transduced cells have been generated and well characterized, they may be used repeatedly to maintain reproducibility between studies. Retroviral-mediated gene transfer has been the most frequently used for inserting genes into cells *in vitro.* One disadvantage of retroviral vector is that it can accommodate ~7 kb of foreign gene. In addition, entry of the retrovirus into a cell depends on the presence of specific viral receptors on the target cells. Since retrovirus-mediated gene transfer requires active cell division for stable integration of the transgenes, postmitotic cells (e.g., neurons) cannot be genetically modified by this method. Both herpes and adenovirus appear to be more promising for gene transfer into such cells, but the cytotoxicity of these vectors remains an issue.

Presently, methods for *in vivo* gene transfer into the brain are under development (see Chapter 9). Herpes simplex and adenovirus vectors are being explored for their potential use. Both of these virus vectors have broad host- and cell-type specificities and remain episomal, thereby excluding the possibility of insertional activation of host genome. Adenovirus vectors can be grown at a high titer (10^{10} PFU/ml) but can accommodate only up to 6 to 8 kb of foreign DNA. Herpes virus vectors can accommodate large pieces of foreign DNA (up to 15 kb) but the titer of the virus is low (10^7 PFU/ml). Transgene expression *in vivo* from both vector systems have been achieved for 1 to 2 months, but the level of gene expression has been found to decline with time. Localized delivery of the virus to the CNS can be cytotoxic, due to the presence of a high amount of virus particles in a small area. The long-term safety of both viruses is unknown.

III. IN VIVO ANIMAL MODELS

There are two neural systems that have been the focus of most research on somatic gene therapy in the CNS: the cholinergic system and the dopaminergic system. One reason for the emphasis on these particular neurotransmitter systems is the well-documented success of intracerebral neurotrophic delivery and/or neurotransmitter-rich tissue grafts in improving the functioning of these systems following experimental injury.[3,81] Second, the well-characterized animal models of dopamine and cholinergic degeneration that have been used to demonstrate the effectiveness of intracerebrally delivered products provide powerful assay systems for assessing the functioning of genetically modified cells. In the following section, the application of somatic gene therapy to these *in vivo* model systems is described.

A. Cholinergic System

Neurons within the diagonal band of Broca, the medial septum, and the nucleus basalis magnocellularis (NBM), collectively within the basal forebrain, provide the major sources of cholinergic input to the hippocampus and neocortex. These pathways have been implicated in learning and memory, primarily based on results obtained following lesions of the basal forebrain.[82] Cholinergic neurons within the basal forebrain show a unique responsiveness to the neurotrophin NGF.[83] These cells express the receptor for NGF and show retrograde transport of NGF from terminal target areas to their somata. In animal models of basal forebrain damage, NGF infused into the ventricle of experimental rats has been shown to enhance the survival of cholinergic neurons and the expression of cholinergic markers.[84-88]

Based on successful results with NGF infusions into the brain, somatic gene therapy directed at the cholinergic system has focused primarily on the genetic modification of cells to produce NGF (Table 1). In all of the studies to date, fibroblasts derived from either established cell lines or from adult animals were selected as the target cell for gene transfer.[10,16,30,89-91] The most striking effects of NGF-producing cells have been in two

Table 1

Studies with Cells Genetically Modified to Produce Nerve Growth Factor (NGF)

Cell Type	Transgene	Model System	Grafted	*In Vivo* Survival	*In Vivo* Measure	Reference
208F fibroblast	Mouse NGF	Septohippocampal lesion	Yes	2 weeks	Septal cholinergic cell savings; cholinergic sprouting	10
NIH 3T3 fibroblast	Rat NGF	Septohippocampal lesion	Yes	6 weeks	Septal cholinergic cell savings; cholinergic sprouting	16
Primary fibroblast	Mouse NGF	Septohippocampal lesion	Yes	6 weeks 8 weeks	Septal cholinergic cell savings; cholinergic sprouting	30
Rat 1 fibroblast	Rodent NGF (?)	Basalo-cortical lesion	Yes	4 weeks	Block atrophy of NBM neurons	90
Rat 1 fibroblast	Rodent NGF (?)	Basalo-cortical lesion	Yes	4 weeks	Block atrophy of NBM neurons	91
208F fibroblast	Rat NGF	Basalo-cortical lesion	Yes	4 weeks	Cholinergic sprouting	89

animal models of retrograde cholinergic cell death: fimbria-fornix transections and cortical devascularization.[10,16,30,90,91] In the fimbria-fornix model, axotomy of the septo-hippocampal pathway results in the loss of over 50% of the cholinergic neurons within the septum.[92] However, when NGF-producing fibroblasts were implanted into the axotomized region, as many as 90% of the septal cholinergic population was sustained.[10,16] In addition to the effect of NGF on cholinergic somata, the genetically modified fibroblasts have been found to elicit a robust sprouting of cholinergic fibers towards the grafts.[10,16,30] Further, some of the cholinergic fibers have been found to traverse the grafts and reinnervate denervated target regions in the hippocampus.[30] These results strongly suggested that genetically modified fibroblasts continue to produce and release NGF *in vivo* and supported the role of NGF in promoting the survival and axonal growth of cholinergic neurons within the adult brain.

Fibroblasts producing NGF have also been found to strongly affect damaged cholinergic neurons within the NBM. In the devascularization model, disruption of the cortical blood supply results in several degenerative changes in the NBM, including shrinkage of cell somata,[93-94] loss of neurites,[88] and a decrease in cholinergic markers.[88,94-96] Mimicking results obtained with NGF infusions, fibroblasts engineered to produce NGF have been found to block completely the atrophy of cholinergic neurons within the NBM[90,91] and maintain neurite density at normal levels when implanted into damaged cortical regions.[91] In addition, NGF-producing fibroblasts induced a marked increase in choline acetyltransferase activity above control levels in the cortex[90,91] and improved KCl-stimulated release of acetylcholine from cortical terminals.[90] These findings were consistent with the results obtained in the fimbria-fornix model and provided additional support for the continued production and release of bioactive NGF from the genetically modified fibroblasts *in vivo*. Further, the potent effect of the NGF grafts on the host brain at doses that were over 40-fold lower than those typically used for intraventricular infusions strongly suggests that the discrete application of NGF to compromised regions of the brain is an exceptionally effective method for preventing neurodegenerative changes in the cholinergic system.

B. Dopaminergic System

The dopaminergic projection from the substantia nigra pars compacta to the striatum can be destroyed easily using the catecholamine neurotoxin 6-hydroxydopamine (6-OHDA). Rats with unilateral lesions of this pathway suffer a loss of dopamine content within the striatum and display a variety of behavioral abnormalities.[97-100] There are two major points of intervention in the dopamine degenerative processes after injury. First, initial damage to the neurons may be restricted by either removing toxic substances from the extracellular environment or by enhancing neural functioning with neurotrophins. Second, the loss of dopamine within the brain may be reversed by supplying an exogenous source of the neurotransmitter. In contrast to the neurotrophic approach in the cholinergic system, virtually all work in the dopamine-depletion model using the intracerebral grafting technique has focused on the replacement of neurotransmitter levels within the striatum. The primary sources of dopamine-rich tissue that has been used for transplantation are fetal substantia nigra neurons, which have been shown to increase dopamine levels within the striatum and reverse many of the behavioral abnormalities of dopamine-depleted rats.[3]

Based on the positive results with dopamine-rich tissues in the 6-OHDA rat, somatic gene therapy in this model system has also focused on a dopamine-replacement strategy. In contrast to the neurotrophins, cells are not genetically modified to produce a neurotransmitter directly from an inserted gene. Rather, cells are engineered to express the synthetic enzyme necessary for neurotransmitter production. Thus, in the case of dopamine, the TH enzyme converts tyrosine to L-dopa, while dopa decarboxylase (DDC) performs the final conversion of L-dopa to dopamine. To date, cells engineered for somatic gene therapy in the dopamine-depleted rat model have been manipulated solely to express TH. The choice of target cells then dictates the final product synthesized by the engineered cells, based on the presence or absence of endogenous DDC.

There is a broad range of target cells that have been genetically manipulated to express various forms of TH (Table 2). Initially, cell lines were chosen for gene transfer based on their favorable *in vitro* characteristics (see above). While the majority of these cells solely engineered to express TH synthesized L-dopa, a small percentage of the immortal target cells (endocrine-derived cells) also had the capacity to manufacture and secrete dopamine. When implanted into the brain of 6-OHDA rats, cells engineered to produce either L-dopa or dopamine were found to continue to synthesize and secrete the catecholamine product for at least 1 week postgrafting.[12,14] In direct comparisons between L-dopa and dopamine-producing cells, both were found to increase dopamine levels within the striatum, indicating that the L-dopa released from the grafts was efficiently converted to dopamine within the host brain.[14] However, only the L-dopa-producing cells appeared to ameliorate some of the behavioral abnormalities of the 6-OHDA rats.[14] Such findings were suggested to reflect the increased diffusion capacity, and thus the larger sphere of influence, of L-dopa prior to its conversion into dopamine.

In subsequent work with cells engineered to express TH, there has been more emphasis on the use of primary cells for gene transfer, since the established cell lines often displayed invasive growth following grafting. Three different cell types have been explored: fibroblasts, Schwann cells, and muscle cells (Table 2). Those genetically modified primary cells that have been grafted into the brain of 6-OHDA rats have shown noninvasive survival for extended periods.[29,33] Prolonged survival of engineered cells within the brain is crucial for establishing somatic gene therapy as a viable strategy for CNS disease. In work with TH-expressing muscle cells, the modified cells have been reported to produce high levels of dopamine both *in vitro* and *in vivo*.[33] These engineered muscle cells were found to induce a profound and stable reduction in behavioral abnormalities of dopamine-depleted rats for at least 6 months postgrafting. Primary fibroblasts modified to express TH have also been shown to reverse some of the behavior abnormalities of dopamine-depleted rats, but this effect appears to diminish with increased time postgrafting.[29] Since the survival of both fibroblasts and muscle cells is qualitatively similar within the brain, the difference in effectiveness of the engineered cells remains unclear. It has been suggested that the use of the Lipofection technique to introduce a TH-containing plasmid into the muscle cells was instrumental in preserving the stability of the plasmid DNA and thus prolonging transgene expression *in vivo*.[33] Whether such a hypothesis is correct remains to be established. Nonetheless,

Table 2

Studies with Cells Genetically Modified to Produce Tyrosine Hydroxylase (TH)

TH Form	Cell Type	Catechol. Produced	Grafted	*In Vivo* Measure	*In Vivo* Survival	Reference
Rat	208F fibroblast	L-dopa	Yes	Behavioral improvement	14 days	11
Rat	1.17 Schwann	L-dopa	No	—	—	103
Rat	Primary fibroblast	L-dopa	Yes	Behavioral improvement	2 months	29
Rat	Primary muscle	L-dopa; dopamine	Yes	Behavioral improvement; increased dopamine levels	6 months	33
Rat	Primary Schwann	L-dopa	No	—	—	103
Human type I	RIN endocrine	L-dopa; dopamine	Yes	Increased dopamine levels; no behavioral effect	9 days	14
Human type I	NIH 3T3 fibroblast	L-dopa	Yes	Behavioral improvement; increased dopamine levels	9 days	14
Human type I	C6 glioma	L-dopa	Yes	TH-labeled cells	10 days	12
Human type I	NS20Y neuroblast.	L-dopa	Yes	Behavioral improvement	9 days	15
Human type I	AtT-20 neuroendo.	L-dopa; dopamine	Yes	Behavioral improvement	15 days	15
Human type II	NRK-49F fibroblast	L-dopa	No	—	—	101
Human type II	C6 glioma	L-dopa	No	—	—	104

the results obtained with TH-expressing cells in the 6-OHDA rat support the use of somatic gene therapy for neurotransmitter replacement strategies.

IV. APPLICATIONS OF SOMATIC GENE THERAPY TO HUMAN DISEASE

The major difficulty in treating human diseases of the CNS is that the etiology and progression of most diseases are poorly understood. Neural dysfunction probably results from a number of factors, including aberrant gene expression, genetic defects that interfere with the production of vital enzymes, developmental defects, programmed changes in cellular development, neoplasia, infections, autoimmunity, toxic factors, or programmed cell death during aging. Due to the complex nature of many diseases and the limitations of available gene transfer methods, somatic gene therapy for the CNS is currently focused on intervening in the disease process and/or ameliorating CNS damage rather than correcting the cause of the disorder.

Presently, human diseases of the CNS that would appear to be most amenable to a somatic gene therapy approach are those in which a palliative strategy has been suggested to be effective. Based on pharmacologic treatments and/or animal models, such diseases would include Alzheimer's disease (cholinergic dysfunction), Parkinson's disease (dopaminergic dysfunction), Huntington's disease (γ-aminobutyric acid dysfunction), and brain cancers. Of the neurodegenerative disorders, Parkinson's disease would be among the first to consider for somatic gene therapy, since the adverse motor functioning has been linked to the loss of the single neurotransmitter dopamine, and this deficit can be effectively reversed with the exogenous administration of dopamine (with drugs and/or intracerebral grafts). In contrast, although cholinergic pathology is a prominent feature of Alzheimer's disease, there are accompanying degenerative changes in several other neurotransmitter systems that make this disorder much more challenging to address. To date, pharmacologic treatments for Alzheimer's disease have focused primarily on supplementing acetylcholine levels within the brain and have achieved minimal success. Based on results from animal models of cholinergic degeneration, a neurotrophic delivery strategy that directly enhances the survival of cholinergic neurons would be much more effective than neurotransmitter replacement in sustaining appropriate cholinergic functioning in the brain. Such an approach has not been pursued vigorously since NGF cannot cross the blood-brain barrier and must be delivered directly to the brain through an intracerebral cannula. Somatic gene therapy could avoid the complications of a cannula delivery approach, such as infections and provide the added benefit of site-specific provision of the neurotrophin.

V. ISSUES AND CONCERNS

In considering somatic gene therapy for human diseases of the CNS, there are several issues to address: cells selected for gene transfer must be able to survive noninvasively for prolonged periods within the CNS; the engineered cells should show stable production of the therapeutic product; the therapeutic product should be easily amenable to external regulation for appropriate dosing; and the cells should contain a safety mechanism that can be activated to halt the growth and/or functioning of the engineered cells *in vivo*. The first of these issues, the prolonged survival of cells within the CNS, is also the most crucial since concerns about gene function would be irrelevant if the engineered cells did not survive within the brain. Similarly, there would be few benefits derived from genetically modified cells that continued to grow after intracerebral grafting. Based on work in animals, the most optimal target cells to use for gene transfer and grafting in the brain would be those obtained from the patients themselves (autologous cells). There are currently two cell types that appear to be viable candidates for consideration: fibroblasts and muscle cells. Both populations have been shown to survive for over 1 year in the brain of rats (unpublished observations) and each would provide an effective vehicle for passively delivering a therapeutic product to the CNS.

Once the survival of engineered cells within the brain is assured, a second concern for human somatic gene therapy is achieving prolonged expression of the inserted gene *in vivo*. In this regard, the use of cell-specific promoters and enhancers, such as the muscle creatin kinase enhancer for myoblasts, may prove more effective in sustaining gene expression *in vivo*.[32] The issues of gene regulation *in vivo* and safety mechanisms are becoming the focus of increased attention. Fisher and colleagues have reported that choline administration to rats implanted with fibroblasts expressing a choline acetyltransferase transgene was able to induce a robust increase in acetylcholine production and release from the cells *in vivo*.[102] Although this route of manipulation demonstrated that the engineered cells were amenable to exogenous control after grafting, it would be preferable to insert or access unique regulatory elements directly within the transgene sequences to restrict the consequences of any external manipulations to the engineered cells. Similarly, in regard to safety, transgene sequences could be constructed to contain an easily activated toxic element. Resolving some of these issues should help contribute to the development of somatic gene therapy for human diseases of the CNS.

VI. FUTURE DIRECTIONS AND CONCLUSIONS

Until recently, it was not possible to consider an *ex vivo* somatic gene therapy approach directed toward primary neurons, since these cells typically survive poorly in culture and show limited mitotic activity. However, bFGF (basic fibroblast growth factor) at a concentration of 20 ng/ml has been found to induce the proliferation of neurons *in vitro*.[43] Using this property of bFGF, primary neuronal cultures have been generated from embryonic rat hippocampus that can be passaged, frozen for long-term storage, thawed, and recultured. Further, the proliferative properties of the cells have made them amenable to standard gene transfer methods (Ray and Gage, unpublished results). The development of these cells for gene therapy will greatly enhance the applications of gene transfer to the exploration and repair of the CNS.

In summary, both *ex vivo* gene transfer followed by implantation and *in vivo* gene transfer models have been developed for gene therapy in the CNS. Retroviral vectors have been used most frequently for gene transfer into both neuronal and non-neuronal cells *in vitro*. For *in vivo* gene transfer, herpes simplex and adenovirus vectors have both been examined. One of the primary concerns regarding clinical trials using gene therapy involves safety. In particular, it is essential to demonstrate the long-term safety of recombinant virus. Retroviral vectors are the only vectors which have been used for targeting cells to be used for clinical trials in patients. So far, no toxicity of this viral vector has been noted. Herpes simplex and adenoviral vectors hold promise, but their long-term safety *in vivo* is unknown. The major technical problem for gene therapy in the CNS is that expression of the transgenes is currently not stable. The continued resolution of these and other technical problems will increase the effectiveness of genetically modified cells for gene therapy in the CNS.

ACKNOWLEDGMENTS

We would like to thank M. L. Gage and Henry Grajeda for help in the preparation of the manuscript. Work in the authors' laboratory has been supported by grants from NIH (PO1 AG10435, RO1 AG06088), the Margaret A. and Herbert Hoover Foundation, and the Hollfelder Foundation.

REFERENCES

1. Ledley, F. D., Are contemporary methods for somatic gene therapy suitable for clinical applications?, *Clin. Invest. Med.*, 16, 78, 1993.
2. Blaese, R. M., Treatment of severe combined immunodefficiency (SCID) due to adenosine deaminase (ADA) with autologous lymphocytes transduced with a human ADA gene, *Hum. Gene Ther.*, 1, 327, 1990.
3. Fisher, L. J. and Gage, F. H., Grafting in the mammalian central nervous system, *Physiol. Rev.*, 73, 583, 1993.
4. Freed, W. J., Substantia nigra grafts and Parkinson's disease: from animal experiments to human therapeutic trials, *Rest. Neurol. Neurosci.*, 3, 109, 1992.
5. Ezzeddine, Z. D., Matuza, R. L., Platika, D., Short, M. P., Malick, A., Choi, B., and Breakefield, X. O., Selective killing of glioma cells in culture and *in vivo* by retrovirus transfer of the herpes simplex virus thymidine kinase gene, *New Biol.*, 3, 608, 1991.
6. Culver, K. W., Ram, Z., Wallbridge, S., Ishii, H., Oldfield, E. H., and Blaese, R. M., *In vivo* gene transfer with retroviral vector-producer cells for treatment of experimental brain tumors, *Science*, 256, 1550, 1992.
7. Mullen, C. A., Kilstrup, M., and Blaese, R. M., Transfer of the bacterial gene for cytosine deaminase to mammalian cells confers lethal sensitivity to 5-flurocytosine: a negative selection system, *Proc. Natl. Acad. Sci. U.S.A.*, 89, 33, 1992.
8. Takamiya, Y., Short, M. P., Ezzeddine, Z. D., Moolten, F. L., Breakefield, X. O., and Martuza, R. L., Gene therapy of malignant brain tumors: a rat glioma line bearing the herpes simplex virus type 1-thymidine kinase gene and wild type retrovirus kills other tumor cells, *J. Neurosci. Res.*, 33, 493, 1992.
9. Shimohama, S., Rosenberg, M. B., Fagan, A. M., Wolff, J. A., Short, M. P., Breakfield, X. O., Friedmann, T., and Gage, F. H., Grafting genetically modified cells into the rat brain: characteristics of *E. coli* β-galactosidase as a reporter gene, *Mol. Brain Res.*, 5, 271, 1989.

10. Rosenberg, M. B., Friedmann, T., Robertson, R. C., Tuszynski, M., Wolff, J. A., Breakefield, X. O., and Gage, F. H., Grafting genetically modified cells to the damaged brain: restorative effects on NGF expression, *Science*, 242, 1575, 1988.
11. Wolff, J. A., Fisher, L. J., Xu, L., Jinnah, H. A., Langlais, P. J., Iuvone, P. M., O'Malley, K. L., Rosenberg, M. B., Shimohama, S., Friedmann, T., and Gage, F. H., Grafting fibroblasts genetically modified to produce L-dopa in a rat model of Parkinson disease, *Proc. Natl. Acad. Sci. U.S.A.*, 86, 9011, 1989.
12. Uchida, K., Takamatsu, K., Kaneda, N., Toya, S., Tsukada, Y., Kurosawa, Y., Fujita, K., Nagatsu, T., and Kohsaka, S., Synthesis of L-3,4-dihydroxyphenylalanine by tyrosine hydroxylase cDNA-transfected C6 cells: application for intracerebral grafting, *J. Neurochem.*, 53, 728, 1989.
13. Horellou, P., Guibert, B., Leviel, V., and Mallet J., Retroviral transfer of a human tyrosine hydroxylase cDNA in various cell lines: regulated release of dopamine in mouse anterior pituitary AtT-20 cells, *Proc. Natl. Acad. Sci. U.S.A.*, 86, 7233, 1989.
14. Horellou, P., Brundin, P., Kalen, P., Mallet, J., and Bjorklund, A., *In vivo* release of dopa and dopamine from genetically engineered cells grafted to the denervated rat striatum, *Neuron*, 5, 393, 1990.
15. Horellou , P., Marlier, L., Privat, A., and Mallet, J., Behavioural effect of engineered cells that synthesize L-DOPA or dopamine after grafting into the rat striatum, *Eur. J. Neurosci.*, 2, 116, 1990.
16. Strömberg, I., Wetmore, C. J., Ebendal, T., Ernfors, P., Persson, H., Olson, L., Rescue of basal forebrain cholinergic neurons after implantation of genetically modified cells producing recombinant NGF, *J. Neurosci. Res.*, 25, 405, 1990.
17. Kawaja, M. D., Fisher, L. J., Schinstine, M., Jinnah, H. A., Ray, J., Chen, L. S., and Gage, F. H., Grafting genetically modified cells within the rat central nervous system: methodological considerations, in *Neural Transplantation: A Practical Approach*, Dunnett, S. B. and Bjorklund, A., Eds., Oxford University Press, 1992, chap. 2.
18. Gage, F. H., Wolff, J. A., Rosenberg, M. B., Xu, L., Yee, J.- K., Shults, C., and Friedmann, T., Grafting genetically modified cells to the brain: possibilities for the future, *Neuroscience*, 23, 795, 1987.
19. St. Louis, D. and Verma, I. M. An alternative approach to somatic cell gene therapy, *Proc. Natl. Acad. Sci. U.S.A.*, 85, 3150, 1988.
20. Palmer, T. O., Thompson, A. R., and Miller, A. D., Production of human factor IX in animals by genetically modified skin fibroblasts: potential therapy for hemophilia B, *Blood*, 73, 438, 1989.
21. Selden, R. F., Skoskiewicz, M. J., Howie, K. B., Russell, P. S., and Goodman, H. M., Implantation of genetically engineered fibroblasts into mice : implications for gene therapy, *Science*, 236, 714, 1987.
22. Kraemer, P. M., Travis, G. L., Ray, F. A., and Cram, L. S., Spontaneous neoplastic evolution of chinese hamster cells in culture: multiple progression of phenotypes, *Cancer Res.*, 43, 4822, 1983.
23. Macieira-Coelho, A. and Azzarone, B., The transition from primary culture to spontaneous immortalization in mouse fibroblast population, *Anticancer Res.*, 8, 669, 1988.
24. Sorge, J., Kuhl, W., West, C., and Beutler, E., Complete correction of the enzymatic defect of type-I Gaucher disease fibroblasts by retroviral-mediated gene transfer, *Proc. Natl. Acad. Sci. U.S.A.*, 84, 906, 1987.
25. Palmer, T. D., Rosman, G. J., Osborne, W. R. A., and Miller, A. D., Genetically modified skin fibroblasts persist long after transplantation but gradually inactivate introduced gene, *Proc. Natl. Acad. Sci. U.S.A.*, 88, 1330, 1991.

26. Scharfmann, R., Axelrod, J. H., and Verma, I. M., Long-term *in vivo* expression of retrovirus-mediated gene transfer in mouse fibroblast implants, *Proc. Natl. Acad. Sci. U.S.A.*, 88, 4626, 1991.
27. Kawaja, M. D., Fagan, A. M., Firestein, B. L., and Gage, F. H., Intracerebral grafting of cultured autologous skin fibroblasts into the rat striatum: an assessment of graft size and ultrastructure, *J. Comp. Neurol.*, 307, 695, 1991.
28. Kawaja, M. D. and Gage, F. H., Reactive astrocytes are substrates for the growth of adult CNS axons in the presence of elevated levels of nerve growth factor, *Neuron,* 7, 1, 1991.
29. Fisher, L. J., Jinnah, H. A., Kale, L. C., Higgins, G. A., and Gage, F. H., Survival and function of intrastriatally grafted primary fibroblasts genetically modified to produce L-dopa, *Neuron*, 6, 371, 1991.
30. Kawaja, M. D., Rosenberg, M. B., Yoshida, K., and Gage, F. H., Somatic gene transfer of NGF promotes the survival of axotomized septal neurons and the regeneration of their axons in adult rats, *J. Neurosci.*, 12, 2849, 1992.
31. Kawaja, M. D. and Gage, F. H., Morphological and neurochemical features of cultured primary skin fibroblasts of Fischer 344 rats following striatal implantation, *J. Comp. Neurol.*, 317, 102, 1992.
32. Dai, Y., Roman, M., Naviaux, R. K., and Verma, I. M., Gene therapy via primary myoblasts: long-term expression of factor IX protein following transplantation *in vivo*, *Proc. Natl. Acad. Sci. U.S.A.*, 89, 10892, 1992.
33. Jiao, S., Gurevich, V., and Wolff, J. A., Long-term correction of rat model of Parkinson's disease by gene therapy, *Nature*, 362, 450, 1993.
34. Dhawan, J., Pan, L. C., Pavlath, G. K., Travis, M. A., Lancetot, A. M., and Blau, H. M., Systemic delivery of human growth hormone by injection of genetically engineered myoblasts, *Science*, 254, 1509, 1991.
35. Barr, E. and Leiden, J. M., Systemic delivery of recombinant proteins by genetically modified myoblasts, *Science*, 254, 1507, 1991.
36. Cepko, C. L., Immortalization of neural cells via oncogene transduction, *TINS*, 11, 6, 1988.
37. Ryder, E. F., Snyder, E. Y., and Cepko, C. L., Establishment and characterization of multipotent neural cell lines using retrovirus vector-mediated oncogene transfer, *J. Neurobiol.*, 21, 356, 1989.
38. Snyder, E. Y., Deitcher, D. L., Walsh, C., Arnold-Aldea, S., Hartwieg, E. A., and Cepko, C. L., Multipotent neural cell lines can engraft and participate in development of mouse cerebellum, *Cell,* 68, 33, 1992.
39. Frederiksen, K., Jat, P. S., Valtz, N., Levy, D., and McKay, R. D. G., Immortalization of precursor cells from the mammalian CNS, *Neuron*, 1, 439, 1988.
40. Renfranz, P. J., Cunningham, M. G., and McKay, R. D. G., Region-specific differentiation of the hippocampal stem cell line HiB5 upon implantation into the developing mammalian brain, *Cell*, 66, 713, 1991.
41. Lendhal, U. and McKay, R. D. G., The use of cell lines in neurobiology, *TINS*, 13, 132, 1990.
42. Cepko, C. L., Immortalization of neural cells via retrovirus-mediated oncogene transduction, *Annu. Rev. Neurosci.*, 12, 47, 1989.
43. Ray, J., Peterson, D. A., Schinstine, M., and Gage, F. H., Proliferation, differentiation and long-term culture of primary hippocampal neurons, *Proc. Natl. Acad. Sci. U.S.A.*, 90, 3602, 1993.
44. Miller, A. D. and Rosman, G. J., Improved retroviral vectors for gene transfer and expression, *BioTechniques*, 7, 980, 1989.

45. Shaw, G. and Kamen, R., A conserved AU sequence from the 3′ untranslated region of GM-CSF mRNA mediates selective mRNA degradation, *Cell,* 46, 659, 1986.
46. Shyu, A. B., Greenberg, M. E., and Belasco, J. G., The c-fos transcript is targeted for rapid decay by two distinct mRNA-degradation pathways, *Genes Dev.,* 3, 60, 1989.
47. Bernstein, P. and Ross, J., Poly (A), poly (A) binding protein, and the regulation of mRNA stability, *TIBS,* 14, 373, 1989.
48. Maxwell, I. H., Brown, J. L., and Maxwell, F., Inefficiency of expression of luciferase reporter from transfected murine leukemia proviral DNA may be partially overcome by providing a strong polyadenylation signal, *J. Gen. Virol.,* 72, 1721, 1991.
49. Palmer, T. D., Hock, R. A., Osborne, W. R. A., and Miller, A. D., Efficient retrovirus-mediated transfer and expression of a human adenosine deaminase gene in diploid skin fibroblasts from an adenosine deaminase-deficient human, *Proc. Natl. Acad. Sci. U.S.A.,* 84, 1055, 1987.
50. Osborne, W. R. A. and Miller, A. D., Design of vectors for efficient expression of human purine nucleoside phosphorylase in skin fibroblasts from enzyme-deficient humans, *Proc. Natl. Acad. Sci. U.S.A.,* 85, 6581, 1988.
51. Hock, R. A., Miller, D. A., and Osborne, W. R. A., Expression of human adenosine deaminase from various strong promoters after gene transfer into human hematopoietic cell lines, *Blood,* 74, 876, 1989.
52. Schinstine, M., Rosenberg, M. B., Routledge-Ward, C., Friedmann, T., and Gage, F. H., Effects of choline and quiesence on Drosophila choline acetyltransferase expression and acetylcholine production by transduced rat fibroblasts, *J. Neurochem.,* 58, 2019, 1992.
53. Schinstine, M., Ray, J., and Gage, F. H., Effect of cytokines on transgene expression from the Moloney murine leukemia virus LTR (MoMLV) and collagen a2(I) promoters in fibroblasts, *Soc. Neurosci. Abstr.,* 17, 571, 1991.
54. Chen, L. S., Ray, J., Fisher, L. J., Kawaja, M. D., Scinstine, M., Kang, U. J., and Gage, F. H., Cellular replacement therapy for neurologic disorders: potential of genetically engineered cells, *J. Cell. Biochem.,* 45, 252, 1991.
55. Mulligan, R. C., The basic science of gene therapy, *Science,* 260, 926, 1993.
56. Gage, F. H., Intracerebral grafting of genetically modified cells acting as biological pumps, *TiPS,* 11, 437, 1991.
57. Anderson, W. F., Prospects of human gene therapy, *Science,* 226, 401, 1984.
58. Friedmann, T., Progress toward human gene therapy, *Science,* 244, 1275, 1989.
59. Miller, A. D., Progress towards human gene therapy, *Blood,* 76, 271, 1990.
60. Verma, I. M., Gene therapy, *Sci. Am.,* 263, 68, 1990.
61. Kriegler, M., *Gene Transfer and Expression. A Laboratory Manual,* W. H. Freeman & Co., New York, 1990.
62. Geller, A. I. and Breakefield, X. O., A defective HSV-1 vector expresses *Escherichia coli* β-galactosidase in cultured peripheral neurons, *Science,* 241, 1667, 1988.
63. Geller, A. I. and Freese, A., Infection of cultured central nervous system neurons with a defective herpes simplex virus vector results in stable expression of *Escherichia coli* β-galactosidase, *Proc. Natl. Acad. Sci. U.S.A.,* 87, 1149, 1990.
64. Geller, A. I., During, M. J., and Neve, R. L., Molecular analysis of neuronal physiology by gene transfer into neurons with herpes simplex virus vectors, *TINS,* 14, 428, 1991.
65. Geller, A. I., A system, using neural cell lines, to characterize HSV-1 vectors containing genes which affect neuronal physiology, or neuronal promoters, *J. Neurosci. Meth.,* 36, 91, 1991.
66. Breakefield, X., Gene delivery into the brain using virus vectors, *Nat. Genet.,* 3, 187, 1993.
67. Johnson, P. A., Yoshida, K., Gage, F. H., and Friedmann, T., Effects of gene transfer into cultured CNS neurons with a replication-defective herpes simplex virus type 1 vector, *Mol. Brain Res.,* 12, 95, 1992.

68. Johnson, P. A., Miyanohara, A., Levine, F., Cahill, T., and Friedmann, T., Cytotoxicity of a replication-defective mutant of herpes simplex virus type 1, *J. Virol.*, 66, 2952, 1992.
69. During, M. J., Geller, A. I., Deutch, A. Y., Neve, R. L., and O'Malley, K. L., Expression of human tyrosine hydroxylase and unregulated signal transduction enzymes in neurons in the mammalian brain from HSV-1 vectors, *Soc. Neurosci. Abstr.*, 17, 140, 1991.
70. Sable, B. A., Vick, A., Hollt, V., and Geller A., HSV-1 vector-mediated gene transfer: genetically modified post-mitotic cells implanted into rat striatum express a foreign gene, *Soc. Neurosci. Abstr.*, 17, 570, 1991.
71. Neve, R. L., Adenovirus vectors enter the brain, *TINS*, 16, 251, 1993.
72. Freese, A., Geller, A. I., and Neve, R., HSV-1 vector mediated neuronal gene delivery, *Biochem. Pharmocol.*, 40, 2189, 1990.
73. Dobson, A. T., Margolis, T. P., Sedarati, F., Stevens, J. G., and Feldman, L. T., A latent, nonpathogenetic HSV-1 derived vector stably express β-galactosidase in mouse neurons, *Neuron*, 5, 353, 1990.
74. Ho, D. Y. and Mocarski, E. S., Herpes simplex virus latent RNA (LAT) is not required for latent infection in the mouse, *Proc. Natl. Acad. Sci. U.S.A.*, 86, 7596, 1989.
75. Akli, S., Caillaud, C., Vigne, E., Stratford-Perricaudet, L. D., Ponaru, L., Perricaudet, M., Kahn, A., and Peschanski, M. R., Transfer of a foreign gene into the brain using adenovirus vectors, *Nat. Genet.*, 3, 224, 1993.
76. Bajocchi, G., Feldman, S. H., Crystal., R. G., and Mastrangeli, A., Direct *in vivo* gene transfer to ependymal cells in the central nervous system using recombinant adenovirus vectors, *Nat. Genet.*, 3, 229, 1993.
77. Davidson, B. L., Allen, E. D., Kozarsky, K. F., Wilson, J. M., and Roessler, B. J., A model system for in vivo gene transfer into the central nervous system using an adenoviral vector, *Nat. Genet.*, 3, 219, 1993.
78. Le Gal La Salle, G., Robert, J. J., Berrard, S., Ridoux, V., Stratford-Perricaudet, L. D., Perricaudet, M., and Mallet, J., An adenovirus vector for gene transfer into neurons and glia in the brain, *Science*, 259, 988, 1992.
79. Holt, C. E., Garlick, N., and Cornel, E., Lipofectin of cDNAs in the embryonic vertebrate central nervous system, *Neuron*, 4, 203, 1990.
80. Ono, T., Fujino, Y., Tsuchiya, T., and Tsuda, M., Plasmid DNAs directly injected into mouse brain with Lipofectin can be incorporated and expressed by brain cells, *Neurosci. Lett.*, 117, 259, 1992.
81. Roback, J. D., Palfrey, H. C., and Wainer, B. H., Nerve growth factor and the neurotrophin family: evolving roles in the central nervous system, *Comments Dev. Neurobiol.*, 6, 311, 1992.
82. Hepler, D. J., Olton, D. S., Wenk, G. L., and Coyle, J. T., Lesions in nucleus basalis magnocellularis and medial septal area of rats produce qualitatively similar memory impairments, *J. Neurosci.*, 5, 866, 1985.
83. Thoenen, H., The changing scene of neurotrophic factors, *TINS*, 14, 165, 1991.
84. Hefti, F., Nerve growth factor promotes survival of septal cholinergic neurons after fimbrial transections, *J. Neurosci.*, 8, 2155, 1986.
85. Gage, F. H., Armstrong, D. M., Williams, L. R., and Varon, S., Morphological response of axotomized septal neurons to nerve growth factor, *J. Comp. Neurol.*, 269, 147, 1988.
86. Haroutunian,V., Kanof, P. D., and Davis, K. L., Attenuation of nucleus basalis of meynert lesion-induced cholinergic deficits by nerve growth factor, *Brain Res.*, 487, 200, 1989.
87. Dekker, A. J. and Thal, L. J., Effect of delayed treatment with nerve growth factor on choline acetyltransferase activity in the cortex of rats with lesions of the nucleus basalis magnocellularis: dose requirements, *Brain Res.*, 584, 55, 1992.
88. Cuello, A. C., Maysinger, D., and Garofalo, L., Trophic factor effects on cholinergic innervation in the cerebral cortex of the adult rat brain, *Mol. Neurobiol.*, 6, 451, 1992.

89. Ernfors, P., Ebendal, T., Olson, L., Mouton, P., Stromberg, I., and Persson, H., A cell line producing recombinant nerve growth factor evokes growth responses in intrinsic and grafted central cholinergic neurons, *Proc. Natl. Acad. Sci. U.S.A.*, 86, 4756, 1989.
90. Maysinger, D., Piccardo, P., Goiny, M., and Cuello, A. C., Grafting of genetically modified cells: effects of acetylcholine release *in vivo*, *Neurochem. Int.*, 21, 543, 1992.
91. Piccardo, P., Maysinger, D., and Cuello, A. C., Recovery of nucleus basalis cholinergic neurons by grafting NGF secretor fibroblasts, *Neuroreport*, 3, 353, 1992.
92. Gage, F. H., Wictorin, K., Fischer, W., Williams, L. R., Varon, S., and Björklund, A., Life and death of cholinergic neurons in the septum and diagonal band region following complete fimbria-fornix transection, *Neuroscience*, 19, 241, 1986.
93. Cuello, A. C., Garofalo, L., Kenigsberg, R. L., and Maysinger, D., Gangliosides potentiate *in vivo* and *in vitro* effects of nerve growth factor on central cholinergic neurons, *Proc. Natl. Acad. Sci. U.S.A.*, 86, 2056, 1989.
94. Maysinger, D., Jalsenjak, I., and Cuello, A. C., Microencapsulated nerve growth factor: effects on the forebrain neurons following devascularizing cortical lesions, *Neurosci. Lett.*, 140, 71, 1992.
95. Maysinger, D., Herrera-Marschitz, M., Goiny, M., Ungerstedt, U., and Cuello, A. C., Effects of nerve growth factor on cortical and striatal acetylcholine and dopamine release in rats with cortical devascularizing lesions, *Brain Res.*, 577, 300, 1992.
96. Stephens, P. H., Cuello, A. C., Sofroniew, M. V., Pearson, R. C. A., and Tagari, P., Effect of unilateral decortication on choline acetyltransferase activity in the nucleus basalis and other areas of the rat brain, *J. Neurochem.*, 45, 1021, 1985.
97. Ungerstedt, U., Postsynaptic supersensitivity after 6-hydroxydopamine induced degeneration of the nigro-striatal dopamine system, *Acta Physiol. Suppl.*, 367, 69, 1971.
98. Ungerstedt, U., Striatal dopamine release after amphetamine or nerve degeneration revealed by rotational behavior, *Acta Physiol. Suppl.*, 367, 49, 1971.
99. Dunnett, S. B., Björklund, A., Stenevi, U., and Iversen, S. D., Grafts of embryonic substantia nigra reinnervating the ventrolateral striatum ameliorate sensorimotor impairments and akinesia in rats with 6-OHDA lesions of the nigrostriatal pathway, *Brain Res.*, 229, 209, 1981.
100. Schmidt, R. H., Ingvar, M., Lindvall, O., Stenevi, U., and Bjorklund A., Functional activity of substantia nigra grafts reinnervating the striatum: neurotransmitter metabolism and (^{14}C)2-deoxy-D-glucose autoradiography, *J. Neurochem.*, 38, 737, 1982.
101. Uchida, K., Ishii, A., Kaneda, N., Toya, S., Nagatsu, T., and Kohsaka, S., Tetrahydrobiopterin-dependent production of L-DOPA in NRK fibroblasts transfected with tyrosine hydroxylase cDNA — future use for intracerebral grafting, *Neurosci. Lett.*, 109, 282, 1990.
102. Fisher, L. J., Schinstine, M., Salvaterra, P., Dekker, A. J., Thal, L., and Gage, F. H., *In vivo* production and release of acetylcholine from primary fibroblasts genetically modified to express choline acetyltransferase, *J. Neurochem.*, 61, 1323, 1993.
103. Owens, G. C., Johnson, R., Bunge, R. P., and O'Malley, K. L., L-3,4-Dihydroxyphenylalanine synthesis by genetically modified schwann cells, *J. Neurochem.*, 56, 1030, 1991.
104. Uchida, K., Takamatsu, K., Kaneda, N., Toya, S., Tsukada, Y., Kurosawa, Y., Fujita, K., Nagatsu, T., and Kohsaka, S., Transfection of tyrosine hydroxylase cDNA into C6 cells, *Proc. Jpn. Acad.*, 64(B), 290, 1988.

11 Direct Gene Transfer *In Vivo*

Hans Herweijer, Jeffery D. Fritz, James E. Hagstrom, and Jon A. Wolff

I. INTRODUCTION

Gene therapy protocols have previously involved the removal of target cells from the body. These cells are then cultured *in vitro* to allow for gene transfer and subsequently reintroduced *in vivo*. This scheme poses severe limits on the applicability of gene therapy. Not every cell type can currently be cultured *in vitro*. Surgical complications may arise while removing or transplanting the target cells. Furthermore, the tissue culture and transplantation elements of the procedure can be quite costly. An alternative approach is transfection or transduction *in vivo*: direct gene transfer. Current direct gene transfer techniques can be divided in three groups: direct injection of naked DNA, injection of complexed DNA, and direct injection of recombinant viruses.

II. DIRECT INJECTION OF NAKED DNA

Direct injection of naked DNA is a simple and attractive approach for the introduction of foreign genes *in vivo*. While exploring the use of liposome-mediated gene transfer *in vivo*, Wolff et al. found that naked DNA is taken up in mouse skeletal muscle.[1] Basically, plasmid DNA is injected in an isotonic solution directly into the muscle tissue. Reporter gene expression can be measured from the first day after injection and is maintained for long periods — at least 1.5 years in mice.[1,2] Cardiac muscle has also been shown to be an excellent host tissue for direct transfection by naked DNA injection.[3,4] No other tissues investigated so far show any significant uptake and expression of naked plasmid DNA after injection.

A. Injection of Naked DNA in Skeletal Muscles

1. Structure of Striated Muscle

Myofibers that constitute skeletal muscle are formed during embryogenesis by fusion of many myoblasts. Mature myofibers can contain up to 1000 nuclei; they do not

0-8493-2440-8/95/$0.00+$.50

divide and show no DNA synthesis beyond DNA repair. Upon damage of a myofiber, which does occur during normal use, repair is provided by muscle stem cells. These satellite cells are resting myoblasts that can divide and fuse into existing fibers. In extensively damaged muscle, fusing satellite cells can actually form a limited number of new myofibers.

Individual myofibers are surrounded by the endomysium, a connective tissue sheath, mainly consisting of collagenous glycoproteins. Other components include fibronectin, entactin, and heparan sulfate. Myofibers are grouped by sheaths of perimysium; the whole muscle is contained within the epimysium. Both the perimysium and epimysium are connective tissue sheaths consisting of collagenous fibers, fibronectin, and other components. Together, these connective tissues form a strong scaffolding, enabling muscle to generate maximal power.

2. Uptake of Naked DNA

The mechanism of DNA uptake into myofibers is poorly understood. Direct intramuscular injection of naked DNA primarily deposits the plasmids in the extracellular space. In order to enter a myofiber, the DNA has to cross at least the endomysium and the myofiber cell (plasma) membrane. Since the injections are usually performed at only one entry and deposit site, plasmid DNA will also have to cross the perimysium to spread throughout the muscle. The plasmid DNA expression cassettes that have been used contain reporter genes under the control of eukaryotic promoters. The plasmid DNA has to enter the nucleus of a cell so that DNA-dependent RNA polymerase II can initiate expression. Therefore, both the cell and nuclear membranes must be crossed. Plasmid DNA crossing of the endomysium and the cellular membrane has been attributed to temporary disruptions of these membranes.[5] Various observations, which will be discussed below, have ruled out physical disruptions as a major means of DNA entry.

So far, only striated muscle has been shown to support *in vivo* plasmid DNA uptake, resulting in sustained expression. It can be assumed that some structural feature unique to striated muscle is responsible for DNA uptake. Directly after injection into mouse muscle, plasmid DNA can be found distributed throughout the whole muscle. Staining of DNA with Hoechst 33258 demonstrates the presence of plasmid DNA in the extramyofiberal space (i.e., within the extracellular space enclosed by epimysium and perimysium) and inside myofibers.[5] It was estimated that in 30 to 80% of the myofibers, plasmid DNA had crossed the external lamina and appeared to be intramyofiberal (i.e., within T-tubules, caveolae, or in the cytoplasm). Spread of high molecular weight (genomic) DNA was severely hindered by the epimysium and the perimysium, thus showing that size is an important factor determining DNA transport. This also suggests that membrane disruption is not an important pathway for DNA entry. Electron microscope studies with colloidal gold-conjugated plasmids showed that the DNA enters T-tubules and caveolae.[5] Transport into the myofiber was not found for gold-conjugated polyethyleneglycol, polylysine, or polyglutamate. Each of these polymers resembles plasmid DNA in molecular weight but differs in total charge. Again, this demonstrates that physical disruption of the myofiber cell membrane is not a very likely mechanism of DNA transport through this membrane. It does suggests a specific transport route for

DNA into the cell that is not available to many other polymers. It may well be that T-tubules are responsible for the entry of plasmid DNA into the myofiber. These structures are found only in skeletal and cardiac muscle tissues. The exact mechanism responsible for this process remains to be elucidated.

3. Persistence of DNA in Striated Muscle

Once plasmid DNA has entered the nucleus, it can remain there in an extrachromosomal state for very long periods of time. *In vivo* data have been published that show stable expression over a period of 19 months.[2] Recently, we have assayed luciferase expression in mice injected 2 years previously and found that expression is still present at approximately the same level as at 1 month after injection (Figure 1). The episomal persistence of plasmid DNA in muscle cells has been investigated in detail. Southern blotting experiments showed the presence of a single band of plasmid DNA after digestion with enzymes that cut once inside the plasmid.[1] If integration occurs, one would expect multiple bands. Since this type of assay is limited in sensitivity, other experiments were performed.[2] PCR analysis showed the presence of plasmid DNA in muscle injected 19 months earlier. These data are extremely important in view of future gene therapy protocols. Considering the persistence of the plasmid DNA, there seems to be no need to repeat introduction of the transgene. This allows for simple, efficient, and cost-effective therapies. The plasmid DNA that was recovered after 19 months still had its original bacterial methylation pattern, thus indicating that the plasmid DNA was not replicated or repaired extensively. Transformation of DNA isolated from the injected muscles into bacteria resulted in the isolation of the injected plasmids. No genomic sequences were detected in these plasmids. Plasmid DNA injected into striated muscle appears to remain unintegrated.

4. Nonhuman Primates

Injection of plasmid DNA into the quadriceps (rectus femoris) resulted in sustained reporter gene expression in mice, rats, cats, and rhesus monkeys.[6] Efficiency of expression, however, was inversely related to the size of the animal. Luciferase expression after injection of 10 μg pRSVL (luciferase expressed from the Rous sarcoma virus promoter) was 50-fold higher in mice than in rhesus monkeys.[6] Since expression does not change much over time for all species, DNA uptake seems to be a more important factor than differences in plasmid stability or expression efficiency. Higher amounts of injected DNA resulted in higher expression levels in all species. Yet, saturation occurs starting at approximately 100 μg of DNA injected into mice. No such effect was observed for rhesus monkeys, even for injections of 500 μg plasmid DNA.

Based on these observations, Jiao et al.[6] reasoned that differences in muscle structure could be the key factor determining DNA uptake efficiency. Two factors stand out so far: (1) Although the length and weight of the rectus femoris is increased in the larger animals, the number of myofibers is approximately the same. Also, in the larger animals, the number of nuclei per myofiber volume is lower. This would increase the average distance that a plasmid molecule has to travel inside the (thicker) myofiber to

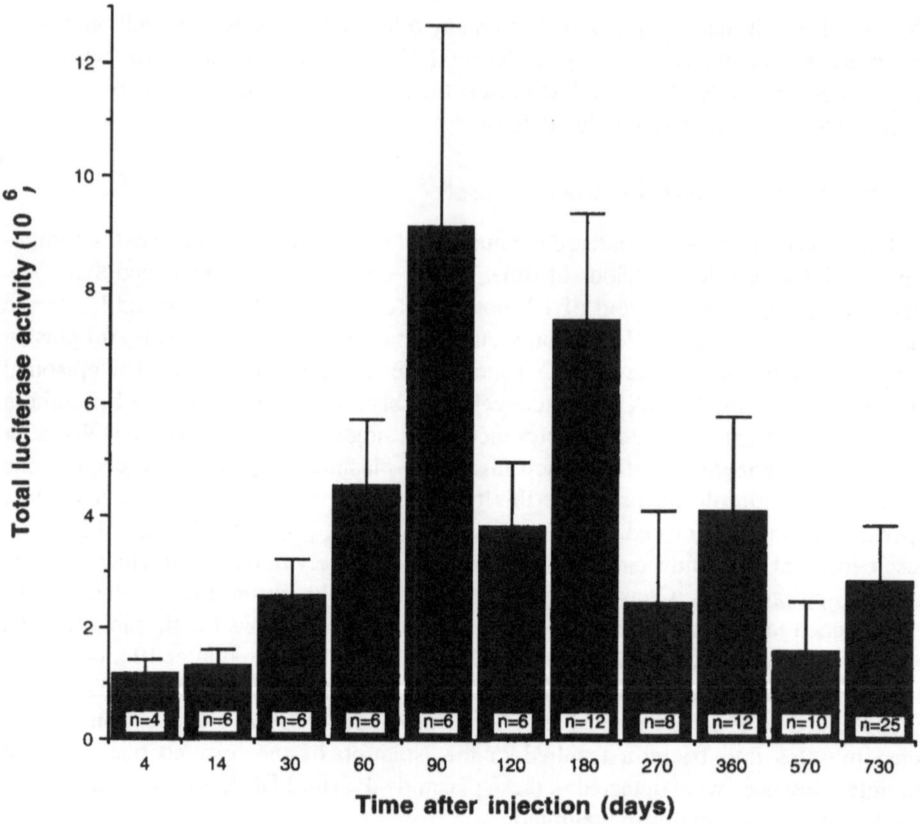

FIGURE 1. Expression of luciferase in the muscles of mice previously injected with 100 μg pRSVL (luciferase under control of the Rous sarcoma virus promoter/enhancer). Plasmid DNA dissolved in normal saline solution was injected in a volume of 100 μl into the rectus femoris of Balb/c mice. Quadriceps were removed at different times after injection, as indicated in the figure. Luciferase activity was measured in muscle lysates as published.[2] The number of muscles analyzed is indicated in each bar. Data are from Wolff et al., 1992[2] and unpublished observations.

reach one of the (more widely scattered) nuclei. As discussed above, it is likely that intracellular DNA transport is the result of slow diffusion. The increase in the average time a plasmid is in the cytoplasm is then responsible for a higher rate of nuclease digestion and thus a lower number of plasmids entering a nucleus. More in-depth studies of intracellular DNA transport are required to strengthen this hypothesis. (2) Connective tissue could well be the most important factor limiting transport of plasmid DNA in the muscle. The thicker perimysium in Rhesus monkey muscle (compared to mouse muscle) may restrict the distribution of DNA to a much smaller area, thus limiting cellular DNA uptake.[6] This explanation is supported by a recent study.[7] Preinjection of mouse muscle with a large volume of a hypertonic sucrose solution slightly increased reporter gene expression from plasmids injected shortly afterward. More importantly, the sample-to-sample variation was decreased from 120 to 25%. The preinjection probably disrupts the connective tissues and separates the myofibers, allowing for a better distribution of the DNA solution. Since expression levels are not much different with or without preinjection, apparently a fairly constant number of plasmid molecules is taken up by the myofibers in the injected muscle. Otherwise, a much greater increase in the mean expression levels in the preinjected group is expected. Therefore, it seems much more likely that the better accessibility of all myofibers results in a more constant uptake of plasmid DNA. Without preinjection, bulk plasmid DNA can be taken up by fewer myofibers or might be lost, resulting in the higher variability. It could be very interesting to test this preinjection method in larger animals. There, increased penetration could still result in much higher DNA uptake and expression. In view of the lower efficiency in larger animals in general, this might be an important improvement toward human gene therapy.

5. Factors Affecting DNA Transfer

A large number of factors do influence the efficiency of *in vivo* transfection by naked DNA injection. In the last few years, several papers were published describing some of these factors. In order to arrive at an "optimal" gene transfer protocol, we will summarize the most important factors here.

a. *Host Animal*

As was discussed above, larger animals tend to express less protein per microgram of injected plasmid DNA. Also, factors like mouse strain, age, and sex influence expression.[8] Expression was highest upon injection of mice at the time of the greatest muscle growth (age 4 to 6 weeks for mice). Females expressed less compared to males of the same age, presumably because of the smaller size of the injected muscle.

b. *Muscle Group*

Expression levels seem to be the same for different striated muscles in a given host. Differences are more likely to reflect the ease of injection in a larger muscle group vs. a smaller muscle group. Also, muscles that can better contain the injected plasmid solution (e.g., tibialis anterior) show higher expression levels. Probably this is due to less leakage of the DNA solution to surrounding tissues.

c. *Plasmid DNA*

Covalently closed circular (CCC) plasmid DNA is superior to linear DNA. Injection of linearized plasmid DNA resulted in 20-[2] to 100-fold[5,9] lower reporter gene expression levels. This probably is a result of a much faster degradation of linear DNA in the extracellular matrix or cytoplasm, resulting in fewer plasmid molecules reaching the nucleus. Moreover, linear DNA could be less well maintained in an extrachromosomal status. The latter hypothesis is consistent with tissue culture experiments where transfection with either CCC or linear DNA can result in stably transfected cells with comparable efficiencies and expression levels. Another important factor is the quality of the injected

DNA. "Cleaner" DNA, i.e., doubly banded on cesium chloride, is more efficient. There is definitely a batch-to-batch variation in expression levels obtained from the same plasmid.[10] We have extensively optimized and standardized our plasmid production, yet occasionally a low-expressing batch of DNA is produced without any apparent reason. Probably, small impurities, such as bacterial proteins or chemicals used during the plasmid preparation, are responsible for this phenomenon. Therefore, it is advisable to test every batch of plasmid DNA by transfection into tissue culture cells. Introduction of pelleted DNA is a viable alternative to direct injection of DNA in solution. In fact, implantation of DNA pellets has appeared to be more efficient, both in rodents and in rhesus monkeys.[6,10]

d. *Amount of DNA*

The expression levels measured for a reporter gene are proportional to the amount of plasmid DNA injected. This has been shown for mice,[7] rats, and rhesus monkeys.[6] This relationship is not always linear, which is probably related to DNA distribution factors (see below).

e. *Gene Regulatory Sequences*

The choice of promoter is important. To obtain high expression levels, a strong, constitutively expressing promoter is necessary. Good examples are the Rous sarcoma virus or the cytomegalovirus promoter/enhancer sequences. Likewise, muscle-specific promoters, such as α-actin or creatine kinase, usually give sustained expression but at somewhat lower levels. The influence of promoter choice will be discussed in more detail in the section on cardiac muscle injections.

f. *Injection Fluid*

In a direct comparison of many different injection fluids, hypertonic saline and sucrose solutions resulted in comparable expression levels as normal saline or sucrose.[6,10] Compared to normal saline, preinjection with hypertonic sucrose was more effective, both in increasing expression levels and in lowering variation.[7]

g. *Injection Volume*

Gal et al. found that the injection volume is an important factor in determining the expression levels after injection into rat cardiac muscle.[11] Luciferase expression increased almost linearly with increasing volume and was four-fold higher for injection in 300 μl than in 50 μl. The dependence on injection volume seems much less in skeletal muscles. Volumes between 25 and 100 μl injected into mouse quadriceps resulted in the same expression values.[10] A study by Davis et al. also showed that skeletal muscle is not very sensitive to variation in injection volumes;[7] yet, the larger volumes gave the highest expression levels. As discussed above, preinjection with a large volume of saline or hypertonic sucrose increased expression and decreased the variation considerably. Also, expression seems to be better correlated with the amount of injected DNA. The optimal time interval between preinjection and DNA injection remains to be determined. All these data indicate that larger injection volumes result in a better DNA distribution throughout the injected muscle and a subsequently higher uptake of the DNA in myofibers.

h. Other Variables

Needle type is not in itself an important factor.[10] Yet, we find that 27-gauge needles are a good size to work with. Different types of anesthetics appear to have no effect on reporter gene expression levels. Anesthetizing of mice with metofane is a fast and easy way to prepare mice for i.m. injections. This anesthetic works long enough for the procedure to be completed, and the mice recover quickly. Fewer operational deaths have been observed with metofane than with avertin or other injectable anesthetics. We generally open the skin in order to observe the injection site. Others choose to inject transcutaneously with good results. The use of a needle collar[10] to standardize the depth of injection can be helpful.

6. An Optimal Protocol for Naked DNA Injections in Mouse Skeletal Muscles

The following is a protocol that currently can be considered as resulting in the highest transfection frequency, highest expression values, and lowest variability in the test animals: use a short-duration anesthetic. Preinject the muscle group with hypertonic saline or sucrose solution. Use a large volume: 100 μl for rectus femoris, 75 μl for tibialis anterior of mice. Inject 10 to 100 μg of very clean plasmid DNA in a large volume identical to that used for the preinjection. Keep the needle in the muscle for 5 to 10 sec after the injection, in order to allow for the fluids to disperse through the muscle.

B. Injection of Naked DNA in Cardiac Muscles

There are major differences in ultrastructure between cardiac and skeletal muscle. Most notable is that intercalated discs connect mononucleated cardiomyocytes into fibers. This network of muscle cells lacks endomysium and epimysium connective tissues. Features considered important in naked DNA uptake, such as T-tubules and caveolae, are larger and more abundant. Cardiac muscle was investigated as a host tissue as a natural next step after the discovery of the uptake of naked plasmid by skeletal striated muscle. Injection of plasmid DNA in cardiac muscle results in DNA uptake and expression.[3,4] The level of reporter gene expression in cardiac muscle is equal to or higher than in skeletal muscle.[3] Yet, expression seems to be restricted to a small area, 2 to 3 mm, surrounding the injection site.[4,9,11] Comparable to skeletal muscle, the total number of cells that can be transfected using currently available techniques seems limited.

Expression of reporter genes could be measured from the first day after injection up to approximately 1 month.[3,4] At that time, expression levels were down by as much as 90%. So far, expression in cardiac muscle appears to be transient. The cause for this appears to be an immune response of the host directed against the expressed transgene.

Immune-suppressed and nude rats showed stable expression in cardiac muscle for periods up to 6 months.[3] It is remarkable that a similar immune response is not directed against transfected skeletal muscle cells. Other groups have reported more stable expression in rat cardiac muscle up to 2 months.[9,11] It is likely that introduction of endogenous genes will result in the absence of an immune response. Then, as in skeletal muscle, long-term stability of expression may be expected.

Conditions affecting DNA uptake and expression in cardiac muscle generally parallel those for skeletal muscle. It has been shown that a larger injection volume increases expression levels,[11] much more than for skeletal muscle injections. It has not been reported yet whether preinjection of cardiac muscle is beneficial.

1. Regulation of Cardiac Gene Expression

Direct injection of plasmid DNA has been used imaginatively to study muscle gene regulation. Measurement of reporter gene expression under control of muscle-specific promoters and/or muscle-specific enhancers can give valuable information on the relative importance of these regulatory sequences.[9,12,13] Using this technique, it was possible to map the regulatory sequences that confer skeletal and cardiac muscle-specific expression of the muscle creatine kinase gene.[12] In another study, it could be demonstrated that a short 18-bp sequence in the β-myosin heavy chain gene enhancer region functions as a repressor on heterologous promoters.[13] Also, the influence of hormones on gene regulation has been studied using this technique.[14,15] In these studies, a reporter gene is again placed under the control of a muscle-specific promoter/enhancer. Direct injection of these plasmids will show the skeletal or cardiac muscle specificity of the combinations. Also, hormonal influences on reporter gene expression can be studied by altering the hormonal status of the animal. Thyroid hormone treatment, for instance, increased expression of a reporter gene under the control of an α-myosin promoter/enhancer two-fold in cardiac muscle.[14] This cassette did not express in skeletal muscle. Using the direct DNA injection technique, these kinds of studies are relatively easy and fast, especially compared to the alternative of generating transgenic animals.

C. Injection of Naked DNA in Other Tissues

Several other tissues have been injected with plasmid DNA. Yet, none of these seems to be able to take up naked DNA in significant quantities that result in reporter gene expression. Very low but unstable expression was found in uterus and stomach (smooth muscle tissues).[3] All other major organs remained negative.

Direct intratumoral injection of naked DNA into mice carrying B16 melanomas resulted in DNA uptake and expression of the reporter gene.[16] Approximately 10% of the tumor cells expressed β-galactosidase 10 days after injection of 1 μg of plasmid DNA. By using a promoter that is preferentially expressed in melanoma cells, the authors claim high tissue specificity of this *in vivo* transfection technique. This gene therapy method may be a valuable new treatment modality for solid tumors. It could be interesting to determine the efficacy of direct naked DNA injection in tumors derived of tissues that normally can be transfected well by this method (i.e., skeletal and cardiac muscle).

D. Gene Gun

Introduction of naked DNA by ballistic methods is an alternative to injection. Small metal particles of tungsten or gold that are coated with DNA are accelerated to very high speeds. When directed at cells *in vitro*, the coated particles move through the cell membrane and deliver the DNA inside the cell. Stable transfection has been observed in a variety of cell types.[17] The efficiency of this particle bombardment method is comparable to other methods used to transfect tissue culture cells (i.e., frequency around 10^{-3}). Particle bombardment also can be used for *in vivo* transfection. Application is limited to those tissues where the applicator can be directed. So far, *in vivo* transfection has been accomplished for skin, liver, kidney, mammary gland, and muscle tissues.[17-19] Up to 25% of the cells can be transfected, yet expression is transient. Expression is most stable in muscle tissues. This probably is a result of stable persistence of plasmid DNA in an episomal state, as has been discussed above for naked DNA injection into muscle. With its properties of high transfection frequency and transient expression, the gene gun could be very useful for *in vivo* immunization schemes.

E. In Vivo Electroporation

Plasmid DNA has been introduced into skin cells of mice by *in vivo* electroporation.[20] A short time after subcutaneous injection of plasmid DNA, a short electric pulse was applied through an applicator clamped onto the skin. Electric pulses of 400 V and 100 to 300 μsec duration resulted in transfection frequencies between 2×10^{-5} and 7×10^{-4}. This was measured by culturing primary fibroblasts in medium containing G418 after transfection with pSV_3-Neo^R plasmids. Electric pulses of 600 V resulted in necrotic damage to the skin, which was repaired during the next 2 weeks. Pulses of less than 200 V did not result in measurable transformation. It is currently not clear if expression is stable. At present, this technique appears to be too inefficient to compete with other techniques for *in vivo* gene delivery.

F. Gene Therapy Using Naked DNA

Efficiency of transfection and maximal obtainable expression levels are of major importance in designing gene therapy protocols for the treatment of human inborn or acquired disorders based on the technique of naked DNA injection in skeletal muscle. Generally, up to 2% of the fibers in a mouse quadricep express β-galactosidase (blue X-Gal staining) after injection of *lac-Z* reporter plasmids. Moreover, injection in larger laboratory animals is much less efficient. Extrapolating this to the human situation predicts an even lower efficiency. The number of stably transfected cells might be improved by using the preinjection technique, but this remains to be investigated. This level of transfection efficiency is currently too low for a gene therapy protocol for most muscle disorders, e.g., Duchenne muscular dystrophy (DMD).[21] To correct the defect in these diseases, it is necessary to introduce a gene in most of the myofibers in the body. An efficiency of 10% seems to be the minimum needed. A significant improvement in the direct injection technique is necessary before this can be attempted for treatment of these genetic diseases. Results in *mdx* mice, a model for human DMD, are encouraging. Expression of a functional dystrophin gene halts the ongoing wasting of myofibers and results in expression of dystrophin-associated proteins. Therefore, (repeated) gene therapy might be more beneficial than is expected based on the relatively low transfection efficiency.

Direct injection of naked DNA can be a very important gene therapy route for the production of proteins that are excreted from the muscle cells. This can ameliorate a disease state in which muscle is not normally involved. Good examples are the production of factor IX and growth hormone. Transplantation of transfected myoblasts has already shown that muscle can be a source for the systemic delivery of these proteins. Other applications can be found in the field of metabolic diseases. Here, muscle could serve as an artificial organ, taking over the task of the affected organ. This is limited to those diseases in which the substrate is circulating and can be taken up by muscle cells. For instance, in phenylketonuria (PKU), a mutation in the phenylalanine hydroxylase (PAH) gene results in the lack of expression of this liver enzyme. High, toxic levels of phenylalanine are the result. Expression of PAH in muscle cells is currently being investigated as gene therapy for PKU. In this case, clearance by PAH-expressing muscle cells could result in a beneficial effect on blood phenylalanine levels.

A possible complication in every gene therapy protocol is the risk of immune reactions. In many genetic diseases, there is no protein expressed at all from the afflicted gene. Therefore, introduction and expression of a normal copy of this gene will probably result in the generation of antitransgene antibodies and cytotoxic T-cells. This did happen for reporter transgenes expressed in cardiac muscle.[3] Expression in skeletal muscle of transgenes that are not secreted does not result in immune responses. Apparently, myofibers are not very efficient in presenting these transgenes in the context of MHC-I. Experiments as described above for PKU treatment should yield more information on the complications of immune responses on excreted proteins. A more definitive way to overcome these problems requires tolerizing the immune system before the introduction of a transgene. Gene therapy methods to this end are being developed.

III. DIRECT INJECTION OF COMPLEXED DNA

A. Liposomes

Liposomes have been used extensively to transfect a wide variety of cells *in vitro.* Coating of DNA with a mixture of lipid molecules results in the creation of a particle that has no charge or is positively charged, which can be endocytosed. Introduction of DNA encapsulated in liposomes can also result in transfection *in vivo.* Injection of DNA-liposome complexes into the portal vein resulted in transfection of liver cells with relatively high efficiency.[22-25] Also *in vivo* transfection of vascular endothelial cells can be performed relatively efficiently.[26-32] This latter application is detailed in Chapter 7.

Currently, one of the major target organs for *in vivo* liposome-mediated transfection is the lung.[33-35] The target cell is ideally the lung stem cell, which could result in long-term persistence and expression of the introduced plasmid DNA. However, primarily the differentiated epithelial cells forming the lining of the lung seem to take up the DNA.[35] This results in loss of expression of the introduced gene, due to the limited lifetime of these cells. This is not a problem in itself, since the *in vivo* liposome-mediated transfection procedure potentially can be repeated frequently.

Other organs that were transfected by liposome complexes are liver[22,36] and kidney.[37] In these studies, the authors included nonhistone chromosomal protein (high mobility group 1) and inactivated Sendai virus in their liposome preparations to increase DNA uptake ten-fold.[22] After introduction of DNA-liposome complexes into the portal vein of rats, approximately 10% of the liver cells stained positive for the presence of the introduced reporter gene (SV40 large T-antigen or human insulin).[22,36] Expression was transient and almost completely absent by day 14. In the kidney, up to 15% of the glomerular cells expressed the introduced reporter gene after injection of the DNA-liposomes into the renal artery. Expression, however, was again transient.[37]

The transient nature of reporter gene expression after *in vivo* liposome-mediated transfection is probably due to nonintegration of the plasmid DNA.[25,38] Plasmid DNA has been demonstrated to exist in an extrachromosomal state; integration into the host genome could not be demonstrated,[25] thus ruling out that integration events occur with high frequency. Apparently, episomal plasmid DNA is not stable in cell types other than muscle. Direct injection of DNA-liposome complexes into muscle does not result in very significant transfection: expression levels were 100-fold lower compared to injection of naked DNA.[5] Still, muscle can take up DNA after intravenous injection of DNA-liposome complexes.[29,38] In these experiments, it appeared that the plasmid DNA was still present in cardiac muscle up to 2 months after injection, again demonstrating that plasmid DNA can persist episomally in muscle much better than in most other tissues. It remains to be tested whether liposome-mediated transfection with linearized plasmid DNA can result in sustained expression. Linear DNA seems to integrate into the host genome more readily. However, transfection efficiencies are lower with linear DNA compared to CCC plasmid DNA.

The intravenous delivery route can result in DNA uptake and reporter gene expression in a variety of cells. Expression has been documented in kidney, lung, spleen, thymus, bone marrow, peripheral and circulating bone marrow-derived cells, uterus, smooth and skeletal muscle, pancreas, and the intestine.[29,38,39] Recently, uptake of liposome-complexed DNA by spinal motor neurons has been described.[40] Motor and sensory neurons are capable of taking up molecules and delivering these into the neuronal perikaryon via retrograde axoplasmic transport. Each of this techniques promises to be valuable in the study of gene expression in specific tissues and in limited gene therapy protocols.

B. Polylysine Complexes

The latest development in *in vivo* gene delivery is the use of DNA conjugated to a ligand that binds the conjugate to complementary receptors present on the intended target cells.[41] Receptor-mediated gene transfer promises a high degree of specificity, something that many of the other *in vivo* gene transfer methods described here lack. The ligand-DNA conjugates contain polylysine to complex the DNA and can be taken up by cells *in vitro* and *in vivo*. Using asialoorosomucoid covalently linked to polylysine, plasmid DNA bound to these molecules could be taken up by asialoglycoprotein receptor-positive liver cells *in vitro*.[42] Later studies also demonstrated the uptake of these DNA complexes by liver cells *in vivo* after intravenous injection.[41,43-45] Expression was found to be transient in these studies. Initially, high copy numbers of episomal plasmid DNA could be found in the liver cells (100 to 10,000 per cells).[46,47] Over time, this plasmid DNA was lost, and in 1 to 2 weeks after injection, plasmid DNA was not detectable. A much slower decrease in plasmid numbers and expression occurred when partial hepatectomy was performed after gene delivery.[47] Still, plasmid DNA was found to be not integrated.

Another well-studied ligand for receptor-mediated gene delivery is transferrin.[48-54] A transferrin-polylysine-DNA complex resulted in efficient reporter gene expression in airway epithelial cells after intratracheal introduction.[52] In a series of experiments, it was shown that one of the major limitations for efficient DNA uptake is the endosomal degradation of the introduced plasmid DNA. Inhibition of this degradative pathway resulted in a very significant enhancement of the transfection efficiency. This can be achieved by addition of chloroquine to the culture medium (*in vitro*) or by the addition of adenovirus particles (*in vitro* and *in vivo*). The use of recombinant viral proteins, which are responsible for destabilizing the endosome, should make this system better defined and much safer to use. The feasibility of this approach has recently been demonstrated using a synthetic oligopeptide derived from the influenza virus hemagglutinin protein.[55]

IV. DIRECT INJECTION OF RECOMBINANT VIRUSES

A. Retroviruses

Most of the currently approved gene therapy protocols use retroviruses for *in vitro* gene transfer. It seemed natural therefore, to explore the use of retroviruses *in vivo*. Recombinant retroviruses can be harvested in titers up to 10^6 infectious particles per ml of culture medium. Compared to other recombinant viruses, this is not very high (see below), yet sufficient for most purposes. Techniques have been developed that enable relatively straightforward construction of vectors and production of viruses. Using these systems, recombinant retroviruses, free of any contaminating helper (i.e., wild-type) virus can be produced. Furthermore, with the proper vector design, sustained expression can be obtained in most target cell types. A very important aspect of the retroviral life cycle is the need for cell division in order for the virus to integrate into the host genome. *In vivo*, three different target organs have been the focus of most studies so far: (1) liver, (2) vascular endothelial cells, and (3) tumors.

1. Liver

Hepatocytes can be transduced efficiently *in vitro*. Yet, isolation of hepatocytes, *in vitro* culture, and reintroduction in the patient is a very complicated and not always successful procedure. It is more attractive to transduce liver cells *in vivo*. The liver is easily accessible for perfusion with retrovirus-containing media via the portal vein. Yet, transduction frequencies are low following this procedure. This can be improved dramatically by performing a partial hepatectomy 1 to 2 days before exposure to the virus.[56] The hepatectomy results in a rapid cycling of the liver cells, making them susceptible for retrovirus replication and integration. Up to 5% of the (regenerated) liver cells express reporter genes long term (more than 6 months).[56-58] This approach is a relatively simple procedure applicable for the treatment of a variety of hepatic disorders.

2. Endothelial Cells

Vascular endothelial cells can be efficiently transduced by retroviruses *in vivo*. After intravenous injection of recombinant retroviruses, the transduction efficiency is equal to liposome-mediated transfection in the same system. Since retroviruses integrate into the host genome, expression is long term and dependent on host cell lifetime and transcription regulation. Retroviral transduction of selected vessels can be applied to the treatment of atherosclerosis, cancer, and the production of recombinant proteins (e.g., blood clotting factors). This is discussed in more detail in Chapter 7.

3. Tumors

Direct injection of retroviruses is being tested as a treatment for brain tumors. Since neuronal cells are postmitotic, only the dividing tumor cells will be susceptible for retroviral transduction. The retrovirus carries a gene for a prodrug-activating enzyme (thymidine kinase activation of ganciclovir). Upon activation, dividing cells are selectively killed, thus achieving a two-step specificity. This approach was effective in a rat glioma model,[59] and will be tested in a forthcoming human trial.[60] Several other tumor sites will be amenable to selective retroviral transduction (e.g., the liver). This application is detailed in Chapter 14. Other approaches for gene therapy for cancer are described in Chapter 15.

B. Adenoviruses

Adenoviral vectors can introduce and express foreign genes in a wide variety of cells *in vivo*. The adenoviral vectors currently in use derive from a system developed in the late 1970s.[61] In these vectors, the E1A, E1B, and E3 genes are replaced with up to 7.5-kb of transgenes. The recombinant adenovirus is propagated in 293 cells (a human kidney cell line) that constitutively express the E1 genes. Titers up to 10^{11} infectious particles per ml can be generated.

In vivo application of adenoviral vectors has focused strongly on the treatment of cystic fibrosis. The most important target cells are the epithelial cells lining the lung. Adenoviruses have a great advantage here, since they can infect nondividing, differentiated cells. Furthermore, adenoviruses are tropic for the respiratory epithelium. Several studies describe the instillation of recombinant adenoviruses into the lungs of cotton rats. These laboratory animals have a human-like sensitivity for adenovirus infection. Infection of lungs of cotton rats with an α1-antitrypsin adenovirus resulted in the transduction of epithelial cells and human α1-antitrypsin could be measured in the serum for 1 week after infection.[62] After intratracheal infection of cotton rats with 10^7 to 10^{10} plaque-forming units of a *lac-Z*-adenovirus, all major categories of airway epithelial cells were found to express β-galactosidase.[63] Expression levels were directly proportional to the instilled dose of virus. After instillation of a hCFTR-adenovirus in cotton rat lungs, expression of human CF could be detected in these lungs by immunocytochemistry for 14 days.[64] Moreover, expression of human CFTR mRNA could be detected in these lungs by PCR for periods up to 6 weeks. This demonstrated the feasibility of an adenoviral gene transduction technique for the treatment of cystic fibrosis. Based on these data, a clinical trial has been proposed. Currently, a dose escalation study is underway to determine tolerance for instillation of large amounts of adenovirus. As discussed above for liposome-mediated transfection of lung endothelial cells, the turnover of these cells probably requires regular introduction of the CF gene. Clinical trials should also give some information on whether repeated instillation of high doses of adenoviral particles will elicit a strong immune response.

Direct intravenous injection of recombinant adenoviruses results in the transduction of cells in many different tissues in addition to the lung, such as liver, intestine, and cardiac and skeletal muscles.[65] Long-term expression (12 months) of β-galactosidase was found in cardiac and skeletal muscles of mice injected as neonates. Analogous to the status of plasmid DNA, adenoviral DNA has been shown to remain extrachromosomally in these tissues. The ability of muscle tissues to maintain episomal DNA well might be an important factor in the sustained expression of reporter genes after adenoviral transduction. Direct intramuscular injection results in highly efficient transduction (up to 50% of the myofibers), but in a limited area surrounding the site of injection.[65,66] Production of a liver enzyme, ornithine transcarbamylase (OTC), has been demonstrated in tissues from deficient mice transduced with an OTC-adenoviral vector.[67] The recombinant virus was administered intravenously into neonates and resulted in a reversal of the disease phenotype. OTC mRNA was found 15 months postinjection, thus showing that in this model, liver is also capable of sustaining expression for long periods. High-efficiency transduction of liver cells has been achieved by direct injection of adenoviral vectors in the portal vein.[68,69] Expression of β-galactosidase could be detected in 90% of the parenchymal cells after delivery of 2×10^9 *lac-Z*-adenoviruses.[69] The same amount of a low-density lipoprotein receptor-expressing adenovirus resulted in expression of this LDL receptor in the liver and a ten-fold increased rate of LDL clearance from the circulation of the injected mice.

Recently, it has been shown that adenoviral vectors can express foreign genes in the central nervous system. A discussion of gene transfer methods for introduction of genes into the brain can be found in Chapter 9. Short-term expression has also been demonstrated in vascular endothelial cells. High expression of β-galactosidase or α1-antitrypsin was measured in the first 2 weeks after infection yet was no longer detectable after 4 weeks.[70] A more in-depth presentation of transduction into endothelial cells can be found in Chapter 7.

C. Herpesviruses

Several groups have been developing herpes simplex virus-derived vectors. Herpesviruses have a natural tropism for neurons but can also infect other postmitotic cells.

Table 1

Examples of Gene Therapy Approaches by Direct Gene Transfer *In Vivo*

Target Tissue	Gene (Defect)	Method	Reference(s)
Muscle	Dystrophin (Duchenne)	Naked DNA injection	21
Liver	LDL receptor	Adenovirus	69
	(hypercholesterolemia)	Polylysine-complexed DNA	45
	α1-Antitrypsin	Adenovirus	68
	(α1-Antitrypsin deficiency)	Retrovirus	58
	Ornithine transcarbamylase (hyperammonemia)	Adenovirus	67
	Albumin (analbuminemia)	Polylysine-complexed DNA	44
Liver → Systemic	Hepatitis B surface antigen	Liposomes	23
	(vaccination)	Herpesvirus	73
	Factor IX (hemophilia)	Herpesvirus	73
	Growth hormone	Retrovirus	56
Lung	CF (cystic fibrosis)	Adenovirus	64
		Liposome-complexed DNA	34
	α1-Antitrypsin	Adenovirus	62
Tumor	Thymidine kinase (glioma)	Retrovirus	59, 60, 74
	Thymidine kinase (liver)	Retrovirus	75
	HLA-B7 (melanoma, tumor vaccination)	Liposomes	76

Recombinant herpes viruses can be grown to high titers and can infect nondividing cells effectively. Experience with *in vivo* application of recombinant herpesviruses is so far limited to the brain[71,72] and the liver.[73] Problems with gene regulation, wild-type virus contamination and vector construction make these viruses currently more difficult to work with than the other viral systems discussed here. It remains unclear whether herpesvirus vectors can express relatively high levels of foreign genes over the long term (at least several months). Still, given their properties, herpesvirus vectors could be important gene therapy vehicles for neuronal cells.

V. SUMMARY

Effective *in vivo* gene transfer leading to high transient expression levels can be accomplished by several of the methods discussed above. For many purposes, this transient solution is sufficient and safe. The actual choice of the gene transfer method is a matter of the intended host tissue, required transfection efficiency, and expression level. A list of *in vivo* gene therapy approaches that were demonstrated to work in animal models is given in Table 1. Sustained expression is still much more difficult to achieve.

Most tissue types do not have the capability to maintain episomal plasmid DNA. Striated muscle does, and therefore a simple technique like direct injection works well, be it with a relatively low transfection efficiency. Being the best investigated viral system so far, retroviruses certainly have specific applications. For other viral vectors, the current problems with transduction efficiency, longevity of expression, toxicity, and general safety issues need to be addressed first. An additional advantage of nonviral transfer of DNA is the lack of cotransfer of viral sequences and proteins. The use of naked DNA is advantageous, because its administration does not elicit any immune responses. With these techniques, there is no need to grow virus, which can be complicated due to the quantity and quality needed for clinical applications.

Injections of complexed DNA, using specific ligands to target the DNA, appear to be the future direction in direct *in vivo* gene therapy research. Yet, an increase in transfection efficiency needs to be obtained. Inclusion of proteins preventing endosomal breakdown of DNA seems very promising in this respect. As is the case for the other introduction methods, sustained expression after receptor-mediated gene delivery depends on the ability of a cell to maintain extrachromosomal plasmid DNA or integrate DNA into the cellular genome. Achieving this in a controlled way will be the next challenge in gene therapy research.

REFERENCES

1. Wolff, J. A., Malone, R. W., Williams, P., Chong, W., Acsadi, G., Jani, A., and Felgner, P. L., Direct gene transfer into mouse muscle in vivo, *Science*, 247, 1456, 1990.
2. Wolff, J. A., Ludtke, J. J., Acsadi, G., Williams, P., and Jani, A., Long-term persistence of plasmid DNA and foreign gene expression in mouse muscle, *Hum. Mol. Genet.*, 1, 363, 1992.
3. Acsadi, G., Jiao, S. S., Jani, A., Duke, D., Williams, P., Chong, W., and Wolff, J. A., Direct gene transfer and expression into rat heart in vivo, *New Biol.*, 3, 71, 1991.
4. Lin, H., Parmacek, M. S., Morle, G., Bolling, S., and Leiden, J. M., Expression of recombinant genes in myocardium in vivo after direct injection of DNA, *Circulation*, 82, 2217, 1990.
5. Wolff, J. A., Dowty, M. E., Jiao, S., Repetto, G., Berg, R. K., Ludtke, J. J., Williams, P., and Slautterback, D. B., Expression of naked plasmids by cultured myotubes and entry of plasmids into T tubules and caveolae of mammalian skeletal muscle, *J. Cell Sci.*, 103, 1249, 1992.
6. Jiao, S., Williams, P., Berg, R. K., Hodgeman, B. A., Liu, L., Repetto, G., and Wolff, J. A., Direct gene transfer into nonhuman primate myofibers in vivo, *Hum. Gene Ther.*, 3, 21, 1992.
7. Davis, H. L., Whalen, R. G., and Demeneix, B. A., Direct gene transfer into skeletal muscle in vivo: factors affecting efficiency of transfer and stability of expression, *Hum. Gene Ther.*, 4, 151, 1993.
8. Wells, D. J. and Goldspink, G., Age and sex influence expression of plasmid DNA directly injected into mouse skeletal muscle, *FEBS Lett.*, 306, 203, 1992.
9. Buttrick, P. M., Kass, A., Kitsis, R. N., Kaplan, M. L., and Leinwand, L. A., Behavior of genes directly injected into the rat heart in vivo, *Circ. Res.*, 70, 193, 1992.
10. Wolff, J. A., Williams, P., Acsadi, G., Jiao, S., Jani, A., and Chong, W., Conditions affecting direct gene transfer into rodent muscle in vivo, *Biotechniques*, 11, 474, 1991.
11. Gal, D., Weir, L., Leclerc, G., Pickering, J. G., Hogan, J., and Isner, J. M., Direct myocardial transfection in two animal models. Evaluation of parameters affecting gene expression and percutaneous gene delivery, *Lab. Invest.*, 68, 18, 1993.

12. Vincent, C. K., Gualberto, A., Patel, C. V., and Walsh, K., Different regulatory sequences control creatine kinase-M gene expression in directly injected skeletal and cardiac muscle, *Mol. Cell. Biol.*, 13, 1264, 1993.
13. Edwards, J. G., Bahl, J. J., Flink, I., Milavetz, J., Goldman, S., and Morkin, E., A repressor region in the human beta-myosin heavy chain gene that has a partial position dependency, *Biochem. Biophys. Res. Commun.*, 189, 504, 1992.
14. Kitsis, R. N., Buttrick, P. M., McNally, E. M., Kaplan, M. L., and Leinwand, L. A., Hormonal modulation of a gene injected into rat heart in vivo, *Proc. Natl. Acad. Sci. U.S.A.*, 88, 4138, 1991.
15. Ojamaa, K. and Klein, I., Thyroid hormone regulation of alpha-myosin heavy chain promoter activity assessed by in vivo DNA transfer in rat heart, *Biochem. Biophys. Res. Commun.*, 179, 1269, 1991.
16. Vile, R. G. and Hart, I. R., In vitro and in vivo targeting of gene expression to melanoma cells, *Cancer Res.*, 53, 962, 1993.
17. Yang, N.-S., Burkholder, J., Roberts, B., Martinell, B., and McCabe, D., In vivo and in vitro gene transfer to mammalian somatic cells by particle bombardment, *Proc. Natl. Acad. Sci. U.S.A.*, 87, 9568, 1990.
18. Zelenin, A. V., Alimov, A. A., Titomirov, A. V., Kazansky, A. V., Gorodetsky, S. I., and Kolesnikov, V. A., High-velocity mechanical DNA transfer of the chloramphenicolacetyl transferase gene into rodent liver, kidney and mammary gland cells in organ explants and in vivo, *FEBS Lett.*, 280, 94, 1991.
19. Yang, N. S., Gene transfer into mammalian somatic cells in vivo, *Crit. Rev. Biotechnol.*, 12, 335, 1992.
20. Titomirov, A. V., Sukharev, S., and Kistanova, E., In vivo electroporation and stable transformation of skin cells of newborn mice by plasmid DNA, *Biochim. Biophys. Acta*, 1088, 131, 1991.
21. Acsadi, G., Dickson, G., Love, D. R., Jani, A., Walsh, F. S., Gurusinghe, A., Wolff, J. A., and Davies, K. E., Human dystrophin expression in mdx mice after intramuscular injection of DNA constructs, *Nature*, 352, 815, 1991.
22. Kaneda, Y., Iwai, K., and Uchida, T., Increased expression of DNA cointroduced with nuclear protein in adult rat liver, *Science*, 243, 375, 1989.
23. Kato, K., Nakanishi, M., Kaneda, Y., Uchida, T., and Okada, Y., Expression of hepatitis B virus surface antigen in adult rat liver. Co-introduction of DNA and nuclear protein by a simplified liposome method, *J. Biol. Chem.*, 266, 3361, 1991.
24. Kato, K., Kaneda, Y., Sakurai, M., Nakanishi, M., and Okada, Y., Direct injection of hepatitis B virus DNA into liver induced hepatitis in adult rats, *J. Biol. Chem.*, 266, 22071, 1991.
25. Leibiger, B., Leibiger, I., Sarrach, D., and Zühlke, H., Expression of exogenous DNA in rat liver cells after liposome mediated transfection in vivo, *Biochem. Biophys. Res. Commun.*, 174, 1223, 1991.
26. Nabel, E. G., Plautz, G., and Nabel, G. J., Site-specific gene expression in vivo by direct gene transfer into the arterial wall, *Science*, 249, 1285, 1990.
27. Lim, C. S., Chapman, G. D., Gammon, R. S., Muhlestein, J. B., Bauman, R. P., Stack, R. S., and Swain, J. L., Direct in vivo gene transfer into the coronary and peripheral vasculatures of the intact dog, *Circulation*, 83, 2007, 1991.
28. Nabel, E. G., Plautz, G., and Nabel, G. J., Transduction of a foreign histocompatibility gene into the arterial wall induces vasculitis, *Proc. Natl. Acad. Sci. U.S.A.*, 89, 5157, 1992.
29. Stewart, M. J., Plautz, G. E., Del, B. L., Yang, Z. Y., Xu, L., Gao, X., Huang, L., Nabel, E. G., and Nabel, G. J., Gene transfer in vivo with DNA-liposome complexes: safety and acute toxicity in mice, *Hum. Gene Ther.*, 3, 267, 1992.

30. Barbee, R. W., Stapleton, D. D., Perry, B. D., Ré, R. N., Murgo, J. P., Valentino, V. A., and Cook, J. L., Prior arterial injury enhances luciferase expression following in vivo gene transfer, *Biochem. Biophys. Res. Commun.*, 190, 70, 1993.
31. Nabel, E. G., Yang, Z., Liptay, S., San, H., Gordon, D., Haudenschild, C. C., and Nabel, G. J., Recombinant platelet-derived growth factor B gene expression in porcine arteries induce intimal hyperplasia in vivo, *J. Clin. Invest.*, 91, 1822, 1993.
32. Nabel, E. G., Yang, Z. Y., Plautz, G., Forough, R., Zhan, X., Haudenschild, C. C., Maciag, T., and Nabel, G. J., Recombinant fibroblast growth factor-1 promotes intimal hyperplasia and angiogenesis in arteries in vivo, *Nature*, 362, 844, 1993.
33. Brigham, K. L., Meyrick, B., Christman, B, Magnuson, M., King, G., and Berry, L. C., *In vivo* transfection of murine lungs with a functioning prokaryotic gene using a liposome vehicle, *Am. J. Med. Sci.*, 298, 278, 1989.
34. Yoshimura, K., Rosenfeld, M. A., Nakamura, H., Scherer, E. M., Pavirani, A., Lecocq, J. P., and Crystal, R. G., Expression of the human cystic fibrosis transmembrane conductance regulator gene in the mouse lung after in vivo intratracheal plasmid-mediated gene transfer, *Nucl. Acids Res.*, 20, 3233, 1992.
35. Bout, A., Valerio, D., and Scholte, B. J., In vivo transfer and expression of the lacZ gene in the mouse lung, *Exp. Lung Res.*, 19, 193, 1993.
36. Kaneda, Y., Iwai, K., and Uchida, T., Introduction and expression of the human insulin gene in adult rat liver, *J. Biol. Chem.*, 264, 12126, 1989.
37. Tomita, N., Higaki, J., Morishita, R., Kato, K., Mikami, H., Kaneda, Y., and Ogihara, T., Direct in vivo gene introduction into rat kidney, *Biochem. Biophys. Res. Commun.*, 186, 129, 1992.
38. Zhu, N., Liggitt, D., Liu, Y., and Debs, R., Systemic gene expression after intravenous DNA delivery into adult mice, *Science*, 261, 209, 1993.
39. Philip, R., Liggitt, D., Philip, M., Dazin, P., and Debs, R., *In vivo* gene delivery: efficient transfection of T lymphocytes in adult mice, *J. Biol. Chem.*, 268, 16087, 1993.
40. Sahenk, Z., Seharaseyon, J., Mendell, J. R., and Burghes, A. H., Gene delivery to spinal motor neurons, *Brain Res.*, 606, 126, 1993.
41. Wu, G. Y. and Wu, C. H., Receptor-mediated gene delivery and expression in vivo, *J. Biol. Chem.*, 263, 14621, 1988.
42. Wu, G. Y. and Wu, C. H., Evidence for targeted gene delivery to Hep G2 hepatoma cells in vitro, *Biochemistry*, 27, 887, 1988.
43. Wu, C. H., Wilson, J. M., and Wu, G. Y., Targeting genes: delivery and persistent expression of a foreign gene driven by mammalian regulatory elements in vivo, *J. Biol. Chem.*, 264, 16985, 1989.
44. Wu, G. Y., Wilson, J. M., Shalaby, F., Grossman, M., Shafritz, D. A., and Wu, C. H., Receptor-mediated gene delivery in vivo. Partial correction of genetic analbuminemia in Nagase rats, *J. Biol. Chem.*, 266, 14338, 1991.
45. Wilson, J. M., Grossman, M., Wu, C. H., Chowdhury, N. R., Wu, G. Y., and Chowdhury, J. R., Hepatocyte-directed gene transfer in vivo leads to transient improvement of hypercholesterolemia in low density lipoprotein receptor-deficient rabbits, *J. Biol. Chem.*, 267, 963, 1992.
46. Wilson, J. M., Grossman, M., Cabrera, J. A., Wu, C. H., and Wu, G. Y., A novel mechanism for achieving transgene persistence in vivo after somatic gene transfer into hepatocytes, *J. Biol. Chem.*, 267, 11483, 1992.
47. Chowdhury, N. R., Wu, C. H., Wu, G. Y., Yerneni, P. C., Bommineni, V. R., and Chowdhury, J. R., Fate of DNA targeted to the liver by asialoglycoprotein receptor-mediated endocytosis in vivo. Prolonged persistence in cytoplasmic vesicles after partial hepatectomy, *J. Biol. Chem.*, 268, 11265, 1993.

48. Curiel, D. T., Agarwal, S., Wagner, E., and Cotten, M., Adenovirus enhancement of transferrin-polylysine-mediated gene delivery, *Proc. Natl. Acad. Sci. U.S.A.*, 88, 8850, 1991.
49. Curiel, D. T., Agarwal, S., Romer, M. U., Wagner, E., Cotten, M., Birnstiel, M. L., and Boucher, R. C., Gene transfer to respiratory epithelial cells via the receptor-mediated endocytosis pathway, *Am. J. Resp. Cell Mol. Biol.*, 6, 247, 1992.
50. Curiel, D. T., Wagner, E., Cotten, M., Birnstiel, M. L., Agarwal, S., Li, C. M., Loechel, S., and Hu, P. C., High-efficiency gene transfer mediated by adenovirus coupled to DNA-polylysine complexes, *Hum. Gene Ther.*, 3, 147, 1992.
51. Wagner, E., Zatloukal, K., Cotten, M., Kirlappos, H., Mechtler, K., Curiel, D. T., and Birnstiel, M. L., Coupling of adenovirus to transferrin-polylysine/DNA complexes greatly enhances receptor-mediated gene delivery and expression of transfected genes, *Proc. Natl. Acad. Sci. U.S.A.*, 89, 6099, 1992.
52. Gao, L., Wagner, E., Cotten, M., Agarwal, S., Harris, C., Romer, M., Miller, L., Hu, P. C., and Curiel, D., Direct in vivo gene transfer to airway epithelium employing adenovirus-polylysine-DNA complexes, *Hum. Gene Ther.*, 4, 17, 1993.
53. Curiel, D. T., Adenovirus facilitation of molecular conjugate-mediated gene transfer, in *Progress in Medical Virology*, Vol. 40, Melnick, J. L., Ed., S. Karger, Basel, 1993, 1.
54. Michael, S. I., Huang, C. H., Romer, M. U., Wagner, E., Hu, P. C., and Curiel, D. T., Binding-incompetent adenovirus facilitates molecular conjugate-mediated gene transfer by the receptor-mediated endocytosis pathway, *J. Biol. Chem.*, 268, 6866, 1993.
55. Midoux, P., Mendes, C., Legrand, A., Raimond, J., Mayer, R., Monsigny, M., and Roche, A. C., Specific gene transfer mediated by lactosylated poly-L-lysine into hepatoma cells, *Nucl. Acids Res.*, 21, 871, 1993.
56. Hatzoglou, M., Lamers, W., Bosch, F., Wynshaw-Boris, A., Clapp, D. W., and Hanson, R. W., Hepatic gene transfer in animals using retroviruses containing the promoter from the gene for phosphoenolpyruvate carboxykinase, *J. Biol. Chem.*, 265, 17285, 1990.
57. Ferry, N., Duplessis, O., Houssin, D., Danos, O., and Heard, J. M., Retroviral-mediated gene transfer into hepatocytes in vivo, *Proc. Natl. Acad. Sci. U.S.A.*, 88, 8377, 1991.
58. Kay, M. A., Li, Q., Liu, T. J., Leland, F., Toman, C., Finegold, M., and Woo, S. L., Hepatic gene therapy: persistent expression of human alpha 1-antitrypsin in mice after direct gene delivery in vivo, *Hum. Gene Ther.*, 3, 641, 1992.
59. Ram, Z., Culver, K. W., Walbridge, S., Blaese, R. M., and Oldfield, E. H., In situ retroviral-mediated gene transfer for the treatment of brain tumors in rats, *Cancer Res.*, 53, 83, 1993.
60. Oldfield, E. H., Ram, Z., Culver, K. W., Blaese, R. M., DeVroom, H. L., and Anderson, W. F., Gene therapy for the treatment of brain tumors using intra-tumoral transduction with the thymidine kinase gene and intravenous ganciclovir, *Hum. Gene Ther.*, 4, 39, 1993.
61. Jones, N. and Shenk, T., Isolation of adenovirus type 5 host range deletion mutants defective for transformation of rat embryo cells, *Cell*, 17, 683, 1979.
62. Rosenfeld, M. A., Siegfried, W., Yoshimura, K., Yoneyama, K., Fukayama, M., Stier, L. E., Pääkkö, P. K., Gilardi, P., Stratford-Perricaudet, L., Perricaudet, M., Jallat, S., Pavirani, A., Lecocq, J.-P., and Crystal, R. G., Adenovirus-mediated transfer of a recombinant α1-antitrypsin gene to the lung epithelium in vivo, *Science*, 252, 431, 1991.
63. Mastrangeli, A., Danel, C., Rosenfeld, M. A., Stratford-Perricaudet, L. D., Perricaudet, M., Pavirani, A., Lecocq, J. P., and Crystal, R. G., Diversity of airway epithelial cell targets for in vivo recombinant adenovirus-mediated gene transfer, *J. Clin. Invest.*, 91, 225, 1993.
64. Rosenfeld, M. A., Yoshimura, K., Trapnell, B. C., Yoneyama, K., Rosenthal, E. R., Dalemans, W., Fukayama, M., Bargon, J., Stier, L. E., Stratford-Perricaudet, L. D., Perricaudet, M., Guggino, W. B., Pavirani, A., Lecocq, J. P., and Crystal, R. G., In vivo transfer of the human cystic fibrosis transmembrane conductance regulator gene to the airway epithelium, *Cell*, 68, 143, 1992.

65. Stratford-Perricaudet, L. D., Makeh, I., Perricaudet, M., and Briand, P., Widespread long-term gene transfer to mouse skeletal muscles and heart, *J. Clin. Invest.*, 90, 626, 1992.
66. Quantin, B., Perricaudet, L. D., Tajbakhsh, S., and Mandel, J.-L., Adenovirus as an expression vector in muscle cells *in vivo*, *Proc. Natl. Acad. Sci. U.S.A.*, 89, 2581, 1992.
67. Stratford-Perricaudet, L. D., Levrero, M., Chasse, J.-F., Perricaudet, M., and Briand, P., Evaluation of the transfer and expression in mice of an enzyme-encoding gene using a human adenovirus vector, *Hum. Gene Ther.*, 1, 241, 1990.
68. Jaffe, H. A., Danel, C., Longenecker, G., Metzger, M., Setoguchi, Y., Rosenfeld, M. A., Gant, T. W., Thorgeirsson, S. S., Stratford-Perricaudet, L. D., Perricaudet, M., Pavirane, A., Lecocq, J.-P., and Crystal, R. G., Adenovirus-mediated in vivo gene transfer and expression in normal rat liver, *Nat. Genet.*, 1, 372, 1992.
69. Herz, J. and Gerard, R. D., Adenovirus-mediated transfer of low density lipoprotein receptor gene acutely accelerates cholesterol clearance in normal mice, *Proc. Natl. Acad. Sci. U.S.A.*, 90, 2812, 1993.
70. Lemarchand, P., Jones, M., Yamada, I., and Crystal, R. G., In vivo gene transfer and expression in normal uninjured blood vessels using replication-deficient recombinant adenovirus vectors, *Circ. Res.*, 72, 1132, 1993.
71. Federoff, H. J., Geschwind, M. D., Geller, A. I., and Kessler, J. A., Expression of nerve growth factor *in vivo* from a defective herpes simplex virus 1 vector prevents effects of axotomy on sympathetic ganglia, *Proc. Natl. Acad. Sci. U.S.A.*, 89, 1636, 1992.
72. Kaplitt, M. G., Pfaus, J. G., Kleopoulos, S. P., Hanlon, B. A., Rabkin, S. D., and Pfaff, D. W., Expression of a functional foreign gene in adult mammalian brain following in vivo transfer via a herpes simplex virus type 1 defective viral vector, *Mol. Cell. Neurosci.*, 2, 320, 1991.
73. Miyanohara, A., Johnson, P. A., Elam, R. L., Dai, Y., Witztum, J. L., Verma, I. M., and Friedmann, T., Direct gene transfer to the liver with herpes simplex virus type 1 vectors: transient production of physiologically relevant levels of circulating factor IX, *New Biol.*, 4, 238, 1992.
74. Culver, K. W., Ram, Z., Wallbridge, S., Ishii, H., Oldfield, E. H., and Blaese, R. M., In vivo gene transfer with retroviral vector-producer cells for treatment of experimental brain tumors, *Science*, 256, 1550, 1992.
75. Caruso, M., Panis, Y., Gagandeep, S., Houssin, D., Salzmann, J.-L., and Klatzmann, D., Regression of established macroscopic liver metastases after *in situ* transduction of a suicide gene, *Proc. Natl. Acad. Sci. U.S.A.*, 90, 7024, 1993.
76. Nabel, G. J., Chang, A., Nabel, E. G., Plautz, G., Fox, B. A., Huang, L., and Shu, S., Immunotherapy of malignancy by in vivo gene transfer into tumors, *Hum. Gene Ther.*, 3, 399, 1992.

12 Nonautologous Somatic Gene Therapy

Patricia L. Chang

I. INTRODUCTION

Since the approval of the many protocols in human gene therapy,[1] a new chapter in medicine has begun in which genes are the key elements for therapy.[2-4] The ultimate goal is to replace a defective or missing gene with the normal copy in the patient's own tissues where the genes are normally expressed. Although this goal has not been reached with current technologies, rapid progress in research within the next decade will likely achieve most, if not all, of the ideal tenets of human gene therapy as envisioned by Anderson.[5] A rather unexpected direction that somatic gene therapy has taken is its application to the treatment of "nongenetic" diseases. These are the multifactorial cardiovascular,[6] neoplastic,[7] and infectious disorders[8] in which genetics only accounts for some of the etiologic factors. In these applications, genes are integrated into the patient's own cells to produce products that ameliorate the disease phenotype. In other words, genes are used as drugs. The therapeutic outcome is achieved by the gene products delivered *in vivo*. Hence, the basis for somatic gene therapy has expanded to include both genetic and multifactorial disorders that can be treated by the delivery of recombinant gene products from genetically-modified host cells.

As with the development of most new technologies, the cost will be high.[9] Unless conscious efforts are made, such costs may not necessarily come down with time, as witness the high costs of recombinant growth hormone for growth enhancement[10] or glucocerebrosidase for Gaucher's disease.[11] Hence, in contrast to current strategies of gene therapy that depend on genetically-modifying autologous cells from the patients, the concept of developing universal recombinant cell lines suitable for treatment of different patients is economically attractive.[12] It is envisioned that once a cell line has been engineered to provide the gene product of interest, it can be kept in frozen storage until needed. At the time of use, the frozen stock can be thawed and propagated, tested for safety and gene expression levels, and then implanted with a simple injection. To avoid immune-rejection, these nonautologous cells can be protected with implantable immuno-isolation devices. Since this form of gene therapy does not require genetic modification of the patient's own cells, it is more appropriately described as *somatic gene therapeutics*, conceptually similar to the provision of a "drug" in the form of

0-8493-2440-8/95/$0.00+$.50

genetically-engineered cells. The therapy obviates the need and hence the cost for patient-specific genetic manipulation. In addition, this strategy allows for verification of the implanted biomaterial to meet industrial standards of safety and other regulations, features that are difficult to comply with if the limited amount of patient-specific cells are the targets of genetic modification and implantation.

II. HISTORICAL REVIEW OF NONAUTOLOGUS TISSUE IMPLANTS

A. Immuno-Isolation

Non-autologous tissue and organ transplants have played an important role in medicine. Although it is a clinically effective treatment for many disorders, life-long immune suppression for the recipients carries many long-term and undesirable effects.[13] Hence, in the past 2 decades, much effort has been devoted to the development of immuno-isolation devices that allow the implantation of allogeneic or xenogeneic tissues without the need for immune-suppressing the recipients. There are three basic designs for such devices: diffusion chamber, vascular shunt, and micro- or macrocapsules.[14] Whether they are intravascular or extravascular devices, they need to fulfill three critical requirements: biocompatibility, permeability to diffusion of oxygen, nutrients, and therapeutic products, and prevention of immune rejection.

Preventing cell-mediated immune rejection is an easily fulfilled requirement for most implantable devices. If the membrane pore size is below 0.1 μm, it will prevent monocyte diapedesis and lymphocyte infiltration.[15] For preventing humoral immune responses, the problem is more complex. Antibody response can occur both within and outside the device enclosing the implanted tissue. Throughout the development of immuno-isolation devices, exclusion of IgG (≈160,000 Da) has been a critical requirement[16] that was thought to be the crucial feature that confers immune protection. If the implantable device has a molecular weight cut-off above the size of IgG, there will be immune reaction from antibodies gaining access and binding to cell surface antigen of the tissue within the implanted device. However, subsequent cell killing requires the binding of a single IgM (mol. wt. 970,000) or two IgG in close proximity. This triggers the C1q complement (mol. wt. 410,000) to initiate the complement cascade, forming the membrane attack complex that causes cell lysis.[17] Therefore, even if the permeability threshold is above 160,000 Da, cell lysis due to the humoral response may not necessarily occur if the IgM or C1q complement has been excluded. On the other hand, even for membranes whose molecular weight cut-off can be kept below the size of IgG, a second type of humoral response can still take place. Chronic antigen leakage from cell secretions or debris could stimulate both antibody response and local cell-mediated immune response. Although the newly synthesized antibody may have no direct effect on the tissue within the implant, cytokines such as IL-1 would be secreted by macrophages in the vicinity of the immuno-isolation chamber. Alone or in concert with other cytokines, IL-1 (mol. wt. 17,500) has cytotoxic effects.[18,19] These cytotoxic lymphokines with their sizes in the 10 to 20 kDa range will be difficult to exclude without impeding diffusion of nutrients or desired products even as small as insulin (mol. wt. 5800). Hence, immuno-protection against humoral response based on exclusion of IgG molecules alone is probably an over-simplified concept.

B. Systemic Delivery

In spite of such a theoretical impasse and imperfect protection against possible immune reactions of the host, various degrees of success have been achieved with the different implantable permselective devices. Most studies relied on the normoglycemic response from diabetic rodents, dogs, and monkeys as the indicator for success by implanting allogeneic or xenogeneic islet cells. The devices used included diffusion chambers,[20,21] hollow-fibers,[22] arteriovenous shunts,[23] and microcapsules fabricated from water-soluble hydrogels[24] or water-insoluble thermoplastics.[25] The development of a hybrid artificial pancreas using immuno-isolation devices has been reviewed in detail by Colton and Avgoustiniatos.[15] Depending on the device, the tissue, and the animal models used, normoglycemia could be maintained with these artificial pancreases from days to even more than a year.[26] However, mechanical and biologic problems could occur, such as rupture of sutures, thrombus formation, and mechanical fragility of the capsule.[26-28]

Of all the permselective immune isolation vehicles, microcapsules (<100 μm in diameter) or macrocapsules (>100 μm in diameter) have had the most success in recent years. They can be fabricated from either a seaweed extract, alginate,[29] or acrylate polymers.[30,31] Alginate is a polysaccharide extracted from the brown algae and composed of mixed blocks of 1,4-linked β-D-mannuronic and α-L-guluronic acid in various proportions.[32] The biocompatibility of microcapsules fabricated from alginate was much improved over the years, primarily through the efforts of Sun and co-workers.[33,34] Encapsulation of allogeneic and xenogeneic islet cells with the alginate type of microcapsules has repeatedly demonstrated the success of this immuno-isolation device in maintaining normoglycemia in rodents for various periods, some even up to more than 2 years.[24,33,35-37] Other cell types, such as hepatocytes,[38-40] parathyroid cells,[41] and hybridoma cells[42] have also been similarly encapsulated with varying degrees of success.

With the thermoplastic type of capsules,[31,43-45] it was shown that proliferative transformed cells, in addition to terminally differentiated tissues,[46] could also survive within these capsules. An advantage of transplanting transformed cell lines within capsules is that tumor formation is prevented.[47] The containment of the tumorogenic cells by the capsule ensures that metastasis does not occur. Even if the membrane of the capsule should rupture, the implanted cells would be eliminated by the host-immune system because of the vast difference in histocompatibility antigens, particularly when xenogeneic implants are used,[48] thus providing a measure of protection from tumorigenesis.

C. CNS Delivery

In addition to peripheral systemic disorders, neurologic diseases have also been targeted for treatment with encapsulated cells, primarily through the pioneering efforts of Aebisher and his co-workers.[47] With the central nervous system, many biologically important molecules are prevented from entry through the systemic circulation because of the blood-brain barrier. Hence, direct implantation into the brain is necessary to deliver the therapeutic products *in situ*. The feasibility of this approach was demonstrated when microencapsulated PC12 cells, a transformed cell line derived from a rat pheochromocytoma, were implanted in the rat brain, they were able to reduce experimentally induced Parkinson's disease due to the release of dopamine from the implanted cells.[49] This is a particularly promising development in new therapies, since neural transplantation of autologous adrenal medullary cells in humans with Parkinson's disease has already shown some therapeutic benefit in controlling the disease symptoms.[50] The high risk associated with the procedure in obtaining autologous tissues would thus be considerably reduced if nonautologous sources can substitute by using immuno-protective implants.

Other neurologic functions, such as pain control, have also been explored for possible therapeutic application with the encapsulation technology. Neurotransmitters, such as enkephalins and endorphins, are released by adrenal chromaffin cells. These pain-modulating substances[51] released from bovine chromaffin cells in thermoplastic capsules implanted in the subarachnoid space of the rat were able to reduce the pain threshold in rats.[46] This strategy of pain control is currently under early clinical trials in patients with terminal cancers, and results are pending further confirmation.[52]

D. Summary

Thus, it is clear that implantation of nonautologous tissue has had a long history of development. Most studies *in vivo* have concentrated on the treatment of diabetes, particularly using the alginate type of polyelectrolyte hydrogel as permselective membrane material. Because of concern about the long-term stability and mechanical strength of such polysaccharide microcapsules, diversification into thermoplastic material considered to be mechanically more stable has provided encouraging results in recent years. In addition to the treatment of systemic disease such as diabetes, neurologic diseases have also become important targets for treatment.[47] Novel cell types, such as transformed cell lines[43,53] or primary cells other than islets,[54] were also explored for other therapeutic applications. However, because of the simplicity in fabrication of the alginate type of microcapsules,[34] it has been the standard with which other permselective devices are compared.[55]

III. APPLICATION TO GENE THERAPY

A. In Vitro Studies

1. Fibroblasts

To apply the technology of immuno-isolation to gene therapy as proposed by Chang et al.,[12] several additional requirements must be met. The most fundamental one is that the proliferating recombinant cell line must adapt to survive within the microenvironment of the implantable device and continue to secrete the gene product of choice. Several types of proliferative cells, such as hybridoma,[56] PC12,[49] CHO,[43] and NGF-transfected PC12 cells[57] have been shown to survive within the alginate or thermoplastic-type capsules. In our experience, transfected fibroblasts from a transformed mouse cell line continued to proliferate and express the transgene after encapsulation in the alginate-polylysine-alginate type of microcapsules (Figure 1). In general, transformed cell lines survive better than primary fibroblasts. However, even within the same transformed cell line, different transfected clones may survive and proliferate with varying degrees of proficiency. Furthermore, such differences in their proliferative capacity are characteristic of each clone (Figure 2). Hence, the ability of a cell line to survive within the microcapsule environment appears to be a genetically determined trait.

When the transformed mouse fibroblast (Ltk$^-$) clones were followed for several weeks in culture after encapsulation in alginate-polylysine-alginate membranes, they increased in cell number by two- to five-fold within a month, but the viability of the cells could decrease from about 90 to 50% (Figure 1). However, these values varied among different cell lines and were highly dependent on the conditions used to fabricate the microcapsules. Several conditions that led to enhanced cell viability were: performing the encapsulation process at low temperatures (e.g. 4 to 10°C); expeditious handling of the cells during the encapsulation process so that they could be returned to normal growth media and incubator conditions as soon as possible; coating with the polycation polylysine appropriately so that the membrane was not too thick; liquefying the alginate core of the microcapsules adequately to provide sufficient space for growth; adjusting the concentration of the cell suspension appropriately during the initial encapsulation procedure to provide optimal density for long-term growth.[58] The longest time that the microencapsulated cells have been kept in culture *in vitro* was about 18 weeks, but by that time, the viability had decreased to only 5 to 6%.[60] Hence, judicious choice of the cell line and attention to the process of microcapsule fabrication will lead to increased survival of cell types such as fibroblasts. However, the long-term survival after 4 to 5 months is poor, highlighting the problem of enclosing a proliferative cell line within the finite space of an immuno-isolation device.

2. Myoblasts

A cell type that offers a potential solution to the problem of limited space for growth is the myoblasts. The mouse myoblast C2C12, a spontaneously transformed cell line,[61] normally grows as adherent proliferative cells in culture. Under special culture condition, such as using 10% horse serum instead of fetal calf serum,[62] these cells are induced to fuse into multinucleated myotubes that exist in a terminally differentiated state. Therefore, a possible strategy is to transfect such a cell line in its myoblast proliferative state to allow for transgene integration, and after selecting for clones that are expressing the desirable transgene at a satisfactory level, the recombinant myoblasts can be encapsulated and induced to differentiate into myotubes that no longer proliferate, thereby solving the problem of space limitation within the implantable device. Our preliminary work showed that myoblasts can be encapsulated similarly as the fibroblasts and that they can differentiate into myotubes after encapsulation.[63] Even after implantation into

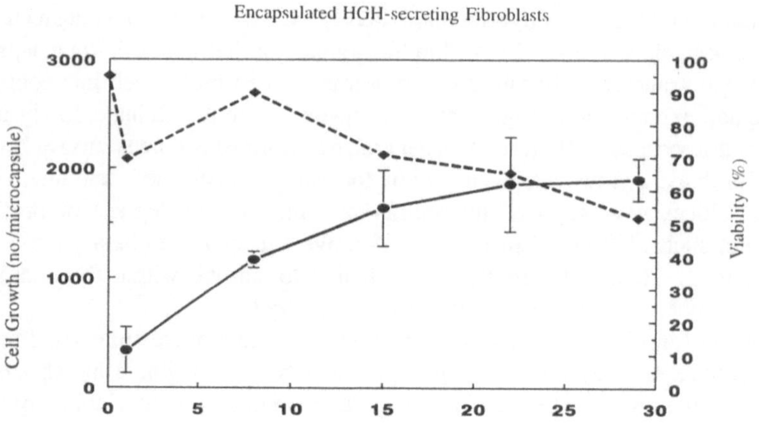

FIGURE 1A

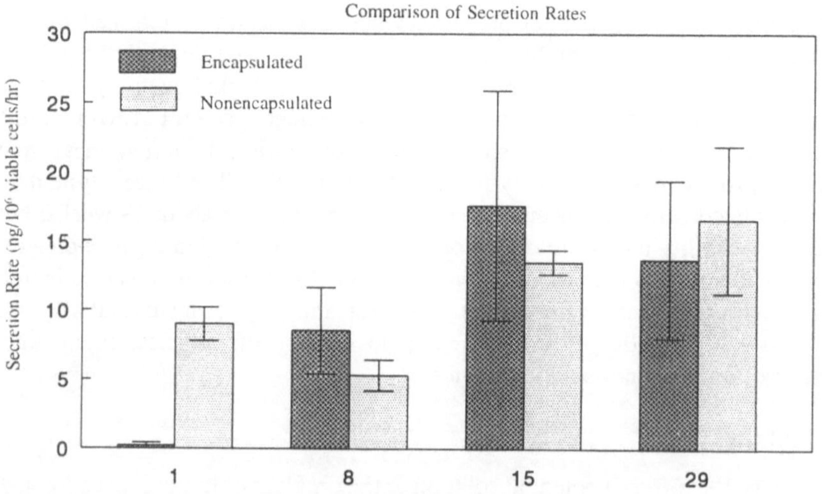

FIGURE 1B

FIGURE 1A–C. Growth of transfected mouse fibroblasts and secretion of recombinant gene product from alginate-polylysine-alginate encapsulated cells. Mouse Ltk⁻ fibroblasts transfected with the *human growth hormone* (HGH) gene were encapsulated at a concentration of 2×10^6 cells/ml potassium alginate. At various days postencapsulation, aliquots of the microcapsules were removed to monitor for (A) cell viability, cell number, (B) HGH secretion rate, and (C) appearance (scale bar = 1000 mm). The proliferation of fibroblasts within the microcapsules with increasing time in culture was evident. The microcapsules were freely permeable to the diffusion of HGH except on day 1, when the encapsulated cells were recovering from the microcapsule fabrication procedure. (From Reference 58.)

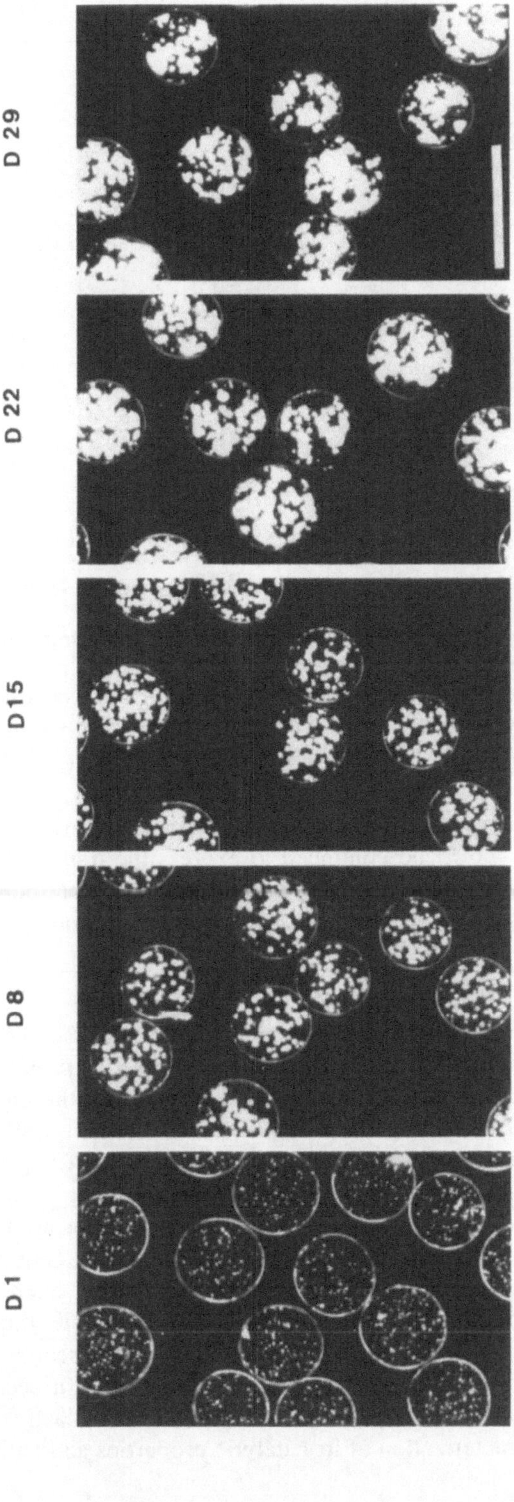

FIGURE 1C

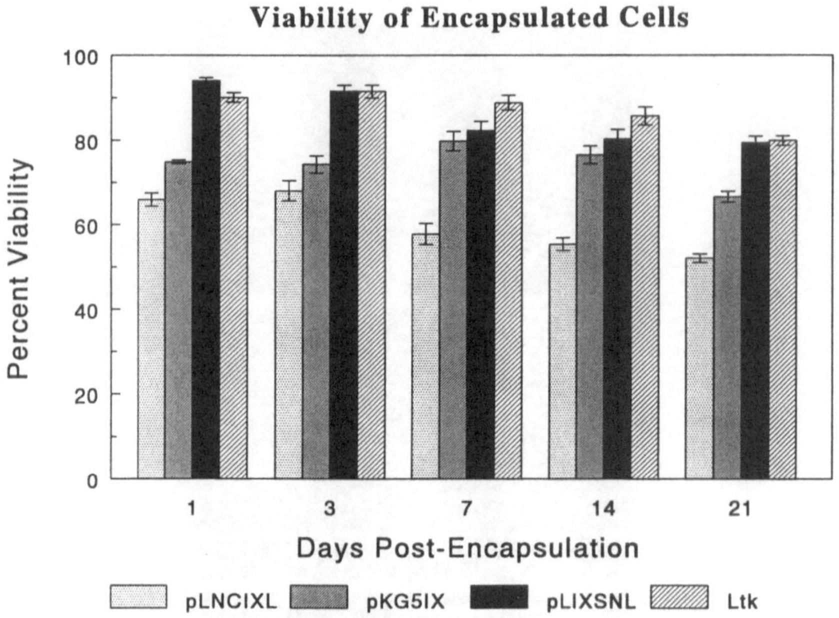

FIGURE 2. Growth characteristics of cell clones isolated from transfected mouse fibroblasts. Clones of mouse fibroblasts transfected with the plasmids pLNCIXL, pKG5IX, and pLIXSNL and parental nontransfected Ltk⁻ cells were isolated and encapsulated in alginate-polylysine-alginate. The viability varied among the different encapsulated clones and appeared to be a genetically-determined trait. (From Reference 59, with permission.)

the mouse peritoneal cavity for 6 months, these encapsulated myoblast-myotubes still retained a viability of ≈60% and continued to express the transgene.[64] Hence, using proliferative cell lines that are inducible to terminal differentiation may solve the problem of long-term survival of recombinant cells within immuno-isolation devices.

3. Secretory Signal

Another primary requirement for recombinant gene products to be delivered with this microcapsule technology is that it must be a secretory gene product, a requirement not always met by the gene of interest. A good example is the enzyme adenosine deaminase that causes one form of severe combined immune deficiency, the first genetic disorder to be treated with somatic gene therapy (see Chapter 3). It is a ≈40 kDa enzyme normally localized in the cytosolic compartment. In order to render it suitable for delivery with nonautologous cell implants, we engineered a 23-amino acid peptide encoding the signal sequence of the bacterial β-lactamase[65] to the amino terminus of adenosine deaminase.[66] Such a signal peptide has been shown to direct nascent polypeptides to dock with the signal recognition particle, thus permitting the fusion protein to traverse the endoplasmic reticulum into the secretory pathway.[67] It was shown that as a result of such molecular engineering, a secretable form of adenosine deaminase was expressed from both mouse fibroblasts and myoblasts.[66] Furthermore, upon characterization of its catalytic properties and antibody reactivity,

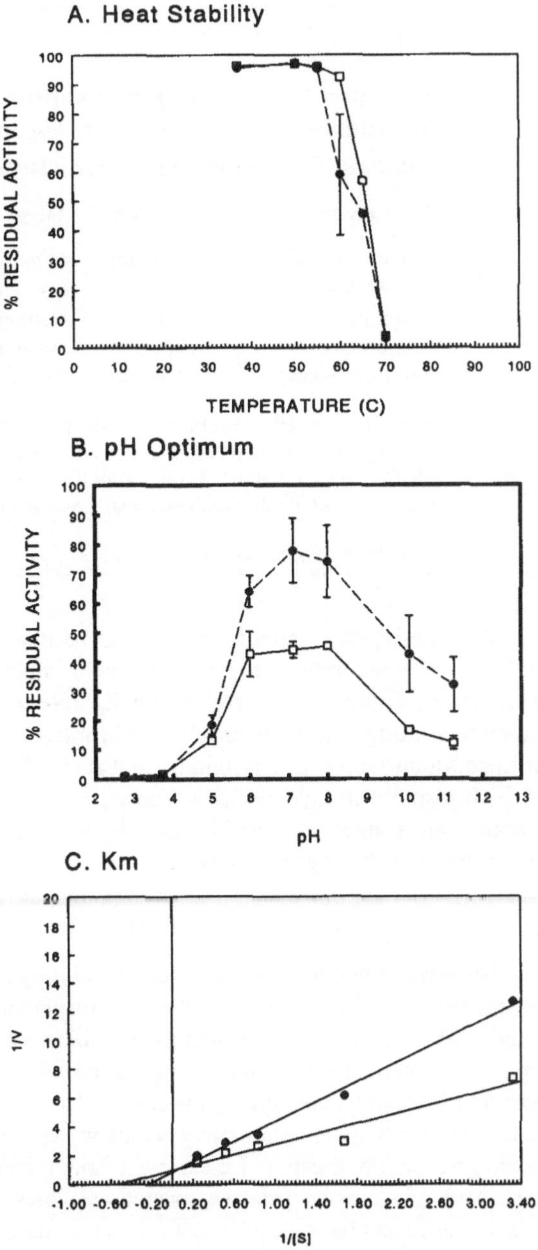

FIGURE 3. Enzymatic characterization of a secreted form of human adenosine deaminase after genetic modification with a signal sequence. The cDNA of the signal sequence from β-lactamase was fused to the 5′ end of the cDNA for human adenosine deaminase, an enzyme that is normally localized in the cytosol. The fusion gene was transfected into mouse Ltk⁻ fibroblasts, and the clones were selected for G418 resistance and screened for adenosine deaminase activity in the media. The secreted (□) adenosine deaminase activity of a transfected clone was compared with the intracellular (●) form of human adenosine adeaminase activity in their (A) heat sensitivity, (B) pH optimum, and (C) K_M. (From Reference 66, with permission.)

the secretable form was indistinguishable from the authentic human enzyme in its structural and enzymatic properties (Figure 3). It appears promising that as our data base expands regarding the structure-function relationship of various DNA motifs, desirable signals for secretion, receptor recognition, catalytic function, and transcription regulation can be engineered into a designer secretory protein for the purpose of nonautologous somatic gene therapeutics.

Table 1

Recombinant Gene Products Expressed from Transfected Mouse Fibroblasts Encapsulated in Alginate-Polylysine-Alginate Microcapsules

Gene Product	Mol. Wt.	Target Disease
Adenosine deaminase	42,000	Severe combined immune deficiency
Growth hormone	48,000	Hypopituitarism
Factor IX	56,800	Hemophilia B
Arylsulfatase A	120,000	Metachromatic leukodystrophy
β-glucuronidase	300,000	Sly Disease, MPS VII

Note: Mouse Ltk⁻ fibroblasts transfected with the various cDNA through calcium phosphate precipitation were encapsulated in alginate-polylysine-alginate microcapsules. Secretion of the various recombinant gene products was monitored either with enzymatic assays or ELISA.

Data from References 68–70.

Potentially, any genetic or somatic disorders that can be treated with systemic delivery of a recombinant gene product is amenable to the proposed somatic gene therapeutics approach. Several clinically relevant gene products have already been expressed through microencapsulated fibroblasts, i.e., growth hormone for the treatment of hypopituitarism, β-glucuronidase and arylsulfatase-A (Table 1)[68-70] for the treatment of the lysosomal storage diseases. These recombinant gene products have the advantage that they are simple polypeptides that do not require unusual post-translational modification for their biologic activity.

4. Post-Translational Modification

However, other gene products such as clotting factor IX is normally expressed only in hepatic tissues and requires extensive post-translational modification to acquire its biologic activity. We therefore investigated the feasibility of expressing biologically active human factor IX through the encapsulated engineered cells. Factor IX normally undergoes complex post-translational γ-carboxylation necessary for its catalytic action in the coagulation cascade.[71] Such modification normally occurs in hepatocytes where γ-carboxylase is available to catalyze the modification. Hence, even though factor IX is naturally secreted, expressing the cDNA of factor IX from engineered cell lines that are nonhepatic in origin may not produce sufficient biologic activity for it to become clinically useful. However, biologically active and hence appropriately processed recombinant factor IX has been expressed from several nonhepatic cell types, such as myoblasts,[72] skin fibroblasts,[73,74] and endothelial cells.[75] Hence, it was not surprising that when mouse fibroblasts (Ltk⁻) were transfected with the cDNA for human factor IX, about 70% of the secreted factor was found to be biologically active, providing appropriate clotting activity when assayed with factor IX-deficient plasma. It was also verified that this biologic activity was derived from properly processed factor IX. Over 98% of the recovered biologic activity was precipitable by barium citrate, indicating appropriate γ-carboxylation of the secreted factor (Table 2). More important, the biologically active factor IX was freely permeable through the alginate-polylysine-alginate microcapsules (Table 2). The demonstration of the secretion of biologically active factor IX from nonhepatic recombinant cells within immuno-isolation devices opens up the possibility of treating diseases that normally require liver-specific proteins, e.g., α_1-antitrypsin, apolipoproteins, and other coagulation factors.

B. Animal Models

1. Delivery of Human Growth Hormone

To test the feasibility of using such encapsulated recombinant cells to deliver a new gene product *in vivo*, we implanted mice with microcapsules enclosing allogeneic mouse fibroblasts transfected with the *human growth hormone* (HGH) gene. HGH was detected in the circulation of the implanted animals within the first 2 weeks, whereas no significant level of circulating HGH was detected in the control mice implanted with only the transfected cells without microcapsules (Figure 4). By about 3 weeks, antibodies against HGH developed in the microcapsule-implanted mice (Figure 4C, D). The immune response was detected only against HGH and no other secretory products from the transfected cells (Figure 5). The antibody titer continued to escalate for more than 3 months, thus demonstrating indirectly the continued delivery of the growth hormone. In contrast, the control mice did not develop detectable antibody titer against the transgene product throughout the same period. The persistent expression of the transgene and survival of the transfected cells were verified when the microcapsules were retrieved periodically to demonstrate that the encapsulated cells remained viable, proliferative (Figure 6), and productive of HGH even by 78 to 111 days.[12]

From the above pilot studies, several salient features of implanting recombinant fibroblasts within alginate-polylysine-alginate microcapsules became clear. First, microencapsulated cells survived longer *in vivo* than *in vitro*. The viability was maintained at about 50%, even by 3 months postimplantation, whereas it would have decreased to 5 to 6% at that time if the microcapsules had been kept *in vitro*.[60] Second, using nonautologous marker transgene such as HGH in mice allows easy detection of transgene product with RIA or ELISA. However, the disadvantage in this case is that such proteins often elicit antibody response after 2 to 3 weeks and precludes any further accurate determination of the level of transgene product delivery. Third, the alginate-polylysine alginate microcapsules eventually showed signs of deterioration by about 3 to 4 months postimplantation. Some of the microcapsules retrieved from the intraperitoneal cavity of the implanted animals appeared fragmented by day 111 (Figure 6, top, f, arrows). Fourth, because growth hormone is species-specific, by delivering HGH to the mouse, its biologic efficacy cannot be assessed.

2. Delivery of Mouse Growth Hormone

To address some of the above problems, we attempted to deliver mouse growth hormone (MGH) to the murine animal model, the Snell dwarf mice, which are deficient in growth hormone production due to a mutation of the transcription factor Pit-1.[76] We also improved the durability of the alginate-polylysine-alginate by using a different formulation of the alginate with a higher viscosity,[77] as well as using the myoblast cell

Table 2

Characterization of Human Factor IX Secreted from Transfected Mouse Ltk⁻ Fibroblasts Encapsulated in Alginate-Polylysine-Alginate Microcapsules

	Plasmid Used	Antigen (ng/ml)	Activity (ng/ml)	% Activity/ Antigen	Barium-Precipitated Fraction: Antigen (ng/ml)	Activity (ng/ml)	% Activity/ Antigen	% Precipitable Activity[a]
Before encapsulation	pLNCIXL	1.75 ± 0.06	1.14 ± 0.05	**65.3 ± 4.56**	0.92 ± 0.04	0.86 ± 0.03	**93.7 ± 5.5**	**100.0 ± 0**
	pKG5IX	16.0 ± 0.23	10.5 ± 0.3	**65.7 ± 2.1**	7.2 ± 0.2	6.1 ± 0.3	**84.0 ± 5.9**	**98.4 ± 0.2**
	pLIXSNL	18.1 ± 0.7	12.2 ± 0.4	**67.4 ± 3.8**	10.6 ± 0.7	8.7 ± 0.3	**82.7 ± 6.9**	**98.7 ± 0.1**
Plasma control		37.1	37.1	**100**	22.8	22.6	**99.4**	**98.5**
After encapsulation	pLNCIXL	3.15 ± 0.5	2.3 ± 0.15	**72.3 ± 1.8**	1.8 ± 0.1	1.17 ± 0.1	**92.0 ± 1.1**	**97.5 ± 0.1**
	pKG5IX	13.7 ± 1.5	9.9 ± 0.8	**72.1 ± 2.6**	9.4 ± 0.3	7.77 ± 0.4	**82.6 ± 3.2**	**99.0 ± 0.0**
	pLIXSNL	17.8 ± 0.3	12.4 ± 1.8	**69.5 ± 9.3**	10.9 ± 0.03	9.0 ± 0.3	**82.4 ± 2.7**	**99.0 ± 0.1**

Note: Mouse Ltk⁻ fibroblasts were transfected with various plasmids encoding the cDNA for the human factor IX. One clone from each transfection was chosen for characterization of factor IX protein antigen with ELISA, biologic activity with clotting assay, and γ-carboxylation with barium precipitation, using normal human plasma as control.

[a] From Reference 59.

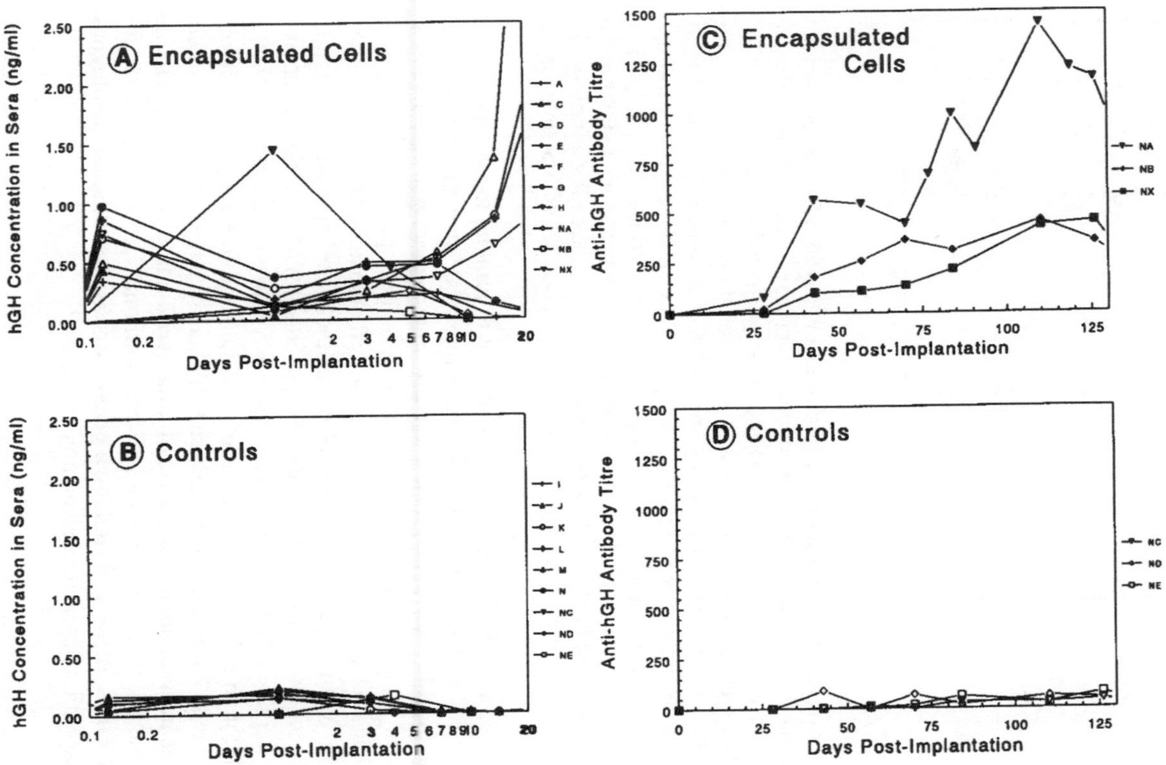

FIGURE 4. Delivery of HGH to mice with transfected mouse Ltk⁻ fibroblasts encapsulated in alginate-polylysine-alginate microcapsules. Mouse Ltk⁻ fibroblasts transfected with the HGH cDNA were encapsulated and implanted intraperitoneally in C57BL/6 mice. Controls were implanted with the same number of transfected cells, but without the immuno-protection of the microcapsules. The level of HGH and mouse anti-human growth hormone IgG were monitored in the circulation with ELISA. (From Reference 12, with permission.)

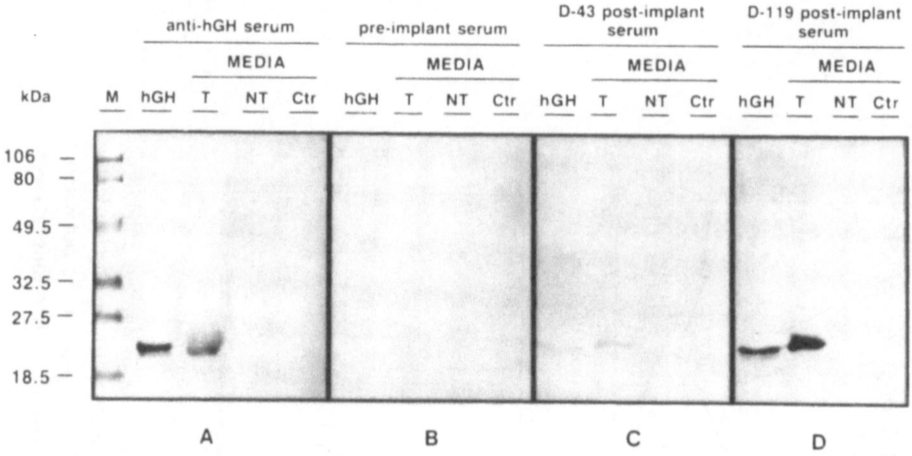

FIGURE 5. Monitoring the immune response in mice implanted with HGH-secreting mouse fibroblasts encapsulated in alginate-polylysine-alginate microcapsules. Culture media collected from nontransfected mouse Ltk⁻ cells (NT) and cells transfected with the HGH cDNA (T) were separated by electrophoresis, transferred to nitrocellulose and blotted with either mouse serum specifically immunized against purified HGH or sera collected from the implanted animals on day 43 or day 119 postimplantation. A negative control consisting of the culture medium without exposure to the cells (Ctr) and a positive control (hGH) of purified human growth hormone were included. M: mol. size markers. Only a single major band of mouse IgG reacting against authentic HGH secreted by the transfected cells was observed in the implanted animals on both days 43 and 119. (From Reference 12, with permission.)

line C2C12 instead of the Ltk⁻ fibroblasts, in an attempt to prolong the survival of the encapsulated cells. As a result, we were able to demonstrate that the encapsulated myoblasts remained viable (≈60%) and continued to secrete the transgene product even after 178 days postimplantation. Furthermore, about 90% of the implanted microcapsules could be retrieved and remained mostly intact with no inflammatory or fibroblast overgrowth at this time. Most important, the pathology of the dwarf mutants was significantly reduced. The animals implanted with the encapsulated MGH-secreting myoblasts increased in body weight and linear growth more than two-fold greater than the controls (Figure 7). In addition, other secondary metabolic responses to the exogenous growth hormone were elicited. These included increased thickness of the epiphyseal growth plate, weights of the internal organs, and level of nonesterified free fatty acid due to the lipolytic effect of the microcapsule-delivered growth hormone.[64] This demonstrates for the first time the clinical efficacy of treating a genetic disease with the nonautologous somatic gene therapeutic approach as originally proposed.[12]

IV. FUTURE DIRECTIONS

In summary, the delivery of recombinant gene products from genetically modified allogeneic cell lines has been proven feasible *in vitro* and *in vivo*. The clinical efficacy

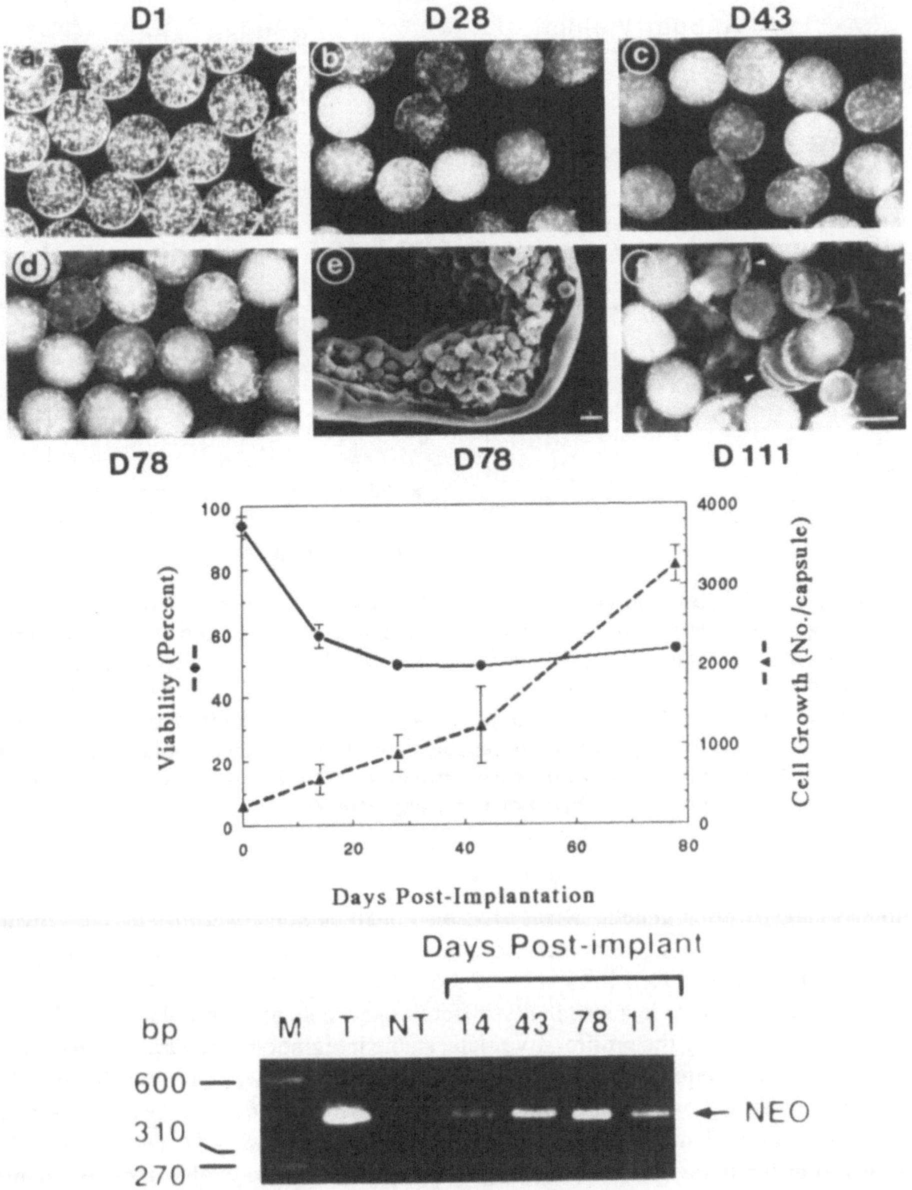

FIGURE 6. Recovery of implanted microcapsules to monitor survival of the recombinant cells. At intervals from day 28 to day 111 postimplantation, microcapsules containing HGH-secreting mouse Ltk⁻ fibroblasts were retrieved from the intraperitoneal cavity of the implanted mice to be (top) photographed under dark-field microscopy (a–d, f: scale = 100 μm) or scanning electron microscopy (e: scale = 10 μm); (middle) monitored for cell viability and cell number; and (bottom) crushed open to release the encapsulated cells to re-establish growth in culture, after which the cells were extracted for DNA to amplify a 415-bp fragment (NEO) specific to the plasmid used in the transfection. M: DNA size marker; T: transfected fibroblast; NT: nontransfected fibroblasts. (From Reference 12, with permission.)

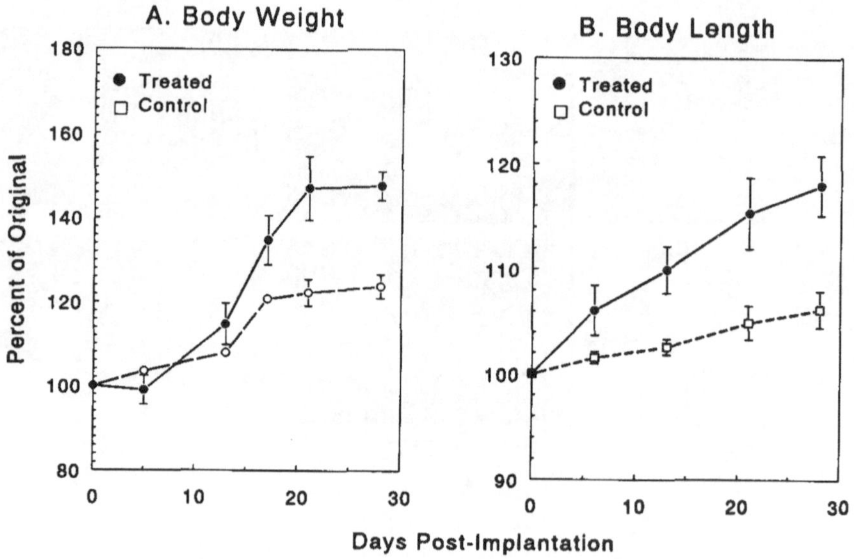

FIGURE 7. Partial correction of the growth deficit in Snell dwarf mice after implantation of mouse-growth-hormone (MGH) secreting myoblasts encapsulated in alginate-poly-lysine-alginate microcapsules. Snell dwarf mice, which were genetically deficient in growth hormone production, were implanted with encapsulated mouse myoblasts engineered to secrete MGH. The treated group (N = 4) was implanted intraperitoneally with 2 ml of alginate-polylysine-alginate microcapsules enclosing mouse myoblasts engineered to secrete MGH at ≈400 ng/day, while the control group (N = 4) received the same amount of encapsulated myoblasts that had not been engineered to secrete MGH. (Data from Reference 64.)

of this strategy to treat genetic disease has also been demonstrated in the Snell dwarf mice. For successful long-term delivery, the cell type used for encapsulation is of paramount importance. Cell lines such as those from myoblasts that can be switched from a proliferative state to a terminally-differentiated state are particularly suitable for several reasons. During the proliferative state, stable integration of the transgene is most likely to occur. Subsequent switching to a differentiated state would relieve the requirement for increasing growth space within the immuno-isolation device. An added advantage of inducing implanted myoblasts into a terminally differentiated state is safety. The risk of a proliferative cell line becoming tumorigenic is considerably reduced, if not entirely eliminated, once the cells have stopped cycling and become differentiated. Together with the fact that the cells are physically contained within the immuno-isolation device and that no muscle tumor has been reported in man,[78] myoblast cell lines appear exceptionally suited for the safety requirements of clinical application. Furthermore, the availability of an almost unlimited supply of the cells for implantation would allow thorough evaluation of short- and long-term safety to meet industrial quality control standards, a requirement that is difficult to meet with existing autologous gene modification protocols.

Our current success with the alginate-polylysine-alginate microcapsules is limited only to allogeneic transplants. Our preliminary effort with xenogeneic cells (encapsulated mouse cells in rats) was met with little success. Massive inflammatory response to the implanted capsules was observed even as early as 14 days postimplantation.[79] It is possible that the intolerance for xenogeneic implants is related to the loss of permeability barrier when cultured cells are encapsulated with the alginate type of microcapsules. During the initial step of encapsulation, the cells had to be mixed with the alginate solution so that some cells can and usually do become lodged in the membrane of the capsules. Consequently, the expected permeability cut-off to molecules >120 kDa cannot be sustained, since the membrane integrity is not uniformly maintained.[68] In fact, molecules as large as 300 kDa are found secreted to the media from the alginate-encapsulated recombinant cells.[70] However, from the success in allogeneic implants *in vivo* that we observed with this type of microcapsule,[12,64] it is apparent that this loss of permeability barrier to immunoglobulin, contrary to accepted belief, is not critical for the host tolerance of encapsulated allogeneic cells.

If this loss of permeability barrier were the cause for the inflammatory response towards encapsulated *xenogeneic* recombinant cells, it would have two important implications. First, to apply this technology to treat disorders requiring the delivery of recombinant gene products of sizes >120 kDa, the permeability threshold would allow the entry of immunoglobulins. This would limit the cell types for implantation to allogeneic sources.[12,64] Otherwise, immune rejection would occur if xenogeneic cells were used.[79] Second, if xenogeneic cells must be used, then only disorders requiring delivery of small recombinant gene products below the threshold of that of immunoglobulins could be treated. Furthermore, it would preclude the use of the alginate-poly-lysine-alginate type of microcapsules, whose permeability integrity is difficult to maintain if cultured cells are encapsulated at a density that is compatible with good cell survival.[58] As an alternative, thermoplastic capsules may be considered. Membranes of the thermoplastic capsules are not perturbed by the presence of the enclosed cells, since the cells are not mixed with the membrane solution during the manufacturing process.[31,53] Consequently, the permeability threshold is readily maintained. Recombinant gene products >120 kDa have been found totally sequestered within the polymer capsules with little detectable leakage.[80] It will be important to verify if xenogeneic recombinant cells can be better tolerated when encapsulated with such thermoplastic capsules. An additional advantage in using thermoplastic capsules is their greater mechanical stability than the water-soluble hydrogel capsules. However, the thermoplastic capsules could elicit a slight foreign-body reaction,[53] and angiogenesis around the implanted capsules has been observed in pilot studies.[81] On the other hand, the increased perfusion around the implanted capsules due to the angiogenic response may confer an added advantage in improving nutrient exchange and gene product delivery between the host and the implanted cells. Hence, the choice of cell types and encapsulation material plays a crucial role in determining the success of the implant.

It is expected that long-term delivery of recombinant products from cells of nonautologous origins requires further improvements on several fronts, e.g., molecular and cell biology and biomaterials sciences. It will be important to create cell types capable of long-term survival and high-level transgene expression, a problem that currently confronts all forms of gene therapy. Furthermore, biomaterials with increased biocompatibility, durability, and membrane integrity will be important to develop. Once these requirements are met, the strategy of somatic gene therapeutics should be applicable to the treatment of genetic disorders, such as enzyme deficiencies in the lysosome, the urea cycle, and other housekeeping metabolic pathways, as well as non-Mendelian disorders such as endocrine deficiencies, neurotransmitter and growth factor imbalance, and multifactorial disorders, such as cancer and cardiovascular diseases. The development of neural implants extends the repertoire of potentially treatable disorders to include those with central nervous system as well as systemic involvements.[47] Furthermore, applications are not only limited to human disorders. Veterinary problems, such as vaccination, growth modulation, and even reproductive control and contraception, have not even begun to exploit this technology. The many potential uses of nonautologous somatic gene therapeutics illustrate well the concept of "genes as drugs".

REFERENCES

1. Morsy, M. A., Mitani, K., Clemens, P., and Caskey, C. T., Progress toward human gene therapy, *JAMA*, 270, 2338, 1993.
2. Mulligan, R. C., The basic science of gene therapy, *Science*, 260, 926, 1993.
3. Morgan, R. A. and Anderson, W. F., Human gene therapy, *Annu. Rev. Biochem.*, 62, 191, 1993.
4. Miller, A. D., Human gene therapy comes of age, *Nature*, 357, 455, 1992.
5. Anderson, W. F., Prospects for human gene therapy, *Science*, 226, 401, 1984.
6. Chien, K. R., Molecular advances in cardiovascular biology, *Science*, 260, 916, 1993.
7. Dorudi, S., Northover, J. M. A., and Vile, R. G., Gene transfer therapy in cancer, *Br. J. Surg.*, 80, 566, 1993.
8. Sarver, N., Gene therapy and ribozyme for HIV infection and immune restoration strategies, *J. Cell. Biochem.*, 17E, 221, 1993.
9. Anderson, C., Research and health care costs, *Science*, 261, 416, 1993.
10. Gibbons, A., Billion-dollar orphans: prescription for trouble, *Science*, 248, 678, 1990.
11. Beutler E., Kay, A. C., Saven, A., Garver, P., Thurston, D. W., and Rosenbloom, B. E., Enzyme-replacement therapy for Gaucher's disease, *N. Engl. J. Med.*, 325, 1809, 1991.
12. Chang, P. L., Shen, N., and Westcott, A., Delivery of recombinant gene products with microencapsulated cells *in vivo*, *Hum. Gene Ther.*, 4, 433, 1993.
13. Bennett, W. M. and Norman, D. J., Action and toxicity of cyclosporine, *Annu. Rev. Med.*, 37, 214, 1986.
14. Reach, G., Bioartificial pancreas. Present state and future prospects, *Biomed. Biochim. Acta*, 43, 569, 1984.
15. Colton, C. K. and Avgoustiniatos, E. S., Bioengineering in development of the hybrid artificial pancreas, *Transact. ASME*, 113, 152, 1991.
16. Goosen, M. F. A., O'Shea, G. M., Gharapetian, H., Chou, S., and Sun, A. M., Optimization of microencapsulation parameters: semipermeable microcapsules as a bioartificial pancreas, *Biotechnol. Bioeng.*, 27, 146, 1985.
17. Whitley, D., Kupiec-Weglinski, J. W., and Tilney, N. L., Antibody mediated rejection of organ grafts, *Curr. Opin. Immunol.*, 2, 864, 1990.
18. Bendtzen, K., Mandrup-Poulsen, T., Nerup, J., Nielsen, J. H., Dinarello, C. A., and Svenson, M., Cytotoxicity of human pI 7 interleukin-1 for pancreatic islets of Langerhans, *Science*, 232, 1545, 1986.
19. Pukel, C., Baquerizo, H., and Rabinovitch, A., Destruction of rat islet cell monolayers by cytokines, synergistic interactions of interferon-gamma, tumor necrosis factor, lymphotoxin, and interleukin 1, *Diabetes*, 37, 133, 1988.

20. Strautz, R. L., Studies of hereditary obese mice (obob) after implantation of pancreatic islets in Millipore filter capsules, *Diabetologia*, 6, 306, 1970.
21. Gates, R. J., Hunt, M. I., Smith, R., and Lazarus, N. R., Return to normal of blood-glucose, plasma-insulin and weight gain in New Zealand obese mice after implantation of islets of Langerhans, *Lancet*, 1972-II, 567, 1972.
22. Chick, W. L., Like, A. A., and Lauris, V., Beta cell culture on synthetic capillaries: an artificial endocrine pancreas, *Science*, 184, 847, 1975.
23. Michaels, A. S., U.S. Patent 3,615,024, 1971.
24. Lim, F. and Sun, A. M., Microencapsulated islets as bioartificial endocrine pancreas, *Science*, 210, 908, 1980.
25. Lacy, P. E., Hegre, O. D., Gerasimidi-Vazeou, A., Gentile, F. T., and Dionne, K. E., Maintenance of normoglycemia in diabetic mice by subcutaneous xenografts of encapsulated islets, *Science*, 254, 1782, 1991.
26. Altman, J. J., Houlbert, D., Callard, P., et al., Long-term plasma glucose normalization in exerimental diabetic rats with macro-encaspulated implants of benign human insulinomas, *Diabetes*, 35, 625, 1986.
27. Sun, A., Parisius, W., Macmorine, H., Sefton, M., and Stone, R., An artificial endocrine pancreas containing cultured islets of Langerhans, *Artif. Endocr. Pancreas*, 4, 275, 1980.
28. Sullivan, S. J., Maki, T., Borland, K. M., Mahoney, M. D., Solomon, B. A., Muller, T. E., Monaco, A. P., and Chick, W. L., Biohybrid artificial pancreas: long-term implantation studies in diabetic, pancreatectomized dogs, *Science*, 252, 718, 1991.
29. Lim, F. and Moss, R. D., Microencapsulation of living cells and tissues, *J. Pharm. Sci.*, 70, 351, 1981.
30. Mallabone, C. L., Crooks, C. A., and Sefton, M. V., Microencapsulation of human diploid fibroblasts in cationic polyacrylates, *Biomaterials*, 10, 380, 1989.
31. Aebischer, P., Wahlberg, L., Tresco, P. A., and Winn, S. R., Macroencapsulation of dopamine-secreting cells by coextrusion with an organic polymer solution, *Biomaterials*, 12, 50, 1991.
32. Smidsrød, O., Molecular basis for some physical properties of alginates in the gel state, *Faraday Disc. Chem. Soc.*, 57, 263, 1974.
33. O'Shea, G. M., Goosen, M. F. A., and Sun, A. M., Prolonged survival of transplanted islets of Langerhans encapsulated in a biocompatible membrane, *Biochim. Biophys. Acta*, 804, 133, 1984.
34. Sun, A. M., Microencapsulation of pancreatic islet cells: a bioartificial endocrine pancreas, *Meth. Enzymol.*, 137, 575, 1988.
35. O'Shea, G. M. and Sun, A. M., Encapsulation of rat islets of Langerhans prolongs xenograft survival in diabetic mice, *Diabetes*, 35, 943, 1986.
36. Sun, A. M., Encapsulated *versus* modified endocrine cells for organ replacement, *Trans. Am. Soc. Artif. Intern. Organs*, 33, 787, 1987.
37. Weber, C. J., Zabinski, S., Koschitzky, T., Wicker, L., Rajotte, R., D'Agati, V., Peterson, L., Norton, J., and Reemtsma, K., The role of $CD4^+$ helper T-cells in destruction of microencapsulated islet xenografts in NOD mice, *Transplantation*, 49, 396, 1990.
38. Sun, A. M., Cai, Z., Shi, Z., Ma, F., O'Shea, G. M., and Gharapetian, H., Microencapsulated hepatocytes as a bioartificial liver, *Trans. Am. Soc. Artif. Intern. Organs*, 32, 39, 1986.
39. Cai, Z., Shi, Z., O'Shea, G. M., and Sun, A. M., Microencapsulated hepatocytes for bioartificial liver support, *Artif. Organs*, 12, 388, 1988.
40. Chang, T. M. S., Living cells and microorganisms immobilized by microencapsulation inside artificial cells, in *Fundamentals of Animal Cell Encapsulation and Immobilization*, Goosen, M. F. A., Ed., CRC Press, Boca Raton, FL, 1993, chap. 8.

41. Fu, X. W. and Sun, A. M., Microencapsulated parathyroid cells as a bioartificial parathyroid, *Transplantation*, 47, 432, 1989.
42. King, G. A., Daugulis, A. J, Faulkner, P., and Goosen, M. F. A., Alginate-polylysine microcapsules of controlled membrane molecular weight cutoff for mammalian cell culture engineering, *Biotechnol. Prog.*, 3, 231, 1987.
43. Dawson, R. M., Broughton, R. L., Stevenson, W. T. K., and Sefton, M. V., Microencapsulation of CHO cells in hydroxyethyl methacrylate-methyl methacrylate copolymer, *Biomaterials*, 8, 360, 1987.
44. Jaeger, C. B., Winn, S. R., Tresco, P. A., and Aebischer, P., Repair of the blood-brain barrier following implantation of polymer capsules, *Brain Res.*, 551, 163, 1991.
45. Uludag, H. and Sefton, M. V., Metabolic activity of CHO fibroblasts in HEMA-MMA microcapsules, *Biotechnol. Bioeng.*, 39, 672, 1992.
46. Wang, H., Tresco, P. A., Aebischer, P., and Sagen, J., Pain reduction by transplants of polymer encapsulated bovine chromaffin cells in the rat spinal subarachnoid space, *Soc. Neurosci. Abstr.*, 17, 235, 1991.
47. Aebischer, P., Goddard, M., and Tresco, P. A., Cell encapsulation for the nervous system, in *Fundamentals of Animal Cell Encapsulation and Immobilization*, Goosen, M. F. A., Ed., CRC Press, Boca Raton, FL, 1993, Chap. 9.
48. Frydel, B. R., Emerich, D. F., McDermott, P. E., Kaplan, F. A., Palmatier, M. A., Christenson, L., Duncan, H., and Sanberg, P. R., Immunocytochemical analysis of allogeneic and xenogeneic PC12 cell implants into striatum, *Soc. Neruosci. Abstr.*, 17, 569, 1991.
49. Winn, S. R., Tresco, P. A., Zielinski, B., Greene, L. A., Jaeger, C. B., and Aebischer, P., Behavioral recovery following intrastriatal implantation of microencapsulated PC12 cells, *Exp. Neurol.*, 113, 322, 1991.
50. Goetz, C. G., Olanow, C. W., Koller, W. C., Penn, R. D., Cahill, D., Morantz, R., Stebbins, G., Tanner, C. M., Klawans, H. L., Shannon, K. M., Comella, C. L., Witt, T., Cox, C., Waxman, M., and Gauger, L., Multicenter study of autologous adrenal medullary transplantation to the corpus striatum in patients with advanced Parkinson's disease, *N. Engl. J. Med.*, 320, 337, 1989.
51. Aebiescher, P., personal communication, 1993.
52. Sagen, J., Pappas, G. D., and Pollard, H. B., Analgesia induced by isolated bovine chromaffin cell transplants in CNS modulatory regions, *Proc. Natl. Acad. Sci. U.S.A.*, 83, 7522, 1986.
53. Sefton, M. V., Kharlip, L., Horvath, V., and Roberts, T., Controlled release using microencapsulated mammalian cells, *J. Contr. Res.*, 19, 189, 1992.
54. Aebischer, P., Winn, S. R., and Galletti, P. M., Transplantation of neural tissue in polymer capsules, *Brain Res.*, 448, 364, 1988.
55. Aebischer, P., Winn, S. R., Tresco, P. A., Jaeger, C. B., and Greene, L. A., Transplantation of polymer encapsulated neurotransmitter secreting cells: effect of the encapsulation technique, *J. Biomech. Eng.*, 113, 178, 1991.
56. Bugarski, B., Jovanovic, G., and Vunjak-Novakovic, G., Bioreactor systems based on microencapsulated animal cell cultures, in *Fundamentals of Animal Cell Encapsulation and Immobilization*, Goosen, M. F. A., Ed., CRC Press, Boca Raton, FL, 1993, Chap. 12.
57. Hoffman, D., Breakefield, X. O., Short, P., and Aebischer, P., Transplantation of a polymer-encapsulated cell line genetically engineered to release NGF, *Exp. Neurol.*, 122, 100, 1993.
58. Chang, P. L., Hortelano, G., Tse, M., and Awrey, D. E., Growth of recombinant fibroblasts in alginate microcapsules, *J. Biotechnol. Bioeng.*, 43, 925, 1994.
59. Liu, H., Ofosu, F. A., and Chang, P. L., Expression of human factor IX by microencapsulated recombinant fibroblasts, *Hum. Gen. Ther.*, 4, 291, 1993.

60. Awrey, D. E., Tse, M., Hortelano, G., and Chang, P. L., Permeability of alginate microcapsules to secretory recombinant gene products, in preparation, 1994.
61. Yaffe, D. and Saxel, O., Serial passaging and differentiation of myogenic cells isolated from dystrophic mouse muscle, *Nature*, 270, 725, 1977.
62. Dhawan, J., Pan, L. C., Pavlath, G. K., Travis, M. A., Lanctot, A. M., and Blau, H. M., Systemic delivery of human growth hormone by injection of genetically engineered myoblasts, *Science*, 254, 1509, 1991.
63. Al-Hendy, A., Hortelano, G., and Chang, P. L., unpublished observation, 1994.
64. Al-Hendy, A., Hortelano, G., Tannenbaum, G. S., and Chang, P. L., Correction of the growth defect in dwarf mice: a novel approach to somatic gene therapy, submitted, 1994.
65. Simon, K., Perara, E., and Lingappa, V. L., Translocation of globin fusion proteins across the endoplasmic reticulum membrane in *Xenopus laevis* oocytes, *J. Cell Biol.*, 104, 1165, 1987.
66. Hughes, M., Delivery of a Secretable Adenosine Deaminase Through Alginate Microcapsules, M.Sc. Thesis, McMaster University, Canada, 1993.
67. von Heijne, G., Signal sequences — the limits of variation, *J. Mol. Biol.*, 184, 99, 1985.
68. Awrey, D. E., Secretion of Marker Proteins from Alginate-poly-L-lysine-alginate Microcapsules, M.Sc. Thesis, McMaster University, Canada, 1993.
69. Liu, H., Delivery of Human Factor IX from Encapsulated Recombinant Fibroblasts, M.Sc. Thesis, McMaster University, Canada, 1993.
70. Tse, M., Secretion of Marker Proteins from Alginate-poly-L-lysine-alginate Microcapsules and Hydroxyethyl Methacrylate-methyl Methacrylate Capsules, M.Sc. Thesis, McMaster University, Canada, 1994.
71. Diuguid, D. L. and Furie, B., Molecular genetics of hemophilia B, in *Hematology — Basic Principle and Practice*, Hoffman, R., Benz, E. J., Shattil, S. J., Furie, B., and Cohen, H. J., Eds., Churchill Livingstone, New York, 1991, 1320.
72. Yao, S. N. and Kurachi, K., Expression of human factor IX in mice after injection of genetically modified myoblasts, *Proc. Natl. Acad. Sci. U.S.A.*, 89, 3357, 1992.
73. St. Louis, D. and Verma, I. M., An alternative approach to somatic cell gene therapy, *Proc. Natl. Acad. Sci. U.S.A.*, 85, 3150, 1988.
74. Palmer, T. D., Thompson, A. R., and Miller, A. D., Production of human factor IX in animals by genetically modified skin fibroblasts. Potential therapy for hemophilia B, *Blood*, 73, 438, 1989.
75. Yao, S. N., Wilson, J. M., Nabel, E. G., Kurachi, S., Hachiya, H. L., and Kurachi, K., Expression of human factor IX in rat capillary endothelial cells: toward somatic gene therapy for hemophilia B, *Proc. Natl. Acad. Sci. U.S.A.*, 88, 8101, 1991.
76. Lin, C., Lin, S.-C., Chang, C.-P., and Rosenfeld, M., Pit-1-dependent expression of the receptor for growth hormone releasing factor mediates pituitary cell growth, *Nature*, 360, 765, 1992.
77. Awrey, D. E. and Chang, P. L., unpublished observation, 1993.
78. Blau, H. M., Dhawan, J., and Pavlath, G. K., Myoblasts in pattern formation and gene therapy, *Trends Genet.*, 9, 269, 1993.
79. Dodge, M. and Chang, P. L., unpublished observation, 1994.
80. Tse, M., Uludag, H., Sefton, M. V., and Chang, P. L., unpublished observation, 1994.
81. Sefton, M. V., personal communication, 1994.

13 Current Gene Marking and Gene Therapy Protocols for Human Bone Marrow Transplantation

Malcolm K. Brenner

I. INTRODUCTION

The attraction of marrow progenitor cells as targets for gene therapy[1-3] (see Chapter 2) is that they are easily accessible, they can be readily manipulated *ex vivo* and, in principle, successful transfer into a single self-renewing stem cell is sufficient to repopulate an entire patient with modified cells. However, most early large-animal models suggested that it would be difficult to transduce novel genes into a significant proportion of marrow stem cells and to express the gene in their progeny.[4-8] Since the course of clinical gene transfer studies has been guided (not always accurately) by feasibility studies in animal models, the first human gene transfer protocols used mature lymphocytes as their targets (see Chapter 3).[2,3,9,10]

II. MARKER GENE STUDIES OF EARLY HEMOPOIETIC CELLS

In general, the transfer of genes into mature hemopoietic cells will not be sufficient for curative gene therapy, which will require instead the transduction of long-lived progenitor cells. The past 2 or 3 years have therefore seen continued effort to develop preclinical models of gene therapy that offer a reasonable chance of introducing the gene into a significant proportion of early hemopoietic cells, with expression maintained into the mature lineages,[11-18] the ultimate aim being translation of the methodology to clinical practice.

In the meantime, interest began to increase in the possibility of using gene transfer simply to mark progenitor cells in marrow removed from patients and subsequently used as autologous bone marrow rescue. These marker studies with marrow cells were to

0-8493-2440-8/95/$0.00+$.50

function as an intermediate step between gene therapy of committed cells and gene therapy of true marrow stem cells.[19-22] Because of the sensitivity of PCR-based detection techniques, even the low level of gene transfer into marrow progenitors that was predicted from animal models would allow these marker studies to provide valuable information about the biology of normal and malignant progenitor cells in patients receiving autologous bone marrow transplants.

III. APPLICATIONS OF MARKER GENES IN AUTOLOGOUS BONE MARROW TRANSPLANTATION

A. Marker Genes to Determine the Source of Relapse

While the dose intensification permitted by autologous bone marrow transplantation has shown promise as effective treatment for neoplasia,[23-29] disease recurrence remains the major cause of treatment failure. When the malignancy originates from or involves the marrow, relapse could originate from malignant cells persisting in the patient, in the rescuing marrow, or in both. The origin of relapse is an important issue to resolve. Concern that the harvested marrow may contain residual malignant cells has led to extensive evaluation of techniques for purging marrow prior to storage and subsequent reinfusion.[30-33] Animal and preclinical human studies have shown that these methods do reduce contamination with malignant cells that have been deliberately added to marrow, but no method has been shown convincingly to reduce the risk of relapse in naturally occurring disease.[32,33]

Marrow for autologous transplants is usually harvested at a time when, by definition, no malignant cells are detectable.[34] It is therefore impossible to undertake any form of quality control after purging to determine whether the putative residual malignant cells have genuinely been eradicated. Unfortunately, these unproven purging techniques almost invariably damage normal progenitor cells, so that engraftment of purged marrow is typically far slower than that of untreated marrow.[33,35,36] Morbidity and mortality from the complications of hemopoietic and immune system failure are correspondingly increased.

If marker genes could be transferred into residual malignant cells in the marrow prior to reinfusion and if gene-marked malignant cells subsequently became detectable in the marrow or peripheral blood in patients who relapsed following autologous transplant, this would be powerful evidence that the harvested marrow contributes to disease recurrence. Moreover, the finding of marked cells at relapse would permit subsequent evaluation of *ex vivo* purging techniques for their ability to eradicate clonogenic cells. Thus, for the first time it would be possible to compare directly the efficacy of the numerous available techniques for marrow purging.

B. Feasibility of Using Gene Marking to Detect Relapse

Gene marking of clonogenic malignant cells in bone marrow can readily be demonstrated *ex vivo*. Recent studies of acute and chronic myeloid leukemia (AML, CML), acute lymphoblastic leukemia (ALL), and neuroblastoma have shown that clonogenic cells can be marked with an efficiency of 0.1 to 40%.[20,21] For safety reasons, only an aliquot of the marrow intended for reinfusion can be transduced, so that between 1 and 10% of any putative malignant cells in the marrow would usually be marked by this approach. The obvious question is whether the marker gene could ever be detected at subsequent relapse, given this low efficiency of marking clonogenic cells. The answer depends on how many malignant cells in the "remission" marrow contributed to the relapse. If fewer than ten cells were involved, the chances of detecting a "marked relapse" are rather small. With larger numbers of cells contributing, the probability of detection improves substantially. The probability of *failing* to detect a marked relapse can be calculated according to the formula $P = (1 - m)^n$, where m is the probability of *not* marking an individual cell and n is the number of malignant cells contributing to relapse. Thus, even if only 1% of malignant cells in the marrow were marked, there would still be a >95% chance of detecting a marked relapse in any one patient, provided that 300 or more cells contributed to the relapse $[P = (1 - 0.99)^{300}]$.

Results of minimal residual disease (MRD) studies indicate that many "remission" marrows will, in fact, contain appreciably more than this minimum number of malignant cells.[34] Detection of MRD is most sensitive in ALL and CML, where unique gene rearrangements allow PCR techniques to detect a single residual malignant cell among 100,000 normal cells. Since a 70-kg patient will receive a minimum of 7×10^9 marrow cells at autologous bone marrow transplantation (ABMT), even in these patients, a reinfused "remission" marrow may in fact contain 70,000 residual malignant cells. For AML, MRD detection techniques are orders of magnitude less sensitive, and thus hundreds of thousands or even millions of blasts may be present in "remission" marrow.

C. Marker Genes to Follow the Fate of Normal Progenitor Cells

Between 1 and 15% of normal progenitor cells are marked in *ex vivo* preclinical studies using retrovirus vectors. This proportion is sufficient to follow the development of normal marked cells *in vivo*, allowing determination of the contribution of infused marrow to short- and long-term hemopoietic recovery and analysis of the factors that modify the growth and development of engrafting marrow.

IV. RESULTS OF MARKER GENE STUDIES TO DETERMINE SOURCE OF RELAPSE

At St. Jude Children's Research Hospital (SJCRH), we have been using the neomycin-resistance gene present in the LNL6 and G1N retroviral vectors to mark the remission marrow of children who receive autologous transplantation for acute myeloblastic leukemia or neuroblastoma.[22]

A. Acute Myeloblastic Leukemia

Twelve AML patients have entered the SJCRH study, and two have relapsed so far. The first evidence of relapse in patient 1 consisted of blast cells found in a marrow

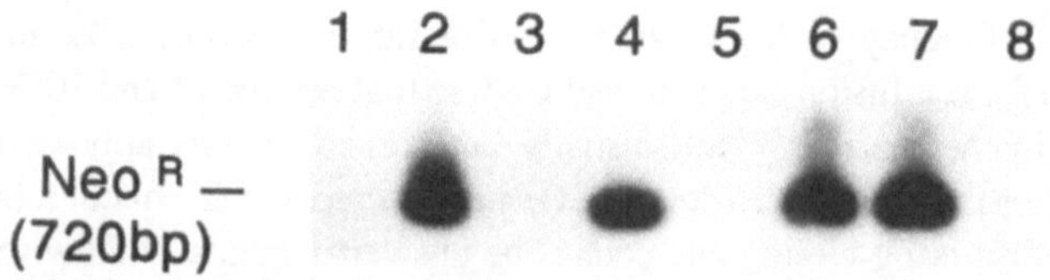

FIGURE 1. Southern blot analysis of Neo^R expression in sorted $CD34^+$ $CD13^+$ peripheral blood blast cells and their colonies. Lane 1 = markers, lane 2 = 10^5 Neo^R-transduced K562 cells (positive control), lane 3 = 10^5 nontransduced K562 cells, lane 4 = 10^5 $CD34^+$ $CD13^+$ blast cells, lane 5 = 10^5 $CD34^-$ $CD13^-$ peripheral blood mononuclear cells, lanes 6 and 7 = two blast colonies grown from $CD34^+$ $CD13^+$ blast cells (23.5 ± 16 G418-resistant blast colonies per 10^5 blast cells). Lane 8 = water control.

aspirate taken routinely on day 180 after ABMT. This aspirate was positive for the neomycin-resistance (Neo^R) marker gene by PCR. Patient 2 had blasts in the peripheral blood at 80 days after transplantation.

Additional studies were used to confirm the leukemic origin of the Neo^R-positive blast cells. Using flow cytometry, we isolated $CD34^+$ $CD56^+$ blasts from the marrow mononuclear cells of patient 1 and $CD34^+$ $CD13^+$ blasts from the circulating mononuclear cells of patient 2, when the blast cell count exceeded $20 \times 10^9/l$. Both subpopulations were positive for the Neo^R marker (Figures 1 and 2), and the $CD34^+$ $CD56^+$ blasts from patient 1 carried the 1;8;21 translocation (Figure 3). Purified blast cells were then cultured in methylcellulose, with and without the neomycin analogue G418, to allow detection of clonogenic malignant cells expressing the transferred neomycin resistance gene. The $CD34^+$ $CD56^+$ blasts formed a mean of 36 ± 12 (SD) blast colonies per 10^5 cells in the presence of G418, a yield that was 2.1% of all blast colonies formed in nonselective medium.

To identify *AML1/ETO* fusion transcripts resulting from the (8;21) translocation, RNA was extracted from leukemic myeloblasts at relapse and from individual $CD56^+$ $CD34^+$ G418-resistant blast colonies, and complementary DNA synthesized with random hexamer primers and reverse transcriptase (RNA PCR kit; Perkin-Elmer Cetus, Norwalk,

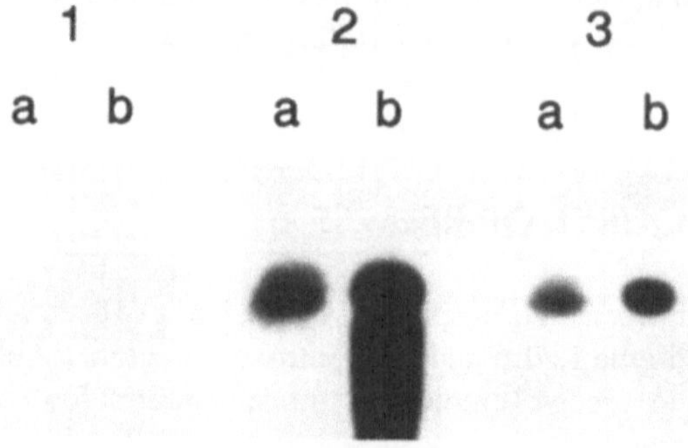

FIGURE 2. Southern blot of PCR analysis of cDNA from $CD34^+$ $CD56^+$ 1;8;21-bearing cells and blast colonies derived from that population. Lane 1a/1b = negative control (10^5 remission marrow cells pretransduction), lane 2a = Neo^R in 10^5 $CD34^+$ $CD56^+$ cells, lane 2b = 8;21 fusion product from the same $CD34^+$ $CD56^+$ cells, lane 3a = Neo^R in single colony grown from $CD34^+$ $CD56^+$ cells, lane 3b = 8;21 fusion product from the same colony.

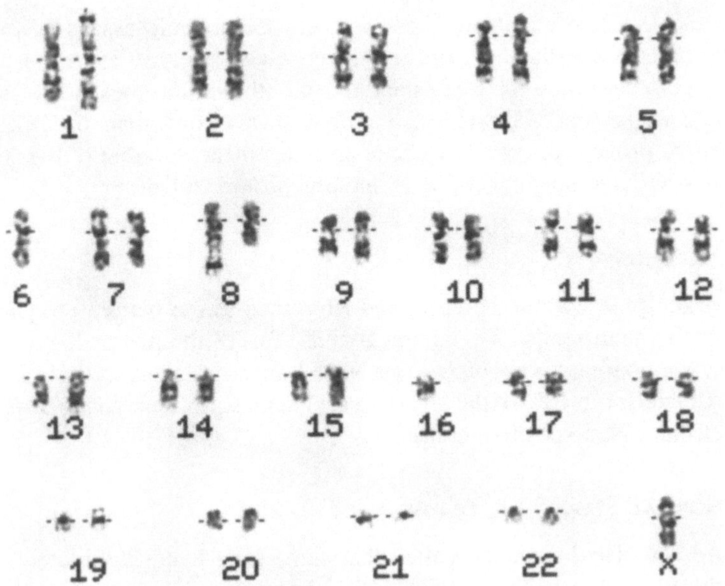

FIGURE 3. Chromosomal analysis of patient 1, showing the 1;8;21 complex translocation.

CT). The 5′ PCR primer AML1-A (5′-TACCACAGAGCCATCAAA-3′) corresponds to nucleotides 1252 to 1269 of the published sequence of *AML1*.[37] The 3′ PCR primer ETO-A (5′-GTTGTCGGTGTAAATGAA-3′) is complementary to nucleotides 119 to 136 of the *ETO* sequence 3′ to the fusion transcript junction.[38] The AML1-A and ETO-A primers are internal to those used to demonstrate *AML1/ETO* fusion transcripts in seven AML patients with the (8;21) translocation.[39] PCR was performed in a similar fashion as described for the Neo^R gene[20,21] except an annealing temperature of 55°C was used.

The cDNA extracted from single colonies was shown to contain the products of both the leukemic translocation and the transferred Neo^R marker (Figure 2). Similar results were obtained with the $CD34^+$ $CD13^+$ blast cells of patient 2 (Figure 1), although a definitive clonal marker was not available for confirmation.

B. Neuroblastoma

Eight patients have been entered into this study at SJCRH,[40] and three have relapsed. In all three, the marker gene was detected in GD2-positive colonies or fluorescence-sorted GD2-positive cells from bone marrow.

V. GENE TRANSFER AND EXPRESSION IN NORMAL PROGENITOR CELLS AND THEIR PROGENY

A. Progenitor Cells

Gene expression was also detected in normal progenitor cells. Semiquantitative PCR analysis of marrow mononuclear cells from 18 patients 1 month after BMT showed between 0 and 0.1 Neo^R copy numbers per cell. Clonogenic assays confirmed that hemopoietic precursor cells had been transduced successfully.[19,20,41] A mean of 6% of hemopoietic colonies were G418 resistant at 1 month and the presence of the marker gene in these individual G418-resistant colonies was confirmed by PCR analysis. Marker-positive progenitor cells have been present for the duration of the study in the two patients who have now exceeded 18 months posttransplant.

B. Mature Cells

The marker gene was also present and expressed in the mature progeny of these progenitor cells (neutrophils, T-cells and B-cells) for 18 months or longer. Levels of expression were comparatively high since Neo^R transcript levels in these mature cells were approximately 1 to 10% of the levels present in an equivalent amount of RNA from K562 cells from a Neo^R-positive clone.

C. Are Normal Stem Cells Transduced?

Several lines of evidence suggested that some long-lived multipotent stem cells were transduced in this study. First, G418-resistant and Neo^R-positive hemopoietic colonies were detected for up to 18 months after ABMT, an unlikely result if only short-lived committed progenitor cells had been transduced. Second, by 6 months after ABMT, multilineage granulocytic-erythroid-monocyte/macrophage (GEMM) colonies were among those found to contain and express the marker gene. Finally, the marker gene is detected in both T- and B-lymphocytes, consistent with the transfection of a primitive lympho-hemopoietic stem cell. Although highly suggestive, these findings do not constitute definitive proof. That will require longer follow-up and the demonstration of shared provirus integrants in cells/colonies derived from different lineages.

VI. CONCLUSIONS OF INITIAL STUDIES

A. Purging Seems Necessary — But is it Effective?

These results directly show that autologous marrow harvested from patients in apparent clinical remission of hematologic and nonhematologic malignancies may in fact harbor malignant cells capable of contributing to relapse. Similar gene marking protocols studying other malignancies have recently begun. If these produce identical results, the implication would be that **effective** marrow purging[28-33] will need to be incorporated into protocols of autologous bone marrow transplantation.

B. Human Marrow Cells May be Suitable Targets for Gene Therapy

Although this study demonstrated only that a marker gene can be transferred *ex vivo* to human hemopoietic progenitor cells and subsequently expressed *in vivo*, it was particularly relevant to the use of human marrow cells as targets for gene therapy. The level and duration of expression of the Neo^R gene in marrow progenitor cells was substantially greater than would have been predicted from virtually all studies using equivalent transduction methods in animals.[4-8,11-17] The overall efficiency of gene transfer into precursor cells *in vivo* at 1 month was about 6% in our clinical system. Little change in that value was seen in the patients who have exceeded 1 year of follow-up. Since only one-third of the marrow sample was transduced, the true transfer rate may be as high as 18%. One explanation for this higher-than-expected efficiency is that patient marrow was harvested during the marrow recovery phase that follows multiple cycles of intensive (marrow-ablative) chemotherapy. During this recovery period, there is profound proliferation of early marrow progenitor cells that may favor integration of the provirus genome.[42] Certainly, animal data suggest that pluripotent progenitor cells are transduced with higher efficiency if the animal is first exposed even to comparatively mild cycle-specific cytotoxic agents, such as 5-fluorouracil.[42] Finally, our results show that the transferred gene is not only present in a large proportion of cells, but it also continues to be expressed, implying that the murine retroviral LTR promoters may remain functional in human marrow progenitor cells and their progeny.

C. Autologous Marrow Produces Long-Term Hemopoietic Repopulation

The demonstration that marked cells of all lympho-hemopoietic lineages are detected beyond 1 year has other implications. Since autologous marrow transplants contribute to long-term hemopoietic recovery rather than simply providing short-term repopulation while residual host stem cells recover, attempts to use hemopoietic growth factors as a substitute for marrow infusion may prove unsuccessful. It will also be possible to use the gene marking to discover whether stem cells harvested from peripheral blood have an equivalent capacity for long-term repopulation (see below).

D. Gene Transfer to Human Marrow Progenitor Cells is Safe — So Far

One of the major concerns about any use of long-lived marrow progenitor cells for retrovirus-mediated gene transfer or therapy is that insertional mutagenesis would occur.[43] This concern has been given extra weight following the discovery of two thymomas in monkeys injected with a vector contaminated with wild-type virus.[44] In all clinical studies, vectors are used that are free of detectable replication-competent retroviruses, and no evidence has yet been obtained for any recombinational events with endogenous retroviral sequences that have generated infectious virus. But although no adverse events due to gene transfer have occurred in the first 140 patient months, all patients will have prolonged follow-up and genetic analysis of any tumors that do appear within the next 15 years.

VII. LIMITATIONS OF ORIGINAL MARKING TECHNIQUES

These original protocols had been submitted for approval early in 1990, when little was known of the safety of gene transfer into human marrow cells. To minimize safety concerns, the protocol used as little marrow manipulation as possible and as simple a method of transduction as was consistent with detectable gene transfer rates.[19] The approach adopted was both extremely wasteful in terms of the quantities of virus used — requiring 5 to 25 l of virus supernatant per patient — and inefficient in terms of gene transfer. As a result of this inefficiency, only limited conclusions could be drawn about the contribution of marrow-based disease to recurrence. The only definitive result was one in which a relapse was gene marked. Unmarked relapses might mean that marrow did not contribute to recurrence, but might also mean that the relapse was generated from only a few marrow-derived malignant cells that escaped being gene marked because of the inefficiency of the process. This particular problem would be accentuated if the marked marrow was also purged, when malignant cell numbers might drop below the detection threshold of the technique. An increased efficiency of marking is one requirement to overcome this limitation.

VIII. FUTURE APPLICATIONS OF GENE MARKING

If the efficiency of the marking techniques could be improved, not only would some of these limitations be overcome, but the technique could be used to address a number of other important scientific and clinical issues in autologous bone marrow transplantation. Some examples follow.

A. Clonality of Malignant and Normal Cell Repopulation

Since retrovirus vectors integrate essentially at random in the host cell genome, each integrant detected must originate from a separate clone of cells. This information may be informative for both malignant and normal cell recovery.

At relapse, analysis of the provirus integration site can reveal whether recurrence is a monoclonal, oligoclonal, or polyclonal event. Analysis of integrants in normal and malignant colonies by restriction fragment length polymorphism (RFLP) or inverted PCR may also indicate whether the malignancy originates in a committed progenitor cell or a more primitive precursor capable of differentiating along a number of separate lineages. This is a crucial issue because the strategy required for cure will likely differ based on this distinction. Should the malignancy prove to be a stem cell disorder, ablation of relatively mature marrow cells by chemotherapy or standard autologous bone marrow transplantation is unlikely to achieve any lasting benefit.

In the normal marrow repopulation, analysis of clonality will provide information on the numbers of stem cells in harvested marrow and can be used to monitor the effects of different regimens intended to expand these progenitors prior to marrow infusion.

B. Efficiency of Purging

As new vectors become available for clinical application, it will be possible to mark distinctively different aliquots of marrow from the same patient and compare several purging technologies simultaneously. Since the end point of such purging studies would be the absence of marked relapse rather than a prolongation of survival or a reduction in relapse risk, even small-scale clinical trials could be informative.

IX. NEW — AND IMPROVED — GENE MARKING PROTOCOLS

The results of the early marking protocols coupled with a realization of the potential of the method have encouraged development of more sophisticated protocols, which are both more efficient and logistically less formidable.[22] Such protocols have opened for patients with CML[45] (M. D. Anderson Cancer Center, University of Texas), breast cancer, and myeloma (National Heart Lung and Blood Institute, National Institutes of Health)[46] and adult ALL and AML (University of Indiana, Indianapolis), while revised protocols for AML and neuroblastoma continue at St. Jude Children's Research Hospital. These protocols incorporate one or more of the modifications described below.

A. Selecting CD34$^+$ Cells

There is now good evidence that selected CD34$^+$ cells will reconstitute patients after autologous marrow transplantation.[46] The dual justification for CD34$^+$ selection in the context of gene marking is that it reduces the volume of vector supernatant required for transduction and that it may be a component of marrow purging in patients in whom the malignant cells are CD34$^-$. The potential disadvantage of CD34$^+$ selection is that immune reconstitution may be delayed, since mature lymphocytes are not transferred with the graft. There is a concern that this, in turn, may increase the risks of subsequent neoplastic change if undetected replication competent retrovirus (RCR) has contaminated the vector — a particular hazard during coculture with producer cell lines (see below). Three of eight primates receiving CD34$^+$-selected autografts transduced with producer lines making RCR — as well as high-titer vector — subsequently developed lymphomas containing wild-type virus.[44] The animals had evidence of multiple cycles of infection with the replication-competent virus, which may only have been possible because their poor immune reconstitution prevented an adequate response to what would normally have been an immunostimulatory inoculum of RCR.

The advantages of CD34 selection may thus be counterbalanced, at least in part, by a decrease in the margin of safety for marker studies. Nonetheless, almost all of the more recently approved marker studies use CD34$^+$ cells as the vector target (see Reference 45 and Table 1).

B. Increasing the Efficiency of Gene Transfer

Three approaches may improve the efficiency of transfer and thus the sensitivity of marking studies. The first is to expose the cells to vector producer cells rather than the supernatants derived therefrom. The second is to manipulate the cells to enter cell cycle to optimize retrovirus-mediated gene transfer. The third is to use alternative gene transfer techniques, which do not require the target cell to be in cycle for transfer to be

Table 1

Current Marrow Marker Studies

Source	Type of Protocol	$CD34^+$ Selection	Primary Interest
St. Jude Children's Research Hospital, Memphis	Neo^R marking of marrow from AML and neuroblastoma patients	No	Source of relapse after BMT
M. D. Anderson Hospital, Houston	Neo^R marking of marrow from CML patients	Yes	1) Source of blast crisis relapse after ABMT to restore chronic phase 2) Contribution of blood and marrow stem cells to recovery
Indiana University, Indianapolis	Neo^R marking of marrow from AML and ALL patients	No	Source of relapse after ABMT
NHLBI/NIH	Neo^R marking of marrow and peripheral blood cells from breast cancer and myeloma patients	Yes	Source of relapse after ABMT Contribution of peripheral blood vs. marrow to reconstitution
Fred Hutchinson Cancer Center, Seattle	Neo^R marking of growth factor-mobilized stem cells	Yes	Contribution of peripheral blood and bone marrow to repopulation

a success. While all these approaches will be tried, only the first two are used in current marking protocols for hemopoietic cells.

1. Coculture with Producer Lines

Retrovirus-mediated gene transfer may be increased in efficiency by coculture of the target cells with the vector producer line rather than with the supernatant derived therefrom.[5,6,17,18] This allows multiple cycles of infection and may catch early progenitor cells as they begin to enter cycle. We do not yet know whether this will modify gene transfer into human pluripotent progenitor cells, particularly since these may have been obtained from patients whose marrow is already undergoing massive regeneration following multiple cycles of chemotherapy. Moreover, a potential benefit is again offset by a potential increase in risk. Coculture with producer cell lines cannot be as closely monitored as coculture with vector supernatant; recombinational events may occur to produce RCR, which infect the target cells and remain undetected by current screening techniques. The thymomas found in monkeys receiving replication-competent retroviruses[44] suggest that such an event would pose a significant risk to the patient. Moreover, cells transduced on producer lines may regularly incorporate more than one copy of the vector, thereby increasing the risk for insertional mutagenesis.

2. Coculture with Growth Factors

While growth factor pretreatment[11-14,44] should not increase the risk of insertional mutagenesis or malignant transformation, it may favor the growth of contaminating malignant cells, many of which have receptors for precisely those growth factors that are most efficient at promoting the growth of normal early progenitor cells. All these potential hazards may be small, but they remind us of the importance for extreme caution when a gene marker protocol is proposed, since these studies cannot benefit the patient directly but must always carry a finite risk. Current proposals try to limit these risks by using IL3, IL6, and kit-ligand in combination to stimulate marrow and peripheral blood $CD34^+$ cells from patients with breast cancer, in whom these agents are unlikely to enhance tumor cell growth (Dunbar et al., NHLBI/NIH clinical protocol).

3. Alternative Vectors

It may be possible to develop vectors that efficiently transduce stem cells even when they are in G_0. The two vectors that have been suggested most commonly for this purpose are adenovirus[47,48] and adeno-associated virus.[49] At present there is no information on whether either of these vectors is capable of transducing human pluripotent progenitor cells. Adenovirus has the additional disadvantage of acting predominantly as an episomal agent, which would almost certainly be rapidly lost during the multiple cycles of expansion and differentiation that are the feature of stem cell development. Nonetheless, suitable alternatives to retrovirus vectors are being pursued actively.

C. Use of Multiple Vectors

Initial marker protocols used a single vector for each patient so that the safety of each vector could be determined individually. As safety data have accumulated, however, it has become permissible to use two distinguishable vectors in a single individual. This enormously increases the power of gene marking, since these vectors can be used to compare the effects of marrow treatment within rather than between individuals. Dual marker studies have been proposed to determine the efficacy of purging, the relative contributions of peripheral blood and marrow-derived stem cells to repopulation after ABMT, and the effects of *ex vivo* marrow pretreatment with cytokines on subsequent engraftment. For example, Deisseroth et al. at M. D. Anderson[45] will study CML accelerated-phase or blast-crisis patients who are reinduced into second chronic phase or cytogenetic remission. The goal of their protocol is to compare preparations of hematopoietic cells collected early in phases of hematopoietic recovery following conventional dose chemotherapy after CD34 selection with those hematopoietic cells collected from the marrow following hematopoietic recovery after conventional dose chemotherapy following CD34 selection. The initial analysis will focus on the reconstitutive capability of the two preparations, and late analysis at the time of relapse will focus on the relative level of representation of the marked retroviruses in the peripheral blood and marrow that appear in the hematopoietic cells of the patient following recovery. Thus, the latter protocol is useful in comparing reconstitutive capability of peripheral blood and marrow and the level of contamination with neoplastic cells in those two settings. At NIH, this dual marker approach is being used for breast cancer and myeloma patients.[46] The primary intent is to compare the relative capacity of progenitor cells obtained from peripheral blood and bone marrow to effect short- and long-term hemopoietic recovery after autologous transplantation. At SJCRH, we are using dual marker studies to compare the efficiency of different purging techniques in patients with AML and neuroblastoma (Figure 4).

X. CONCLUSION: WHAT IS THE VALUE OF GENE MARKING IN AUTOLOGOUS BONE MARROW TRANSPLANTATION?

It is likely that gene marking will continue to be used to help define the role of autologous bone marrow transplantation in cancer therapy and to help guide purging methodologies. In this regard, it should be a valuable complement to conventional randomized clinical efficacy trials. Finally, data from gene marking studies have provided encouragement to proposed gene therapy studies targeting bone marrow cells.

XI. GENE THERAPY PROTOCOLS

While hemopoietic cell gene marking protocols are valuable tools for analyzing and evaluating available therapies, it is obviously in the gene therapy of these cells that most interest lies. Many of the techniques explored in the early marking protocols can be translated to improve the probability of success in human gene therapy protocols, and several potential applications have already been described (see Chapters 2 and 3). This chapter only briefly outlines those clinical protocols that are currently approved.

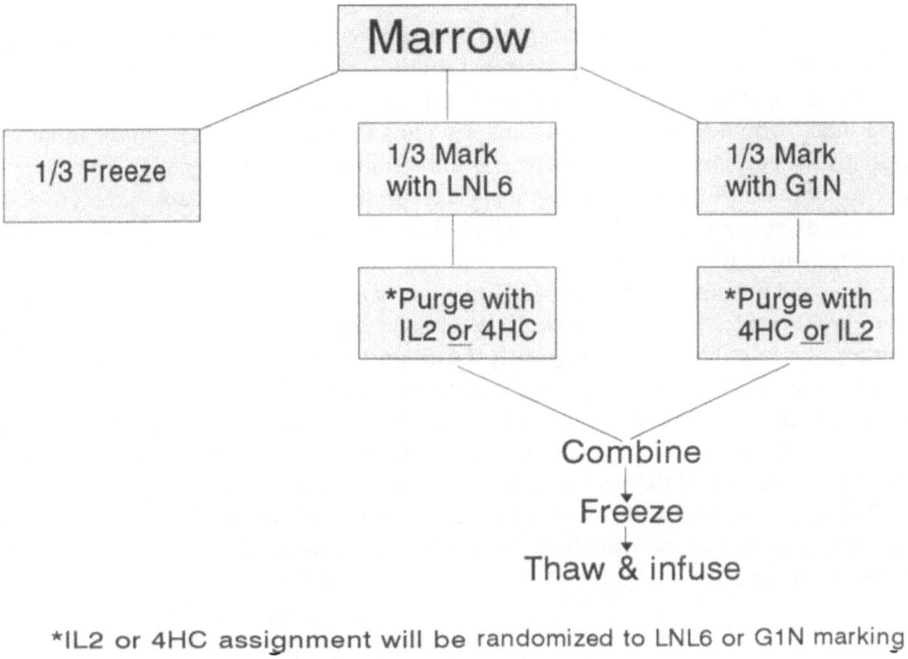

FIGURE 4. Outline of double marker studies for comparing efficiency of purging techniques in individual patients.

A. Single Gene Defects

1. Adenosine Deaminase (ADA) Deficiency Protocols

These protocols have been discussed in detail in Chapter 2 and 3. To date, patients have been treated on three protocols that use ADA gene transfer into hemopoietic progenitor cells. Blaese et al. (NIH/U.S.) use $CD34^+$ cells from peripheral blood or umbilical cord and expose them to retrovirus (G1Na.SaVd) supernatant. Valerio, Hoogerbruge, et al.[50] use $CD34^+$-selected cells from bone marrow stimulated *ex vivo* with IL3 and then cultured on vector producer cell lines, while Bordignon et al. (Milan, Italy) have used a similar bone marrow gene transfer protocol.[13]

2. Gaucher Disease

A protocol to introduce the glucocerebrosidase gene in the G1N vector has recently been approved for use in children with severe Gaucher disease. Bone marrow and peripheral blood $CD34^+$ cells will be the target and will be exposed to retrovirus vectors in the presence of the hemopoietic growth factors IL3, IL6, and stem cell factor. The cells will be returned to patients who have received no preinfusion conditioning.

B. Multidrug Resistance Gene Therapy

Three protocols have now been approved to insert the multidrug resistance (MDR)-1 gene into marrow progenitor cells.[51,52] The gene product forms part of a membrane-localized transport system that is capable of pumping a wide variety of cytotoxic drugs out of the cell. All these MDR gene transfer protocols have two aims. Introduction of the gene into marrow stem cells in patients with cancer may allow these cells to express higher than normal levels of the protein, providing increased resistance to the effects of cytotoxic agents. The resultant increase in the therapeutic index may allow tolerance of more intensive cytotoxic drug regimens and produce a higher cure rate. There is, however, dispute about the degree to which dose intensification will be possible before other organ toxicities supervene.

A second benefit may be that MDR-transfer will allow *in vivo* selection, analogous to the *in vitro* selection made possible by transferring the Neo^R gene.[53,54] In mice, for example, it is possible to increase greatly the proportion of hemopoietic cells expressing a transferred MDR-1 gene simply by treating the animal with low dose Taxol®, a drug that kills marrow cells that do not express the MDR-1 gene product. If this approach works in humans, even low levels of therapeutic gene transfer into hemopoietic cells would be beneficial, if the vector also encoded the selectable MDR-1 gene.

The approved protocols use $CD34^+$ marrow cells, stimulated by culture with hemopoietic growth factors (usually including IL-6 and kit-ligand) and cocultivated with producer cell lines.

XII. GENERAL CONCLUSIONS

As with other tissues, marrow gene therapy will only reach fruition when there are substantial improvements both in vector design and in our capacity to regulate transferred genes. Nonetheless, even the crude methodologies currently available continue to provide us with unique information about the biology of autologous marrow transplantation and are likely to provide therapeutic benefits in selected patients.

ACKNOWLEDGMENTS

I would like to thank my colleagues in the Division of Bone Marrow Transplantation, the Departments of Hematology/Oncology and Biochemistry and Genetic Therapy, Inc. for their help. I would also like to thank Nancy Parnell for word processing.

REFERENCES

1. Smith, C., Retroviral vector-mediated gene transfer into hematopoietic cells: prospects and issues, *J. Hematother.*, 1, 155, 1992.
2. Anderson, W. F., The ADA human gene therapy clinical protocol, *Hum. Gene Ther.*, 1, 327, 1990.
3. Miller, A. D., Human gene therapy comes of age, *Nature*, 357, 455, 1992.
4. Eglitis, M. A., Kantoff, P. W., McLachlin, J. R., Gillio, A., Flake, A. W., Bordignon, C., Moen, R. C., Karson, E. M., Zwiebel, J. A., Kohn, D. B., et al., Gene therapy: efforts at developing large animal models for autologous bone marrow transplant and gene transfer with retrovirus vectors, *Ciba Found. Symp.*, 130, 229, 1987.

5. Schuening, F. G., Kawahara, K., Miller, A. D., To, R., Goehle, S., Stweard, S., Mullally, K., Fisher, L., Graham, T. C., Appelbaum, F. R., Hackman, R., Osborne, W. R. A., and Storb, R., Retrovirus-mediated gene transduction into long-term repopulating marrow cells of dogs, *Blood*, 78, 2568, 1991.
6. Van Beusechem, V. W., Kukler, A., Heidt, P. J., and Valerio, D., Long-term expression of human adenosine deaminase in rhesus monkeys transplanted with retrovirus-infected bone-marrow cells, *Proc. Natl. Acad. Sci. U.S.A.*, 89, 7640, 1992.
7 Ekhterae, D., Crumbleholme, T., Karson, E., Harrison, M. R., Anderson, W. F., and Zanjani, E. D., Retroviral vector-mediated transfer of the bacterial neomycin resistance gene into fetal and adult sheep and human hematopoietic progenitors in vitro, *Blood*, 75, 365, 1990.
8. Lothrop, C. J., Jr., Allebban, Z. S., Niemeyer, G. P., Jones, J. B., Peterson, M. G., Smith, J. R., Baker, J. H., Morgan, R. A., Eglitis, M. A., and Anderson, W. F., Expression of a foreign gene in cats reconstituted with retroviral vector infected autologous bone marrow, *Blood*, 78, 237, 1991.
9. Culver, K., Cornetta, K., Morgan, R., Morecki, S., Aebersold, P., Kasid, A., Lotze, M., Rosenberg, S. A., Anderson, W. F., and Blaese, R. M., Lymphocytes as cellular vehicles for gene therapy in mouse and man, *Proc. Natl. Acad. Sci. U.S.A.*, 88, 3155, 1991.
10. Rosenberg, S. A., Aebersold, P., Cornetta, K., Kasid, A., Morgan, R. A., Moen, R., Karson, E. M., Lotze, M. T., Yang, J. C., Topalian, S. L., Merino, M. J., Culver, J., Miller, D., Blaese, R. M., and Anderson, W. F., Gene transfer into humans — immunotherapy of patients with advanced melanoma, using tumor-infiltrating lymphocytes modified by retroviral gene transduction, *N. Engl. J. Med.*, 323, 570, 1990.
11. Bodine, D. M., Karlsson, S., Papayannopoulou, T., and Nienhuis, A. W., Introduction and expression of human beta globin genes into primitive murine hematopoietic progenitor cells by retrovirus mediated gene transfer, *Prog. Clin. Biol. Res.*, 319, 589, 1989.
12. Gelinas, R. E., Bender, M. A., Miller, A. D., and Novak, U., Long-term expression of human beta-globin gene after retrovirus transfer into pluripotent hematopoietic stem cells of the mouse, *Adv. Exp. Med. Biol.*, 271, 135, 1989.
13. Bordignon, C., Yu, S. F., Hanzopoulos, P., Ungers, G. E., Keever, C. A., O'Reilly, R. J., and Gilboa, E., Retrovirus-mediated vector high efficiency expression of adenosine deaminase in hematopoietic long term cultures of ADA deficient marrow cells, *Proc. Natl. Acad. Sci. U.S.A.*, 86, 6748, 1989.
14 Correll, P. H., Fink, J. K., Brady, R. O., Perry, L. K., and Karlsson, S., Production of human glucocerebrosidase in mice after retroviral gene tranfer into multipotential hematopoietic progenitor cells, *Proc. Natl. Acad. Sci. U.S.A.*, 86, 8912, 1989.
15 Hogge, D. E. and Humphries, R. K., Gene transfer to primary normal and malignant human hemopoietic progenitors using recombinant retroviruses, *Blood*, 69, 611, 1987.
16. Wilson, J. M., Danos, O., Grossman, M., Raulet, D. H., and Mulligan, R. C., Expression of human adenosine deaminase in mice reconstituted with retrovirus-transduced hematopoietic stem cells, *Proc. Natl. Acad. Sci. U.S.A.*, 87, 439, 1990.
17. Lim, B., Apperley, J. F., Orkin, S. H., and Williams, D. A., Long-term expression of human adenosine deaminase in mice transplanted with retrovirus-infected hematopoietic stem cells, *Proc. Natl. Acad. Sci. U.S.A.*, 86, 8892, 1989.
18. Kaleko, M., Garcia, J. V., Osborne, W. R. A., and Miller, A.D., Expression of human adenosine deaminase in mice after transplantation of genetically-modified bone marrow, *Blood*, 75, 1733, 1990.
19. Rill, D. R., Moen, R. C., Buschle, M., Bartholomew, C., Foreman, N. K., Mirro, J., Jr., Krance, R. A., Ihle, J. N., and Brenner, M. K., An approach for the analysis of relapse and marrow reconstitution after autologous marrow transplantation using retrovirus-mediated gene transfer, *Blood*, 79, 2694, 1992.

20. Rill, D. R., Buschle, M., Foreman, N. K., Bartholomew, C., Moen, R. C., Santana, V. M., Ihle, J. N., and Brenner, M. K., Retrovirus mediated gene transfer as an approach to analyze neuroblastoma relapse after autologous bone marrow transplantation, *Hum. Gene Ther.*, 3, 129, 1992.
21. Brenner, M., Mirro, J., Jr., Hurwitz, C., Santana, V., Ihle, J., Krance, R., Ribeiro, R., Roberts, W. M., Mahmoud, H., Schell, M., and Garth, K., Autologous bone marrow transplant for children with AML in first complete remission: use of marker genes to investigate the biology of marrow reconstitution and the mechanism of relapse, *Hum. Gene Ther.*, 2, 137, 1991.
22. Brenner, M. K., Rill, D. R., Moen, R. C., Krance, R. A., Mirro, J., Jr., Anderson, W. F., and Ihle, J. N., Gene-marking to trace origin of relapse after autologous bone marrow transplantation, *Lancet*, 341, 85, 1993.
23. Appelbaum, F. R. and Buckner, C. D., Overview of the clinical relevance of autologous bone marrow transplantation, *Clin. Haematol.*, 15, 1, 1986.
24. Burnett, A. K., Tansey, P., Watkins, R., Alcorn, M., Maharaj, D., Singer, C. R., McKinnon, S., McDonald, G. A., and Robertson, A. G., Transplantation of unpurged autologous bone-marrow in acute myeloid leukaemia in first remission, *Lancet*, 2, 1068, 1984.
25. Goldstone, A. H., Anderson, C. C., Linch, D. C., Franklin, I. M., Boughton, B. J., Cawley, J. C., and Richards, J. D., Autologous bone marrow transplantation following high dose chemotherapy for the treatment of adult patients with acute myeloid leukaemia, *Br. J. Haematol.*, 64, 529, 1986.
26. Chopra, R., Goldstone, A. H., McMillan, A. K., Powles, R., Smith, A. G., Prentice, H. G., Reid, C., Marcus, R., Bell, A., Milligan, D., McCarthy, D., Morgenstern, G., and Barnard, D., Successful treatment of acute myeloid leukemia beyond first remission with autologous bone marrow transplantation using busulfan/cyclophosphamide and unpurged marrow: the British Autograft Group experience, *J. Clin. Oncol.*, 9, 1840, 1991.
27. Shuster, J. J., Cantor, A. B., McWilliams, N., Graham-Pole, J., Castleberry, R. P., Marcus, R., Pick, T., Smith, E. I., and Hayes, F. A., The prognostic significance of autologous bone marrow transplant in advanced neuroblastoma, *J. Clin. Oncol.*, 9, 1045, 1991.
28. De Fabritiis, P., Ferrero, D., Sandrelli, A., Tarella, C., Meloni, G., Pulsoni, A., Pregno, P., Badoni, R., DeFelice, L., Gallo, E., et al., Monoclonal antibody purging and autologous bone marrow transplantation in acute myelogenous leukemia in complete remission, *Bone Marrow Transpl.*, 4, 669, 1989.
29. Gambacorti-Passerini, C., Rivoltini, L., Fizzotti, M., Rodolfo, M., Sensi, M. L., Castelli, C., Orazi, A., Polli, N., Bregni, M., Siena, S., et al., Selective purging by human interleukin-2 activated lymphocytes of bone marrows contaminated with a lymphoma line or autologous leukaemic cells, *Br. J. Haematol.*, 78, 197, 1991.
30. Gorin, N. C., Aegerter, P., Auvert, B., Meloni, G., Goldstone, A. H., Burnett, A., Carella, A., Korbling, M., Herve, P., Maraninchi, D., et al., Autologous bone marrow transplantation for acute myelocytic leukemia in first remission: a European survey of the role of marrow purging, *Blood*, 75, 1606, 1990.
31. Santos, G. W., Yeager, A. M., and Jones, R. J., Autologous bone marrow transplantation, *Annu. Rev. Med.*, 40, 99, 1989.
32. Petersen, F. B. and Buckner, C. D., Allogeneic and autologous bone marrow transplantation for acute leukemia and malignant lymphoma: current status, *Hematol. Oncol.*, 5, 233, 1987.
33. Yeager, A. M., Kaizer, H., Santos, G. W., Saral, R., Colvin, O. M., Stuart, R. K., Braine, H. G., Burke, P. J., Ambinder, R. F., Burns, W. H., et al., Autologous bone marrow transplantation in patients with acute nonlymphocytic leukemia, using ex vivo marrow treatment with 4-hydroperoxycyclophosphamide, *N. Engl. J. Med.*, 315, 141, 1986.
34. Campana, D., Coustan-Smith, E., and Behm, F. G., The definition of remission in acute leukemia with immunologic techniques, *Bone Marrow Transpl.*, 8, 429, 1991.

35. Kaizer, H., Stuart, R. K., Brookmeyer, R., Beschorner, W. E., Braine, H. G., Burns, W. H., Fuller, D. J., Korbling, M., Mangan, K. F., Saral, R., et al., Autologous bone marrow transplantation in acute leukemia: a phase I study of in vitro treatment of marrow with 4-hydroperoxycyclophosphamide to purge tumor cells, *Blood*, 65, 1504, 1985.
36. Gorin, N. C., Douay, L., Laporte, J. P., Lopez, M., Mary, J. Y., Najman, A., Salmon, C., Aegerter, P., Stachowiak, J., David, R., et al., Autologous bone marrow transplantation using marrow incubated with Asta Z 7557 in adult acute leukemia, *Blood*, 67, 1367, 1986.
37. Miyoshi, H., Shimizu, K., Kozu, T., et al., t(8;21) breakpoints on chromosome 21 in acute myeloid leukemia are clustered within a limited region of a single gene, AML1, *Proc. Natl. Acad. Sci. U.S.A.*, 88, 10431, 1991.
38. Erickson, P., Gao, J., Chang, K.-S., et al., Identification of breakpoints in t(8;21) AML and isolation of a fusion transcript with similarity to Drosophila segmentation gene runt, *Blood*, in press, 1992.
39. Nucifora, G., Birn, D. J., Erickson, P., Gao, J., LeBeau, M. M., Drabkin, H. A., and Rowley, J. D., Detection of DNA rearrangements in the AML1 loci and of an AML1/ETO fusion mRNA in patients with t(8;21) AML, *Blood*, in press, 1992.
40. Santana, V. M., Brenner, M. K., Ihle, J., Krance, R., Furman, W., Bowman, L., Ribeiro, R., Hurwitz, C., Mahmoud, H., Moen, R. C., and Anderson, W. F., A phase I trial of high-dose carboplatin and etoposide with autologous marrow support for treatment of stage D neuroblastoma in first remission: use of marker genes to investigate the biology of marrow reconstitution and the mechanism of relapse, *Hum. Gene Ther.*, 3, 257, 1991.
41. Morgan, R. A., Cornetta, K., and Anderson, W. F., Applications of the polymerase chain reaction in retroviral mediated gene transfer and the analysis of gene marked human TIL cells, *Hum. Gene Ther.*, 1, 135, 1990.
42. Wieder, R., Cornetta, K., Kessler, S. W., and Anderson, W. F., Increased efficiency of retrovirus-mediated gene transfer and expression in primate bone marrow progenitors after 5-fluorouracil-induced hematopoietic suppression and recovery, *Blood*, 77, 448, 1991.
43. Cornetta, K., Morgan, R. A., and Anderson, W. F., Safety issues related to retrovirus-mediated gene transfer in humans, *Hum. Gene Ther.*, 2, 5, 1991.
44. Donahue, R. E., Kessler, S. W., Bodine, D., McDonagh, K., Dunbar, C., Goodman, S., Agricola, B., Byrne, E., Raffeld, M., Moen, R., Bacher, J., Zsebo, K. M., and Nienhuis, A. W., Helper virus induced T cell lymphoma in nonhuman primates after retroviral mediated gene transfer, *J. Exp. Med.*, 176, 1125, 1992.
45. The University of Texas M. D. Anderson Cancer Center, Autologous bone marrow transplantation for CML in which retroviral markers are used to discriminate between relapse which arises from systemic disease remaining after preparative therapy versus relapse due to residual leukemia cells in autologous marrow: a pilot trial, *Hum. Gene Ther.*, 2, 359, 1991.
46. O'Shaughnessy, J. A., Cowan, K. H., Wilson, W., Bryant, G., et al., Pilot study of high dose ICE (ifosfamide, carboplatin, etoposide) chemotherapy and autologous bone marrow transplant (ABMT) with neo^R-transduced bone marrow and peripheral blood stem cells in patients with metastatic breast cancer, *Hum. Gene Ther.*, 4, 331, 1993.
47. Lemarchand, P., Jaffe, H. A., Danel, C., Cid, M. C., Kleinman, H. K., Stratford-Perricaudet, L. D., Perricaudet, M., Pavirani, A., Lecocq, J. P., and Crystal, R. G., Adenovirus-mediated transfer of a recombinant human alpha 1-antitrypsin CDNA to human endothelial cells, *Proc. Natl. Acad. Sci. U.S.A.*, 89, 6482, 1992.
48. Quantin, B., Perricaudet, L. D., Tajbakhsh, S., and Mandel, J. L., Adenovirus as an expression vector in muscle cells in vivo, *Proc. Natl. Acad. Sci. U.S.A.*, 89, 2581, 1992.
49. Walsh, C. E., Liu, J. M., Xiao, X., Young, N. S., Nienhuis, A. W., and Samulski, R. J., Regulated high level expression of a human gamma-globin gene introduced into erythroid cells by an adeno-associated virus vector, *Proc. Natl. Acad. Sci. U.S.A.*, 89, 7257, 1992.

50. Hoogerbrugge, P. M., Vosssen, J. M. J. J., Van Beusechem, V. W., and Valerio, D., Treatment of patients with severe combined immunodeficiency due to adenosine deaminase (ADA) deficiency by autologous transplantation of genetically modified bone marrow cells, *Hum. Gene Ther.*, 3, 553, 1992.
51. Hock, R. A. and Miller, A. D., Retrovirus mediated transfer and expression of drug resistance genes in human haematopoietic progenitor cells, *Nature*, 320, 275,1986.
52. Chin, K. V., Pastan, I., and Gottesman, M. M., Function and regulation of the human multidrug resistance gene, *Adv. Cancer Res.*, 60, 157, 1993.
53. Sorrentino, B. P., Brandt, S. J., Bodine, D., Gottesman, M., Pastan, I., Cline, A., and Nienhuis, A. W., Selection of drug-resistant bone marrow cells in vivo after retroviral transfer of human MDR1, *Science*, 257, 99, 1992.
54. Podda, S., Ward, M., Himelstein, A., Richardson, C., de la Flor-Weiss, E., Smith, L., Gottesman, M., Pastan, I., and Bank, A., Transfer and expression of the human multiple drug resistance gene into live mice, *Proc. Natl. Acad. Sci. U.S.A.*, 89, 9676, 1992.

14 Gene Therapy for Brain Tumors

Kenneth W. Culver, John Van Gilder, Thomas Carlstrom, Michael Prados, and Charles J. Link, Jr.

I. INTRODUCTION

Gene therapy as a potential treatment of cancer has exploded onto the forefront of medicine. To date, more than 30 human gene therapy clinical trials are approved for cancer treatment in the U.S. While none of these trials has yet reported significant antitumor efficacy, these efforts clearly signal the dawning of a new era in the evolution of cancer therapy. In this chapter, we will discuss the molecular biology of gene therapy and its application to the clinical research on treatments for primary and metastatic brain tumors.

Primary brain tumors represent a significant health problem in the U.S. National surveys estimate the number of new primary brain tumors at about 15,000 per year in the U.S.[1,2] These 15,000 tumors represent more than 25 different histologic types of primary brain tumors. Of these, glioblastoma multiforme (GBM) comprises the largest single group (20 to 30%), depending upon the study. The etiology of primary tumors is unclear, as there are no known, documented environmental risk factors. Genetics are thought to play a role, since 16% of patients with primary brain tumors have a positive family history of cancer.[3]

For many patients, GBM evolves from a diffuse astrocytoma.[4] It is slightly more prevalent (55%) in males. It has the oldest mean (60.2 years) and median (62.0 years) ages of onset of all brain tumors. Recent data suggest that the incidence of GBM is increasing, especially in the elderly.[5,6] Five-year survival of GBM is poor, 5.5%. Primary brain tumors, including GBM, do not regularly metastasize outside the CNS.[7,8] Therefore, management of the primary tumor is directed to the local tumor area. Traditional therapy for GBM typically consists of an attempt at complete surgical resection of the tumor and postoperative high dose irradiation (usually 5500 to 6000 cGy over 5 to 6 weeks).[9,10] The median survival of GBM patients with this regimen is 9 to 10 months.[11] Although controversial, the length of survival is increased by 2 months when chemotherapy is added to surgical resection and irradiation.[12-13] At the diagnosis of a recurrence of GBM, there is a 100% mortality within weeks to a few months. Harsh et al. reported a mean survival of only 36 weeks in patients with recurrent GBM who underwent a second operation.[2] A reasonable quality of life in these patients was limited to 10 weeks following the diagnosis of recurrent GBM.

0-8493-2440-8/95/$0.00+$.50

Table 1

Experimental Approaches to the Treatment of Brain Tumors

Chemotherapy [110-113]	Radiation[3,114,115]
Systemic	External beam
Regional (intravascular, intratumoral)	Stereotaxic
Immunotherapy[87,88,115-117]	Brachytherapy (radioactive implants)
Injection with irradiated autologous tumor cells	Hypoxic radiosensitizers
Lymphokine-activated killer cells (LAK)	Antibody-guided
Hypothermia[3]	

A variety of experimental treatments for malignant brain tumors has been attempted in humans (Table 1). Nontraditional experimental modalities have not resulted in an improvement in the overall survival with GBM. New treatment strategies are needed for GBM, and we believe that human gene therapy offers novel opportunities for the development of effective therapies for brain tumors.

II. MOLECULAR ASPECTS OF PRIMARY BRAIN TUMORS

Numerous molecular mechanisms are utilized by tumors to resist destruction. These include invasion into the brain, preventing complete surgical resection, evasion of destruction by the immune system, and the development of resistance to radiation and chemotherapy. All of the details have not been clearly elucidated, but several clues exist to these mechanisms. A more detailed understanding of these mechanisms may allow the development of novel therapies.

A. Invasion

GBM tumors infiltrate into the surrounding brain parenchyma, successfully avoiding complete surgical resection and complicating radiotherapy. Efforts to surgically excise an entire temporal lobe involved with tumor are typically followed by recurrence in another area of the brain. Biopsy studies have demonstrated that at the time of the initial diagnosis, most GBM tumors have spread more than 1 cm beyond the area of enhancement noted by MRI scan.[14,15] Nonetheless, 80 to 90% of patients die from complications related to local tumor recurrence. While the application of local therapies is rarely curative, these therapies can substantially improve the duration and quality of life.[12]

B. Molecular Abnormalities

The degree of tumor infiltration varies between patients. The variable invasive properties of these tumors are associated with specific genetic abnormalities occurring within a cell. The wild-type (WT) p53 tumor suppressor gene has multiple functions within the cell, including suppression of tumor cell growth and maintenance of genetic stability. Progression from a low grade (astrocytoma) to a high grade (GBM) has been associated with p53 gene mutations.[16] Introduction of a WTp53 gene into human tumor cells can result in suppression of tumor cell growth *in vitro*.[17-19] Another example is the up-regulation of the endothelial cell-specific mitogen, vascular endothelial growth factor (VEGF) as a tumor progresses from an astrocytoma to a GBM.[20] VEGF appears to be a potent inducer of angiogenesis, which is essential to solid tumor growth. Treatment of animals with a monoclonal antibody to VEGF inhibited tumor growth, suggesting that VEGF is another potential therapeutic target.[21] Overexpression of the platelet-derived growth factor (PDGF) receptor may also result in an autocrine stimulation of endothelial cell proliferation and neovascularization.[22] There are probably many other molecular abnormalities involved in carcinogenesis and invasion in GBM that need to be elucidated.

Other than the abnormalities in p53 (chromosome 17p) noted above, chromosomal abnormalities in astrocytomas and GBM tumors have been identified in chromosomes 1, 2, 6, 7, 9, 10, 11, 13, 17, 19, and 22.[23-25] These cytogenetic anomalies include an array of deletions, rearrangements, and amplifications. The most commonly amplified gene is the epidermal growth factor receptor (EGFr) gene, which is located on chromosome 7.[26] EGFr gene amplification is much more common in GBM than astrocytomas, suggesting that this genetic abnormality may be associated with a late stage glioma. The molecular mechanisms of tumorigenesis involved likely includes several oncogenes (e.g., c-*myc*, c-*sis*, N-*ras*, *Gli*, c-*erb*-B) and several tumor suppressor genes other than p53, but all the specific chromosomal abnormalities have not yet been identified.[27-30]

C. Immunosuppressive Aspects of GBM Tumors

Human gliomas depress host systemic immunity. One mechanism is believed to result from tumor cell production of the immunosuppressive compound TGF-β (transforming growth factor-β), which results in both local and systemic immunosuppression. TGF-β is thought to down-regulate IL-2 secretion and diminish expression of high-affinity IL-2 receptors on T-lymphocytes.[31] Other immune-modulating factors include the modulation of tumor antigens.[32,33]

D. Other Resistance Mechanisms in GBM Tumors

The development of resistance to chemotherapeutic agents due to multiple drug resistance-1 (MDR-1) gene overexpression has been identified in GBM.[34] Investigators have identified multidrug-resistant gliomas that overexpress MDR-1 mRNA. MDR-1 overexpression is also common in tumors that metastasize to the brain (e.g., lung, breast). Attempts at blocking MDR-1 expression through the administration of systemic drugs such as verapamil or cyclosporine A have not been optimal. Other molecular and cellular mechanisms of brain tumor resistance to DNA damaging agents (e.g., radiation, chemotherapy) undoubtedly occur but have not been identified specifically.

III. METASTATIC BRAIN TUMORS

The number and type of metastatic brain tumors generally follow the incidence of systemic cancers in the general population, with lung, breast, and colorectal cancers

Table 2

Important Features of the Murine Retroviral Vector System

- Stable vector gene insertion into the genome of proliferating cells
- Can be genetically engineered to carry genes
- Insertion of viral genome into the target cell DNA is random
- Can be genetically disabled to prevent the induction of disease
- Infect a wide range of cell types
- Low titer

being the most common.[35,36] While the incidence of CNS metastasis in patients with cancer is thought to be about 25% (~250,000/yr) of all systemic malignancies, most of these tumors are asymptomatic and detected only at autopsy. The actual incidence of cancer patients with symptomatic metastatic CNS tumors is thought to be around 10% (~100,000/yr), making metastases the most frequent type of intracranial neoplasm.[37] Traditional treatments for CNS metastases involve the surgical resection of solitary metastasis, if accessible, and/or whole brain irradiation for single or multiple lesions.[38] Multiple lesions are common with 70% of patients having one or two lesions by CT scan, and 30% of patients with three or more. While there are patients who die from the CNS disease, with no other evidence of systemic malignancy, most patients have progressive systemic disease. Therefore, the use of gene therapy as a treatment for metastatic CNS cancer would probably need to induce systemic antitumor effects or be used in conjunction with other systemic therapies to have any chance of effecting complete remission.

IV. GENE THERAPY APPROACHES TO BRAIN TUMORS

Depending upon the mode of action of the gene that is inserted into the tumor, a different treatment approach may be indicated. For example, the use of immune modulating agents may require only transient expression in T-cytotoxic cells or tumor cells to induce a vigorous antitumor immune response. In contrast, the insertion of a p53 gene for prevention or therapy of cancer may require restoration of permanent p53 function to deficient tumor cells. Therefore, the type of delivery system and the gene transferred is critical in the development of a genetic therapy for CNS tumors.

A. Gene Delivery Methods

1. Moloney Murine Leukemia Virus (MoMLV) Vectors

These are the most commonly used gene transfer methods in current clinical trials. Table 2 highlights critical features of this gene delivery system. These vectors are produced by replacing the viral genes required for the production of replication-competent viruses with the genes desired for transfer, creating replication-defective or replication-incompetent viruses.[39] This switch of genetic material renders the virus unable to produce a productive viral infection while maintaining the ability of the vector to bind to the cell surface. Once the virus binds to the cell, it is internalized, and the vector genes are inserted into the target cell chromosomes. This technique has the advantages of high-efficiency gene transfer, approaching 95% in some cell types *in vitro* and stable integration into the target cell genome. Once the vector is integrated, the vector genes become a stable part of the inheritance of that cell, being passed along to all daughter cells.

Murine retroviral vectors integrate and subsequently express vector genes only in proliferating cells.[39] In the brain, the tumor is the most mitotically active cell; macrophage-derived cells, blood cells, and endothelial cells are only minimally mitotic. Therefore, the probability of specific transduction of the tumor is enhanced. One potential disadvantage of retroviral vectors is that they randomly integrate into the host cell genome. This random integration means that vector gene expression may vary in each cell due to differences in the local chromosomal environment, where neighboring genes may influence the level of vector gene expression. For many genes, a very wide range of gene expression from cell to cell will be tolerated so long as sufficient protein is produced.

Random integration could occur within a gene that is absolutely required for normal cellular function or survival. In such a case, the transduced cell might be killed. However, a more serious theoretical problem with random integration may result if vector insertion results in oncogene activation or tumor suppressor gene inactivation. Either event could potentially lead to cellular transformation to a more malignant phenotype. Induction of oncogenesis by this mechanism is termed "insertional mutagenesis". Such events remain a theoretical possibility only, since the use of replication-incompetent retroviral vectors has never been reported to result in malignant transformation in any *in vivo* system.[40]

Many investigators have utilized murine retroviral vectors for *ex vivo* gene transfer. The *ex vivo* approach may be suitable for the stable transduction of hematopoeitic stem cells, but tumors typically cannot be removed from the body, so the transfer of genes into cancer cells *in vivo* would be greatly preferred. *In vivo* strategies also obviate the need to grow cells in culture. The first *in vivo* approach considered was the direct injection of retroviral supernate into tumor masses. The transduction frequency with direct supernate injection is low (1 to 3%) due to three facts: low vector titers (generally 10^5 to 10^6 cfu/ml), tumor cells are not all proliferating at the same time, and the vector particles do not penetrate solid tumor. Retroviral supernate may suffice in some circumstances, such as the treatment of meningeal metastases, which may be more efficiently transduced by injection of retroviral vector supernate into the CSF since penetration of the vector particles into a solid mass is not required.[41] As a result, we and others have been investigating the implantation of the retroviral vector producer cells (VPC) as a method for improved *in vivo* gene transfer into solid tumor masses.[42-44]

The direct implantation of VPC has been shown to result in superior gene transfer efficiency, ranging from 10 to 60%. Using β-galactosidase as a marker for gene transfer, we have observed selective tumor cell transduction, since gene integration occurs only within proliferating cells. Serial sections through the brain and a variety of visceral organs in rodents and monkeys have demonstrated no evidence of gene transfer outside the immediate area of VPC inoculation into the tumor.[45,46] The transduced cells were nearly all tumor cells except for an occasional endothelial cell. This finding is not surprising since the proliferative rate of endothelial cells within the tumor is generally higher than in normal tissues due to tumor angiogenesis.

2. Adenovirus Vectors

This vector system is based upon a common respiratory virus that produces infections of the respiratory tract. Adenoviruses have a natural tropism for respiratory epithelium. In order to improve their safety for use as a gene therapy vehicle, the E1A and E1B structural genes are deleted to render the virus replication-deficient. However, one of the risks of this gene transfer system is that wild-type adenoviruses may infect the same cell with their own E1A and E1B genes, allowing the production of infectious recombinant replication-competent vector particles. Applications of this vector system have focused primarily on the transfer of the cystic fibrosis transmembrane conductance regulator (CFTR) gene into airway cells as a treatment for cystic fibrosis. However, adenovirus vectors have also been shown to have the ability to transfer genes into most tissues of the body, including the brain.[47-50] The primary advantages of this vector system are high-efficiency gene transfer (approaching 100%, depending on the tissues and the mode of delivery), regardless of the proliferative state of the tissue, and the fact that the vectors can be produced at very high titer ($10^{11\text{-}12}$ cfu/ml) vs. retroviral vectors ($10^{5\text{-}6}$ cfu/ml).

There are several disadvantages to the use of adenoviral vectors in the brain. First, these vectors appear to be able to infect nearly all cell types, expressing their genes in each infected cell regardless of the cell's proliferative state. This lack of discrimination could result in significant toxicity to normal tissues. Tissue-specific promoters may be required to overcome this problem. One example is glial fibrillary acidic protein (GFAP), a gene that is naturally active in glial cells. Transfer of a vector containing a gene under the control of a GFAP promoter should result in expression only within the cells that normally synthesize GFAP. This principle has been demonstrated with tissue-specific promoters in animal models.[51] Second, this vector does not integrate. Therefore, long-term expression without repeated administration of the vector is a problem, especially in proliferating cell types where the gene frequency is diluted out as the tumor cells divide. The duration of vector gene expression in nonreplicating cells is under investigation. Third, as noted above, these vectors may recombine *in vivo* with endogenous adenoviruses to produce new, recombinant, replication-competent viruses that could cause disease.

3. Herpes Simplex Virus (HSV) Vectors

These viruses have the unique advantage of being tropic for the central nervous system, establishing life-long latent infections in neurons. Genetic attenuation of these viruses to prevent lytic destruction of infected cells can theoretically be made by deleting virulence genes, such as the thymidine kinase or IF-3 genes. These vectors could then potentially be used to deliver genes into neurons for the treatment of neurologic diseases such as Parkinson's disease or for the treatment of CNS tumors.[52]

Researchers have injected attenuated HSV-1 vectors directly into the brains of animals, demonstrating gene transfer into the surrounding neurons. HSV-1 vectors can induce an *in vivo* immune response following their injection into murine CNS tumors.[53] However, significant collateral damage to the surrounding normal brain tissue occurred, presumably due to the immune response to the experimental viral infection. Another major concern is that these vectors may activate latent wild-type herpes simplex virus in the brain and induce herpes encephalitis. Until these safety concerns have been adequately resolved, these vectors are unlikely to be used in human clinical trials for the treatment of CNS malignancy.

4. Direct, Nonviral-Mediated DNA Transfer

Gene transfer may be accomplished by direct injection of DNA without modification, inside artificially generated cationic lipid vesicles (liposomes) or conjugated to a soluble carrier to target the DNA to certain cell-specific ligands. The direct injection of DNA has shown promise for immunization.[54-57] Genes from the influenzae virus directly injected into skeletal muscle elicit a strong immune response. Perhaps, it may be possible to use this type of method to vaccinate the patients against the occurrence or the recurrence of GBM once the genes encoding brain tumor antigens have been specifically identified.

Liposomes are advantageous in that they will allow the DNA to survive *in vivo* for longer periods and bind to cells.[58] This *in vivo* gene delivery method is being used in one human clinical trial as a means to deliver a foreign HLA gene (B7) into human melanoma tumor deposits.[56] Other approaches include the direct injection of DNA conjugated to a carrier (e.g., transferrin) that targets specific cell surface receptors or the bombardment of cells with DNA-coated gold particles.[59,60] The latter is accomplished by coating metal pellets with DNA and shooting the pellets into tissue. The advantage of these direct DNA transfer systems is their simplicity compared to the production of recombinant virus vectors, which diminishes the potential risks and expense of the procedure. The main disadvantage is the low level of stable integration of the injected DNA. Therefore, long-term expression without repeated administration of the exogenous DNA may be a problem, especially in proliferating tumor cells. However, in cell types that do not regularly proliferate (e.g., neurons), the new DNA may continue to express its genes for months.

V. THERAPEUTIC GENES FOR THE TREATMENT OF BRAIN TUMORS

A. Sensitivity Genes (Nonimmune-Mediated)

1. Herpes Simplex-Thymidine Kinase (HS-tk)

The HS-tk gene catalyzes the phosphorylation (monophosphorylation) of the antiherpes drugs acyclovir (ACV) and ganciclovir (GCV).[61-63] Cellular kinases then convert the monophosphate form of the drugs to a triphosphate form that competitively inhibits the association of deoxynucleoside triphosphates with DNA polymerase and is incorporated into DNA, markedly reducing the rate of DNA synthesis (GCV) or acts as a chain terminator (ACV). As a result, when a herpes simplex virus-infected cell is exposed to ACV or GCV, the cells are destroyed indirectly by ACV-TP or GCV-TP.[64,65] Host cellular thymidine kinases do not phosphorylate ACV or GCV, only the MP and DP forms.[66]

The same principle can be used for cancer therapy, where the transfer of the HS-tk gene into tumor cells *in vitro* results in their destruction with treatment by ACV or GCV either *in vitro* or after reimplantation of the tumor *in vivo*.[67-74] GCV is more potent than ACV in our animal model systems. At least two factors may be involved. The uptake of GCV into herpes simplex-infected cells is greater than for ACV, and GCV appears to be a better substrate for both viral and host kinase enzymes involved in nucleoside phosphorylation with minimal GCV phosphorylation in HS-tk negative cells.[75]

In order to study the potential application of the HS-tk gene for the treatment of cancer, we have pursued an *in vivo* gene transfer method using the implantation of VPCs. In a series of murine experiments, we demonstrated that the transfer of the HS-tk gene into a syngeneic fibrosarcoma *in vivo* resulted in sufficient gene transfer to allow complete tumor ablation with GCV treatment.[43] Surprisingly, complete tumor ablation occurred in the subcutaneous murine tumor models when less than 100% of the tumor cells contained the gene. Additional studies in mice suggest that if at least 10% of the tumor cells in a mixture contain the HS-tk gene, more than 50% of the cancer cells can be completely eliminated. The killing of adjacent, non-HS-tk-containing cells is termed the "bystander" effect. The etiology of the bystander effect is not completely understood. The leading hypothesis is that the phosphorylated forms of GCV pass via gap junctions into neighboring tumor cells.[76] This type of intracellular communication has been described previously with phosphorylated thioguanine derivatives.[77] *In vivo*, the bystander effect is limited to the tumor. Since many tumors do not have gap junctions communicating with surrounding normal tissues, this may explain the tumor-specific bystander killing seen *in vivo*.[78] A variety of investigators are pursuing the mechanism of action of the bystander effect.

The combination of *in vivo* gene transfer and the bystander effect might be a useful therapy for localized tumors. There is as yet no evidence that the "bystander" factor can act systemically. Unfortunately, most patients with cancer have systemic disease at diagnosis. Of the tumors that do not metastasize early, brain tumors seem to have several advantages for the application of this technique (Table 3). Cells of the CNS are not actively growing in number. The direct injection of VPC into the tumor should limit gene transfer to the tumor, since the surrounding normal brain is not proliferating, in contrast to other organs such as the liver where a higher proportion of cells are dividing. There are receptors for the vector on nondividing cells (e.g., neurons), but the vector cannot integrate and is lost. This should allow selective transduction of the tumor, eliminating undesirable side effects to normal brain. The CNS is also relatively immunologically privileged. The HS-tk/GCV system can produce significant tumor destruction in immunodeficient animals, which suggests that the immune system plays a minor role in either the direct killing or the bystander effect. Therefore, this method of tumor destruction is also useful outside the blood-brain barrier. In addition, this immunologically privileged feature might allow prolonged survival of allogeneic or xenogeneic cells. If these cells are VPC, this permits a longer period of time for gene transfer to

Table 3

Advantages of *In Vivo* Gene Transfer in the Brain

The Brain
- Immunologically privileged
- Brain cells are not generally growing in number
- Brain tumors do not generally spread outside the CNS

Murine Retroviral Vectors
- Integrate only into dividing cells

Delivery System
- The producer cells continuously produce vector particles for 7–14 days

Herpes Simplex Thymidine Kinase (HS-tk)
- Ganciclovir can destroy HS-tk containing cells
- Ganciclovir crosses the blood-brain barrier
- Tumor destruction is nonimmune-mediated
- HS-tk and ganciclovir produce "bystander" tumor killing

Safety Aspects
- Vast excess of receptors in the area of injection
- Vector particles are inactivated by complement
- Producer cells are destroyed by ganciclovir

occur, as well as providing the possibility for repeated treatments without the development of a rapidly destructive immune response.

In order to study these features, an *in vivo* rat brain tumor model was established following the stereotaxic implantation of the 9L rat gliosarcoma.[79] These experiments demonstrate that the injection of HS-tk VPC and subsequent GCV treatment can result in complete histologic destruction of the gliosarcoma in 80% of rats. In survival experiments, 50 to 60% of the rats survived long term, suggesting that this procedure may be curative in this model. No associated systemic toxicity or evidence of systemic spread of the retroviral vectors was noted with this form of *in vivo* gene transfer using the injection of β-galactosidase as a marker gene. Additional studies performed in monkeys demonstrated no evidence of acute or long-term toxicities after the direct injection of VPC into the brain followed by intravenous GCV.[45]

The only FDA-approved clinical indication for GCV administration is in the treatment of immunocompromised patients with CMV retinitis. Clinical toxicities of GCV in humans, at the recommended doses, have been primarily hematologic with 40% developing neutropenia and 15% thrombocytopenia.[80] These side effects have been reversible with discontinuation of GCV. Since the toxicities seem to involve tissues with a higher rate of proliferation (e.g., stem cells), other side effects have included aspermatogenesis and abnormalities of the intestinal mucosa. GCV will cross the blood-brain barrier and was present in the CSF at a mean of about 38% of the plasma concentration in the three patients studied.[81]

2. Cytosine Deaminase (CD)

CD is a bacterial gene that converts 5-fluorocytidine (5-FC) to the antimetabolite 5-fluorouracil (5-FU).[82] When murine retroviral vectors are used to insert the CD gene into tumor cells, the genetically-altered cells can be killed *in vitro* and in animals. Not only might the HS-tk and CD sensitivity genes be used as a primary means of tumor cell destruction, they may also be used as a safety mechanism. For use as a safety, one or more "sensitivity" genes would be transferred along with the therapeutic genes. If the genetically-altered cell type resulted in undesirable consequences such as uncontrolled proliferation, the transplanted cells could be destroyed *in vivo* with the appropriate form of systemic chemotherapy (ganciclovir or 5-FC).

B. Tumor Suppressor Genes and Antioncogenes

Several *in vivo* human gene transfer trials for cancer have been approved using the direct injection of retroviral vector supernate into tumor deposits. One involves the direct injection of supernate repeatedly into endobronchial lung cancers.[83] In this experiment, the retroviral vectors will carry genes that target the genetic mechanisms responsible for the malignancy.[84,85] If lung tumors are deficient in the p53 tumor suppressor gene, a p53 vector will be used to transfer a normal copy of the gene. In lung cancers that overexpress the K-ras oncogene, a vector containing an antisense K-ras gene will be used. The antisense K-ras vector will produce mirror image RNA molecules that will bind those being produced by the oncogene. The RNA:RNA hybrids are then degraded by the cell. Experiments in animals have demonstrated that the insertion of an antisense oncogene can result in *in vivo* destruction of injected tumors.[86] Another trial proposes to inject retroviral vector supernate containing an interferon-γ vector directly into melanoma tumor deposits. At this time, neither of these research groups has begun human experimentation.

C. Immune-Mediated Methods

1. In Vitro Manipulation of T-cells for Antitumor Therapy

In vitro IL-2-activated T-lymphocytes have been used for the experimental therapy of brain tumors.[87-91] Unfortunately, the direct *in situ* injection of activated T-lymphocytes into gliomas has not resulted in a significant antitumor effect in most patients. The genetic alteration of T-lymphocytes with cytokine genes is under investigation. *In vitro* studies with allogeneic human T-lymphocytes have noted an enhanced cytoxicity against human glioma cells in culture following transfection with a human TNF-α vector.[92] This approach has been problematic for several reasons, including an inability of the T-lymphocytes to penetrate into the tumor mass, a risk of graft vs. host disease, and the potential toxicities of the cytokine production of the T-lymphocytes in the brain.

2. In Vitro Gene Transfer of Human Glioma Cells

a. *Insertion of Immunostimulatory Genes*

Tumor cells containing immune-modulating cytokine genes either do not grow or grow transiently with subsequent regression. These gene-modified cells may result in the development of systemic immunity against the tumor cell upon rechallenge, suggesting an approach for a cancer vaccine or therapy. Examples include IL-2, IL-7, TNF-α, IL-4, IL-6, γ-interferon, and GM-CSF.[93-100] Such tumor vaccines may make more effective use of cytokines, since the systemic infusion of cytokines (e.g., IL-2) has been quite toxic, limiting their therapeutic usefulness. Whether or not the local expression of these factors by tumor cells in the brain will be less toxic to the CNS than more toxic cytokines alone is not currently known. Some cytokines have minimal CNS toxicities (e.g., IL-4) and may be useful in the treatment of brain tumors.[101]

The expression of a variety of cytokine genes in human glioma cells as a method for vaccination is under evaluation. When the human interferon-β (IFN-β) or tumor necrosis factor-α (TNF-α) genes are transfected with liposomes into human glioma cells, there is a growth-inhibitory effect.[102,103] This inhibitory effect is not seen by adding similar amounts of cytokines into the culture medium. *In vitro* growth inhibition can be augmented with the addition of exogenous dexamethasone to IFN-β-transfected cells or with the addition of exogenous IFN-γ to TNF-α-transfected cells.

Two signals are required for effective T-lymphocyte stimulation: antigen presentation to the T-cell receptor and binding to a costimulatory molecule. Two costimulatory cell surface antigens have been identified in humans: B7-1 and B70 or B7-2.[104-106] The expression of these molecules on tumor cells results in a significant enhancement of tumor immunogenicity. The concurrent expression of membrane costimulatory molecules and cytokine genes may permit a greater cell-mediated, antitumor immune response.

b. *Down-Regulation of Tumor-Derived Immunosuppressive Factors*

Another approach is to suppress or eliminate the expression of genes suspected of causing tumor evasion of the immune system. Insulin-like growth factor-1 (IGF-1) is over-expressed in many tumors, including GBM.[107] Investigators have noted that the transfer of an antisense IGF-1 gene into C6 rat glioma cells blocked IGF-1 production *in vitro* and resulted in a loss of tumorigenicity. Rejection of these genetically-modified tumor cells was mediated by an infiltration of CD8+ T-lymphocytes over a 2 to 3 week period. This suggested that the production of IGF-1 by the C6 glioma cells resulted either directly or indirectly in an inability of T-lymphocytes to recognize and respond to the tumor cells. Strikingly, if the genetically-altered cells were injected SQ into a rat bearing a wild-type C6 glioma in the brain, both the brain tumor and the SQ gene-modified tumors regressed.[108] The mechanism by which IGF-1 blocks immune-mediated destruction of the tumor remains an unknown.

VI. APPROVED HUMAN GENE THERAPY CLINICAL TRIALS FOR BRAIN TUMORS

There are five Recombinant DNA Advisory Committee (RAC)-approved protocols in the U.S. Four of those involve the *in vivo* gene transfer of the HS-tk gene (Table 4). These initial applications of gene therapy have focused on recurrent adult GBM and

Table 4

Approved Gene Therapy Trials for Brain Tumor Treatment

Gene Transfer Method	Gene	Study Locations	Tumor Type(s)	No. of Patients to be Enrolled
In vivo	HS-tk	NIH	GBM/metastasis	20
In vivo	HS-tk	IMMC, UI, UCSF	GBM	30
In vivo	HS-tk	CHLA	Astrocytoma	30[a]
Ex vivo	Antisense IGF-1	CWRU	GBM	12
In vivo	HS-tk	St. Jude	Astrocytoma	6[a]

Note: NIH: National Institutes of Health, Bethesda, MD; IMMC: Iowa Methodist Medical Center, Des Moines; UI: University of Iowa, Iowa City; UCSF: University of California, San Francisco; CHLA: Children's Hospital Los Angeles; CWRU: Case Western Reserve University, Cleveland, OH; St. Jude: St. Jude Children's Cancer Research Hospital, Memphis, TN.

[a] Clinical trials involving children and young adults (2–18 years old).

recurrent childhood astrocytoma, because these disease circumstances are considered incurable. In addition, the use of retroviral vectors require a relatively localized tumor that is rapidly proliferating. Other common tumors, such as meningioma and medulloblastoma, tend to be too diffuse at recurrence and/or proliferate too slowly to expect localized therapy to have any significant antitumor effect.

A. In Vivo Gene Transfer of the Herpes Simplex-Thymidine Kinase Gene

The first trial was initiated at the National Institutes of Health in 1992.[109] In the Phase I NIH trial, murine fibroblast cells (NIH 3T3) producing HS-tk-containing retroviral vectors (VPC) are directly implanted into growing brain tumors in human patients. Implantation of VPC within multiple sites in the tumor mass are achieved with an MRI-guided stereotactic apparatus. After a period of 7 days, during which the VPC transduce the HS-tk gene into surrounding tumor cells, the patients are treated with GCV at a dose of 5 mg/kg/dose twice daily i.v. for 14 days.

The trial at NIH will comprise patients with recurrent GBM and recurrent metastatic disease.[109] As of September 1993, eight patients had been treated, seven with recurrent GBM and one with metastatic melanoma. Each patient received multiple injections of HS-tk VPC into a portion of the tumor. Early results have shown no biochemical or clinical evidence of acute or subacute toxicity. Five of the eight patients had changes in tumor consistency (i.e., cystic changes) based upon MRI scans (four GBM and one melanoma). Three of the five responders had a decrease in tumor size greater than 50% (three GBM). These findings are consistent with the preclinical animal studies that suggested that this method of gene delivery would result in toxicity limited to the tumor. Efforts are underway at NIH to improve the stereotactic delivery of the VPC throughout the body of the tumor.

The initial NIH protocol focused on the use of stereotactic delivery of the VPC. This surgical delivery method is advantageous for patients in whom a surgical resection cannot be attempted. However, this approach has the potential limitation that previously irradiated tumors often contain a large amount of reactive gliosis and necrosis, which may limit the efficiency of gene transfer and the bystander effect. None of the currently used brain scans (i.e., CT, MRI, PET) can accurately differentiate viable tumor within the areas of dead tumor. Therefore, two additional protocols have been developed to circumvent this disadvantage.

A second approach involves adult patients with recurrent GBM and will focus on the direct injection of HS-tk VPC into the walls of the tumor bed at the time of a surgical resection of the tumor. An Ommaya reservoir will then be inserted that will allow administration of additional courses of VPC into the tumor bed without further surgery. This trial is expected to begin in early 1994 as a cooperative trial, including the Iowa Methodist Medical Center in Des Moines, the University of Iowa in Iowa City, and the University of California in San Francisco.

A third trial has been developed at Children's Hospital Los Angeles. This trial is aimed at the treatment of children, ages 2 to 18 years, who have a recurrent astrocytoma. This trial offers an opportunity to ask several important questions. First, do recurrent childhood astrocytomas respond differently than adult GBM tumors to this type of therapy? Second, children with astrocytomas are not generally given radiotherapy before the age of 6 years. Will those children with nonirradiated tumors have an improved response since less gliosis and fibrosis should be present compared to irradiated GBM? This protocol has two treatment groups: children who have a tumor in a location that precludes surgery will receive the VPC by stereotactic injection (similar to the NIH trial), while the other half will have a surgical resection and injection of VPC at surgery and via the Ommaya reservoir, as in the Iowa/UCSF trial. This pediatric trial is expected to begin in November 1993. A second trial using this stereotactic delivery system in children has recently been approved for clinical trials at the St. Jude Children's Cancer Research Hospital.

B. Ex Vivo Gene Transfer of Antisense IGF-1

In this protocol, scientists at Case Western Reserve University will obtain glioma cells from the initial surgical resection. The cells will then be placed in tissue culture. Tumors that grow *in vitro* and produce IGF-1 will be transfected with an episomal-based vector containing an antisense IGF-1 under the control of a metallothionein (MT)-inducible promoter. MT can be induced to initiate transcription of the antisense IGF-1 gene by the administration of heavy metals such as zinc. The genetically altered tumor cells are irradiated and injected subcutaneously into the patient from whom the cells were removed in an effort to generate an enhanced systemic immune response to GBM. The patients will be treated with zinc sulfate by mouth to induce transcription of the antisense IGF-1 gene *in vivo*. Patients may undergo as many as three successive treatments with this therapy. A clinical trial using this approach to treat GBM is expected to begin in early 1994.

VII. SUMMARY

The early applications of gene therapy to genetic disease have rapidly expanded to the treatment of cancer. While brain tumors are clearly molecular-based disorders, the initial applications are attempting to either destroy the malignancy directly through genetic manipulation or by inducing an immune response against the tumor. Neither of these approaches is targeting the actual molecular basis of the tumor. Over the next decade, our knowledge of the genetic basis of brain tumors will grow, along with the number of options for selective gene delivery *in vivo*. These events will provide an opportunity to attack the tumor with gene therapy and perhaps prevent the development of these genetically based malignancies.

REFERENCES

1. Mahaley, M. S., Jr., Mettlin, C., Natarajan, N., Lewis, E. R., Jr., and Peace, B. B., National survey on patterns of care for brain-tumor patients, *J. Neurosurg.*, 71, 826, 1989.
2. Harsh, G. R., IV, Levin, V. A., Gutin, P. H., Seager, M., Silver, P., and Wilson, C. B., Reoperation for recurrent glioblastoma and anaplastic astrocytoma, *Neurosurgery*, 21, 615, 1987.
3. Black, P. McL., Brain tumors, *N. Engl. J. Med.*, 324, 1471, 1991.
4. Russell, D. S. and Rubinstein, L. J., Eds., *Pathology of Brain Tumors of the Central Nervous System*, E. Arnold, London, 1989, 219.
5. Salcman, M., Epidemiology and factors affecting survival, in *Malignant Cerebral Glioma*, Apuzzo, M. L. J., Ed., Am. Assoc. of Neurological Surgeons, Park Ridge, 1990, 95.
6. Greig, N. H., Reis, L. G., Yanick, R., and Rapoport, S. I., Increasing annual incidence of primary malignant brain tumors in the elderly, *J. Natl. Cancer Inst.*, 82, 1621, 1990.
7. Sadik, A. R., Port, R., Garfinkel, B., and Bravo, J., Extracranial metastasis of cerebral glioblastoma multiforme: case report, *Neurosurgery*, 15, 549, 1984.
8. Trattnig, S., Schindler, E., Ungersbock, K., Schmidbauer, M., Heimberger, K., Hubsch, P., and Stiglbauer, R., Extra-CNS metastases of glioblastoma: CT and MR studies, *J. Comp. Assisted Tomography*, 14, 294, 1990.
9. Nazzaro, J. M. and Neuwelt, E. A., The role of surgery in the management of supratentorial intermediate and high-grade astrocytomas in adults, *J. Neurosurg.*, 73, 331, 1990.
10. Wilson, C. B., Glioblastoma: the past, the present and the future, *Clin. Neurosurg.*, 38, 32, 1992.
11. Ammirati, M., Galicich, J. H., Arbit, E., and Liao, Y., Reoperation in the treatment of recurrent intracranial malignant gliomas, *Neurosurgery*, 21, 607, 1987.
12. Walker, M. D., Alexander, E., Hunt, W. E., MacCarty, C. S., Mahaley, M. S., Jr., Mealey, J., Norrell, H. A., Owens, G., Ransohoff, J., Wilson, C. B., Gehan, E. A., and Strike, T. A., Evaluation of BCNU and/or radiotherapy in the treatment of anaplastic astrocytoma. A cooperative clinical trial, *J. Neurosurg.*, 49, 333, 1978.
13. Green, S. B., Byar, D. P., Walker, M. D., Pistenmaa, D. A., Alexander, E., Jr., Batzdorf, U., Brooks, W. H., Hunt, W. E., Mealy, J., Jr., Odom, G. L., Paoletti, P., Ransohoff, J., II, Robertson, J. T., Selker, R. G., Shapiro, W. R., Smith, K. R., Jr., Wilson, C. B., and Strike, T. A., Comparisons of carmustine, procarbazine, and high-dose methylprednisolone as additions to surgery and radiotherapy for the treatment of malignant glioma, *Cancer Treat. Rep.*, 67, 121, 1983.

14. Greene, G. M., Hitchon, P. W., Schelper, R. L., Yuh, W., and Dyste, G. N., Diagnositic yield in CT-guided stereotactic biopsy of gliomas, *J. Neurosurg.*, 71, 494, 1989.
15. Kelly, P. J., Daumas-Duport, C., Kispert, D. B., Kall, B. A., Scheithauer, B. W., and Illig, J. J., Imaging-based stereotaxic serial biopsies in untreated intracranial glial neoplasms, *J. Neurosurg.*, 66, 865, 1987.
16. Harris, C. C. and Hollstein, M., Clinical implications of the p53 tumor-suppressor gene, *N. Engl. J. Med.*, 329, 1318, 1993.
17. Sidransky, D., Mikkelsen, T., Schwechheimer, K., Rosenblum, M. L., Cavavee, W., and Vogelstein, B., Clonal expansion of p53 mutant cells is associated with brain tumor progression, *Nature*, 355, 846, 1992.
18. Baker, S. J., Markowitz, S., Fearon, E. R., Wilson, J. K., and Vogelstein, B., Suppression of human colorectal carcinoma cell growth by wild-type p53, *Science*, 249, 912, 1990.
19. Mercer, W. E., Shields, M. T., Amin, M., Suave, G. J., Appella, E., Romano, J. W., and Ullrich, S. J., Negative growth regulation in a glioblastoma tumor cell line that conditionally expresses human wild-type p53, *Proc. Natl. Acad. Sci. U.S.A.*, 87, 6166, 1990.
20. Plate, K. H., Breier, G., Weich, H. A., and Risau, W., Vascular endothelial growth factor is a potential tumour angiogenesis factor in human gliomas in vivo, *Nature*, 359, 845, 1992.
21. Kim, K. J., Li, B., Winer, J., Armanini, M., Gillett, N., Phillips, H. S., and Ferrara, N., Inhibition of vascular endothelial growth factor-induced angiogenesis suppresses tumour growth in vivo, *Nature*, 362, 841, 1993.
22. Hermansson, M., Nister, M., Betsholtz, C., Heldin, C.-H., Westermark, B., and Funa, K., Endothelial cell hyperplasia in human glioblastoma: coexpression of mRNA for platelet-derived growth factor (PDGF) B chain and PDGF receptor suggests autocrine growth stimulation, *Proc. Natl. Acad. Sci. U.S.A.*, 85, 7748, 1988.
23. James, C. D., Carlbom, E., Dumanski, P. L., Hansen, M., Nordenskjold, M., Collins, V. P., and Cavenee, W. K., Clonal genomic alterations in glioma malignancy stages, *Cancer Res.*, 48, 5546, 1988.
24. Bigner, S. H., Mark, J., Burger, P. C., Mahaley, M. S., Bullard, D. E., Muhlbaier, L. H., and Bigner, D. D., Specific chromosomal abnormalities in malignant human gliomas, *Cancer Res.*, 88, 405, 1988.
25. El-azouzi, M., Chung, R. Y., Farmer, G. E., Martuza, R. L., Black, P. McL., Rouleau, G. A., Hettlich, C., Hedley-Whyte, E. T., Zeervas, N. T., Panagopoulos, K., Nakamura, Y., Gusella, J. F., and Seizinger, B. R., Loss of distinct regions on the short arm of chromosome 17 associated with tumorigenesis of human astrocytomas, *Proc. Natl. Acad. Sci. U.S.A.*, 86, 7186, 1989.
26. Bigner, S. H. and Vogelstein, B., Cytogenetics and molecular genetics of malignant gliomas and medulloblastoma, *Brain Pathol.*, 1, 12, 1990.
27. Gerosa, M. A., Talarico, D., Fognani, C., Raimondi, E., Colombatti, M., Tridente, G., De Carli, L., and Della Valle, G., Overexpression of N-ras oncogene and epidermal growth factor receptor gene in human glioblastomas, *J. Natl. Cancer Inst.*, 81, 63, 1989.
28. Libermann, T. A., Nusbaum, H. R., Razon, N., Kris, R., Lax, I., Soreq, H., Whittle, N., Waterfield, M. D., Ullrich, A., and Schlessinger, J., Amplification, enhanced expression and possible rearrangement of EGF receptor gene in primary human brain tumours of glial origin, *Nature*, 313, 144, 1985.
29. Trent, J., Meltzer, P., Rosenblum, M., Rosenblum, M., Harsh, G., Kinzler, K., Mashal, R., Feinberg, A., and Vogelstein, B., Evidence for rearrangement, amplification, and expression of c-myc in human glioblastoma, *Proc. Natl. Acad. Sci. U.S.A.*, 83, 470, 1986.
30. Kinzler, K. W., Bigner, S. H., Bigner, D. D., Trent, J. M., Law, M. L., O'Brien, S. J., Wong, A. J., and Vogelatein, B., Identification of an amplified, highly expressed gene in a human glioma, *Science*, 233, 70, 1987.

31. Roszman, T., Elliott, L., and Brooks, W., Modulation of T-cell function by gliomas, *Immunol. Today*, 12, 370, 1991.
32. Colombatti, M., Dipasquale, B., Del-L'Arciprete, L., Gerosa, M., and Tridente, G., Heterogeneity and modulation of tumor-associated antigens in human glioblastoma lines, *J. Neurosurg.*, 71, 388, 1989.
33. Pfreundschuh, M., Shiku, H., Takahashi, T., Ueda, R., Ransohoff, J., Oettgen, H. F., and Old, L. J., Serological analysis of cell surface antigens of malignant human brain tumors, *Proc. Natl. Acad. Sci. U.S.A.*, 75, 5122, 1978.
34. Matsumoto, T., Tani, E., Kaba, K., Kochi, N., Shindo, H., Yamamoto, Y., Sakamoto, H., and Furuyama, J., Amplification and expression of a multidrug resistance gene in human glioma cell lines, *J. Neurosurg.*, 72, 96, 1990.
35. Walker, A. E., Robbins, M., and Weinfeld, F. D., Epidemiology of brain tumors: the national survey of intracranial neoplasms, *Neurology*, 35, 219, 1985.
36. Delattre, J. Y., Krol, G., Thaler, H. T., and Posner, J. B., Distribution of brain metastases, *Arch. Neurol.*, 45, 741, 1988.
37. Wright, D. C. and Delaney, T. F., Treatment of metastatic cancer: treatment of metastatic cancer to the brain, in *Cancer: Principles and Practice of Oncology*, DeVita, V. Y., Jr., Hellman, S., and Rosenberg, S. A., Eds., Lippincott, Philadelphia, 1989, 2245.
38. Patchell, R. A., Tibbs, P. A., Walsh, J. W., Dempsey, R. J., Maruyama, Y., Kryscio, R. J., Markesbery, W. R., MacDonald, J. S., and Young, B., A randomized trial of surgery in the treatment of single metastases to the brain, *N. Engl. J. Med.*, 322, 494, 1990.
39. Miller, A. D., Retroviral vectors, *Curr. Top. Microbiol. Immunol.*, 158, 1, 1992.
40. Donahue, R. E, Kessler, S. W., Bodine, D., McDonagh, K., Dunbar, C., Goodman, S., Agricola, B., Byrne, E., Raffeld, M., Moen, R., Bacher, J., Zsebo, K. M., and Nienhius, A. W., Helper virus induced T cell lymphoma in nonhuman primates after retroviral mediated gene transfer, *J. Exp. Med.*, 176, 1125, 1992.
41. Yamada, M., Shimizu, K., Miayo, Y., Hayakawa, T., Ikenaka, K., Nakahira, K., Nakajima, K., Kagawa, T., and Mikoshiba, K., Retrovirus-mediated gene transfer targeted to malignant glioma cells in murine brain, *Jpn. J. Cancer Res.*, 83, 1244, 1992.
42. Short, M. P., Choi, J. K., Lee, A., Malick, A., Breakefield, X. O., and Martuza, R. L., Gene delivery to glioma cells in rat brain by grafting of a retrovirus packaging cell line, *J. Neurosci. Res.*, 27, 427, 1990.
43. Culver, K. W., Ram, Z., Walbridge, S., Ishii, H., Oldfield, E. H., and Blaese, R. M., In vivo gene transfer with retroviral vector producer cells for treatment of experimental brain tumors, *Science*, 256, 1550, 1992.
44. Caruso, M., Panis, Y., Gagandeep, S., Houssin, D., Salzmann, J.-L., and Klatzmann, D., Regression of established macroscopic liver metastases after in situ transduction of a suicide gene, *Proc. Natl. Acad. Sci. U.S.A.*, 90, 7024, 1993.
45. Ram, Z., Culver, K. W., Walbridge, S., Frank, J. A., Blaese, R. M., and Oldfield, E. H., Toxicity studies of retroviral-mediated gene transfer for the treatment of brain tumors, *J. Neurosurg.*, 79, 400, 1993.
46. Ram, Z., Culver, K. W., Walbridge, S., Blaese, R. M., and Oldfield, E. H., In situ retroviral-mediated gene transfer for the treatment of brain tumors in rats, *Cancer Res.*, 53, 83, 1993.
47. Akli, S., Caillaud, C., Vigne, E., Stratford-Perricaudet, L. D., Poenaru, L., Perricaudet, M., Kahn, A., and Peschanski, M. R., Transfer of a foreign gene into the brain using adenovirus vectors, *Nat. Genet.*, 3, 224, 1993.
48. Bajocchi, G., Feldman, S. H., Crystal, R. G., and Mastrangeli, A., Direct in vivo gene transfer to ependymal cells in the central nervous system using recombinant adenovirus vectors, *Nat. Genet.*, 3, 229, 1993.

49. Davidson, B. L., Allen, E. D., Kozarsky, K. F., Wilson, J. M., and Roessler, B. J., A model system for in vivo gene transfer into the central nervous sytem using an adenoviral vector, *Nat. Genet.*, 3, 219, 1993.
50. La Salle, G. L., Robert, J. J., Berrard, S., Ridoux, V., Stratford-Perricaudet, L. D., Perricaudet, M., and Mallet, J., An adenovirus vector for gene transfer into neurons and glia in the brain, *Science*, 259, 988, 1993.
51. Vile, R. G. and Hart, I. R., In vitro and in vivo targeting of gene expression to melanoma cells, *Cancer Res.*, 53, 962, 1993.
52. Coen, D. M., Kosz-Vnenchak, M., Jacobson, J. G., Leib, D. A., Bogard, C. L., Schaffer, P. A., Tyler, K. L., and Knife, D. M., Thymidine kinase-negative herpes simples virus mutants establish latency in mouse trigeminal ganglia but do not reactivate, *Proc. Natl. Acad. Sci. U.S.A.*, 86, 4736, 1989.
53. Martuza, R. L., Malick, A., Markert, J. M., Ruffner, K. L., and Coen, D. M., Experimental therapy of human glioma by means of a genetically engineered virus mutant, *Science*, 252, 854, 1991.
54. Ulmer, J. B., Donnelly, J. J., Parker, S. E., Rhodes, G. H., Felgner, P. L., Dwarki, V. J., Gromkowski, S. H., Deck, R. R., DeWitt, C. M., Friedman, A., Hawe, L. A., Leander, K. R., Martinez, D., Perry, H. C., Shiver, K. W., Montgomery, D. L., and Liu, M. A., Heterologous protection against influenza by injection of DNA encoding a viral protein, *Science*, 259, 1745, 1993.
55. Tang, D., DeVit, M., and Johnston, S. A., Genetic immunization is a simple method for eliciting an immune response, *Nature*, 356, 1, 1992.
56. Nabel, G. J., Chang, A., Nabel, E. G., and Plautz, G., Immunotherapy of malignancy by in vivo gene transfer into tumors, *Hum. Gene Ther.*, 3, 399, 1992.
57. Wu, G. Y. and Wu, C. H., Receptor-mediated in vitro gene transformation by a soluble DNA carrier system, *J. Biol. Chem.*, 262, 4429, 1987.
58. Zhu, N., Liggitt, D., Liu, Y., and Debs, R., Systemic gene expression after intravenous DNA delivery into adult mice, *Science*, 261, 209, 1993.
59. Wagner, E., Zenke, M., Cotten, M., Beug, H., and Birnstiel, M. L., Transferrin-polycation conjugates as carriers for DNA uptake into cells, *Proc. Natl. Acad. Sci. U.S.A.*, 87, 3410, 1990.
60. Jiao, S., Cheng, L., Wolff, J. A., and Yang, N.-S., Particle bombardment-mediated gene transfer and expression in rat brain tissues, *Bio/technology*, 11, 497, 1993.
61. McKnight, S., The nucleotide sequence and transcript map of the herpes simplex thymidine kinase gene, *Nucl. Acids Res.*, 8, 5949, 1980.
62. Elion, G. B., The chemotherapeutic exploitation of virus-specified enzymes, *Adv. Enz. Regul.*, 18, 53, 1980.
63. Faulds, D. and Heel, R. C., Ganciclovir, *Drugs*, 39, 597, 1990.
64. Terry, B. J., Cianci, C. W., and Hagen, M. E., Inhibition of herpes simplex virus type 1 DNA polymersase by [1R(1a,2b,3a)]-9-[2,3-Bis(hydroxymethyl)cyclobutyl]guanine, *Mol. Pharm.*, 40, 591, 1991.
65. Smee, D. F., Martin, J. C., Verheyden, J. P. H., and Matthews, T. R., Anti-herpesvirus activity of the acyclic nucleoside 9-(1,3-dihydroxy-2-propoxymethyl)guanine, *Antimicrob. Agents Chemother.*, 23, 676, 1983.
66. Field, A. K., Davies, M. E., DeWitt, C., Perry, H. C., Liou, R., Germershausen, J., Karkas, J. D., Ashton, W. T., Johnston, D. B. R., and Tolman, R. L., 9-{[2-hydroxy-1-(hyroxymethyl)ethoxy]-methyl}guanine: a selective inhibitor of herpes group virus replication, *Proc. Natl. Acad. Sci. U.S.A.*, 80, 4139, 1983.
67. Moolten, F. L. and Wells, J. M., Curability of tumors bearing herpes thymidine kinase genes transferred by retroviral vectors, *J. Natl. Cancer Inst.*, 82, 297, 1990.

68. Golumbek, P. T., Hamzeh, F. M., Jaffee, E. M., Levitsky, H., Leitman, P. S., and Pardoll, D. M., Herpes simplex-1 virus thymidine kinase gene is unable to completely eliminate live, non-immunogenic tumor cell vaccines, *J. Immunother.*, 12, 224, 1992.
69. Gutierrez, A. A., Lemoine, N. R., and Sikora, K., Gene therapy for cancer, *Lancet,* 339, 715, 1992.
70. Borrelli, E., Heyman, R., Hsi, M., and Evans, R. M., Targeting of an inducible phenotype in animal cells, *Proc. Natl. Acad. Sci. U.S.A.*, 85, 7572, 1988.
71. Moolten, F. L., Tumor sensitivity conferred by inserted herpes thymidine kinase genes: paradigm for a perspective cancer control strategy, *Cancer Res.*, 46, 5276, 1986.
72. Heyman, R. A., Borrelli, E., Lesley, J., Anderson, D., Richman, D. D., Baird, S. M., Hyman, R., and Evans, R. M., Thymidine kinase obliteration: creation of transgenic mice with controlled immune deficiency, *Proc. Natl. Acad. Sci. U.S.A.*, 86, 2698, 1989.
73. Ezzedine, Z. D., Martuza, R. L., Platika, D., Short, M. P., Malick, A., Choi, B., and Breakefield, X. O., Selective killing of glioma cells in culture and in vivo by retrovirus transfer of the herpes simplex virus thymidine kinase gene, *New Biol.*, 3, 608, 1991.
74. Moolten, F. L., Wells, J. M., Heyman, R. A., and Evans, R. M., Lymphoma regression induced by ganciclovir in mice bearing a herpes thymidine kinase transgene, *Hum. Gene Ther.*, 1, 125, 1990.
75. Cheng, Y.-C., Grill, S. P., Dutschman, G. E., Nakayama, K., and Bastow, K. F., Metabolism of 9-(1,3-dihydroxy-2-propoxymethyl)guanine, a new anti-herpes virus compound, in herpes-simplex virus-infected cells, *J. Biol. Chem.*, 258, 12460, 1983.
76. Bi, W. L., Parysek, L. M., Warnick, R., and Stambrook, P. J., In vitro evidence that metabolic cooperation is responsible for the bystander effect observed with HSV tk retroviral gene therapy, *Hum. Gene Ther.*, 4, 725, 1993.
77. Hooper, M. L. and Subak-Sharpe, J. H., Metabolic cooperation between cells, *Int. Rev. Cytol.*, 69, 45, 1991.
78. Plautz, G., Nabel, E. G., and Nabel, G. J., Selective elimination of recombinant genes in vivo with a suicide retroviral vector, *New Biol.*, 7, 709, 1991.
79. Weisacker M., Deen, D. F., Rosenblum, M. L., Hoshino, T., Gutin, P. H., and Barker, M., The 9L rat brain tumor model: description and application of the model, *J. Neurol.*, 224, 183, 1981.
80. Stankus, B. J., *Cytovene Product Monograph*, Syntex Laboratories, Palo Alto, 1992.
81. Shepp, D. H., Dandliker, P. S., de Miranda, P., Burnette, T. C., Cederberg, D. M., Kirk, E., and Meyers, J. D., Activity of 9-[2-hydroxy-1-(hydroxy-methyl)ethoxymethyl]guanine in the treatment of cytomegalovirus pneumonia, *Ann. Int. Med.*, 103, 368, 1985.
82. Mullen, C. A., Kilstrup, M., and Blaese, R. M., Transfer of the bacterial gene for cytosine deaminase to mammalian cells confers lethal sensitivity to 5-fluorcytosine: a negative selection system, *Proc. Natl. Acad. Sci. U.S.A.*, 89, 33, 1992.
83. Roth, J. A., Molecular surgery for cancer, *Arch. Surg.*, 127, 1298, 1992.
84. Bishop, J. M., Molecular themes in oncogenesis, *Cell*, 64, 235, 1991.
85. Weinberg, R. A., Tumor suppresser genes, *Science*, 254, 1138, 1991.
86. Zhang, Y., Mukhopadhyay, T., Donehower, L. A., Georges, R. N., and Roth, J. A., Retroviral vector-mediated transduction of K-ras antisense RNA into human lung cancer cells inhibits expression of the malignant phenotype, *Hum. Gene Ther.*, 4, 451, 1993.
87. Jacobs, S. K., Wilson, D. J., Kornblith, P. L., and Grimm, E. A., In vitro killing of human glioblastoma by interleukin-2-activated autologous lymphocytes, *J. Neurosurg.*, 64, 114, 1986.
88. Jacobs, S. K., Wilson, D. J., Kornblith, P. L., and Grimm, E. A., Interleukin-2 and autologous lymphokine-activated killer cells in the treatment of malignant glioma, *J. Neurosurg.*, 64, 743, 1986.

89. Ingram, M., Shelden, C. H., Jacques, S., Skillen, R. G., Bradley, W. G., Techy, G. B., Freshwater, D. B., Abts, R. M., and Rand, R. W., Preliminary clinical trial of immunotherapy for malignant glioma, *J. Biol. Respir. Mod.*, 6, 489, 1987.
90. Merchant, R. E., Merchant, L. H., and Cook, S. H. S., Intralesional infusion of lymphokine-activated killer (LAK) cells and recombinant interleukin-2 (rIL-2) for the treatment of patients with malignant brain tumor, *Neurosurgery*, 23, 725, 1988.
91. Barba, D., Saris, S. C., Holder, S. C., Rosenberg, S. A., and Oldfield, E. H., Intratumoral LAK cell and interleukin-2 therapy of human gliomas, *J. Neurosurg.*, 70, 175, 1989.
92. Tashiro, T., Yoshida, J., Mizuno, M., and Sugita, K., Reinforced cytotoxicity of lymphokine-activated killer cells toward glioma cells by transfection with the tumor necrosis factor-a gene, *J. Neurosurg.*, 78, 252, 1993.
93. Fearon, E. R., Pardoll, D. M., Itaya, T., Golumbek, P., Levitsky, H. I., Simons, J. W., Karasuyama, H., Vogelstein, B., and Frost, P., Interleukin-2 production by tumor cells bypasses T helper function in the generation of an antitumor response, *Cell*, 60, 397, 1990.
94. McBride, W. H., Thacker, J. D., Comora, S., Economou, J. S., Kelley, D., Hogge, D., Dubinett, S. M., and Dougherty, G. J., Genetic modification of a murine fibrosarcoma to produce interleukin 7 stimulates host cell infiltration and tumor immunity, *Cancer Res.*, 52, 3931, 1992.
95. Asher, A. L., Mule, J. J., Kasid, A., Restifo, N. P., Salo, J. C., Reichert, C. M., Jaffe, G., Fendly, B., Kriegler, M., and Rosenber, S. A., Murine tumor cells transduced with the gene for tumor necrosis factor-a, *J. Immunol.*, 146, 3227, 1991.
96. Tepper, R. I., Pattengale, P. K., and Leder, P., Murine interleukin-4 displays potent anti-tumor activity in vivo, *Cell*, 57, 503, 1989.
97. Mullen, C. A., Coale, M. M., Levy, A. T., Stetler-Stevnson, W. G., Liotta, L. A., Brandt, S., and Blaese, R. M., Fibrosarcoma cells transduced with the IL-6 gene exhibit reduced tumorigenicity, increased immunogenicity and decreased metastatic potential, *Cancer Res.*, 52, 6020, 1992.
98. Gansbacher, B., Bannerji, R., Daniels, B., Zier, K., Cronin, K., and Gilboa, E., Retroviral vector-mediated g-interferon gene transfer into tumor cells generates potent and long lasting antitumor immunity, *Cancer Res.*, 50, 7820, 1990.
99. Esumi, N., Hunt, B., Itaya, T., and Frost, P., Reduced tumorigenicity of murine tumor cells secreting g-interferon is due to nonspecific host responses and is unrelated to class I major histocompatibility complex expression, *Cancer Res.*, 51, 1185, 1991.
100. Dranoff, G., Jaffee, E., Lazenby, A., Golumbek, P., Levitsky, H., Brose, K., Jackson, V., Hamada, H., Pardoll, D., and Mulligan, R. C., Vaccination with irradiated tumor cells engineered to secrete murine granulocyte-macrophage colony-stimulating factor stimulates potent, specific, and long-lasting anti-tumor immunity, *Proc. Natl. Acad. Sci. U.S.A.*, 90, 3539, 1993.
101. Yu, J. S., Wei, M. X., Chiocca, E. A., Martuza, R. L., and Tepper, R. I., Treatment of glioma by engineered interleukin-4 secreting cells, *Cancer Res.*, 53, 3125, 1993.
102. Yoshida, J., Mizuno, M., and Yagi, K., Cytotoxicity of human b-interferon produced in human glioma cells transfected with its gene by means of liposomes, *Biochem. Int.*, 28, 1005, 1992.
103. Mizuno, M., Yoshida, J., Oyama, H., and Sugita, K., Growth inhibition of glioma cells by liposome-mediated cell transfection with tumor necrosis factor-a gene, *Neurologia Medico-chirurgica*, 32, 873, 1992.
104. Townsed, S. E. and Allison, J. P., Tumor rejection after direct co-stimulation of CD8+ T cells by B7-transfected melanoma cells, *Science*, 259, 368, 1993.
105. Azuma, M., Ito, D., Yagita, H., Okumura, K., Phillips, J. H., Lanier, L. L., and Somoza, C., B70 antigen is a second ligand for CTLA-4 and CD28, *Nature*, 366, 76, 1993.

106. Freeman, G. J., Gribben, J. G., Boussiotis, V. A., Ng, J. W., Restivo, V. A., Lombard, L. A., Gray, G. S., and Nadler, L. M., Cloning of B7-2: a CTLA-4 counter-receptor that costimulates human T cell proliferation, *Science*, 262, 909, 1993.
107. Trojan, J., Blossey, B. K., Johnson, T. R., Rudin, S. D., Tykocinski, M., Ilan, J., and Ilan, J., Loss of tumorigenicity of rat glioblastoma directed by episome-based antisense cDNA transcription of the insulin-like growth factor-1, *Proc. Natl. Acad. Sci. U.S.A.*, 89, 4874, 1992.
108. Trojan, J., Johnson, T. R., Rudin, S. D., Ilan, J., Tykocinski, M. L., and Ilan, J., Treatment and prevention of rat glioblastoma by immunogenic C6 cells expressing antisense insulin-like growth factor 1 RNA, *Science*, 259, 94, 1993.
109. Oldfield, E. H., Culver, K. W., Ram, Z., and Blaese, R. M., A clinical protocol: gene therapy for the treatment of brain tumors using intra-tumoral transduction with the thymidine kinase gene and intravenous ganciclovir, *Hum. Gene Ther.*, 4, 39, 1993.
110. Levin, V. A., Wara, W. M., Davis, R. L., Silver, P., Resser, K. J., Yatsko, K., Nutik, S., Gutin, P. H., and Wilson, C. B., Northern California oncology group protocol 6G91: response to treatment with radiation therapy and seven-drug chemotherapy in patients with glioblastoma multiforme, *Cancer Treat. Rep.*, 70, 739, 1986.
111. Shapiro, G. H., Green, S. B., Burger, P. C., Mahaley, M. S., Selker, R. G., Van Gilder, J. C., Robertson, J. T., Ransohoff, J., Mealey, J., Strike, T. A., and Pistenmaa, D. A., Randomized trial of three chemotherapy regimens and two radiotherapy regimens in postoperative treatment of malignant glioma. Brain tumor cooperative group trial 8001, *J. Neurosurg.*, 71, 1, 1989.
112. Bashir, R., Hochberg, F. H., Lingood, R. M., and Hottleman, K., Pre-irradiation internal carotid artery BCNU in treatment of glioblastoma multiforme, *J. Neurosurg.*, 68, 917, 1988.
113. Gutin, P. H., Wara, W. M., Phillips, T. L., and Wilson, C. B., Hypoxic cell radiosensitizers in the treatment of malignant brain tumors, *Neurosurgery*, 6, 567, 1980.
114. Kalofonos, H. P., Pawlikowska, T. R., Hemingway, A., Courtenay-Luck, N., Dhokia, B., Snook, D., Sivolapenko, G. B., Hooker, G. R., McKenzie, C. G., Lavender, P. J., Thomas, D. G. T., and Epenetos, A. A., Antibody guided diagnosis and therapy of brain gliomas using radiolabeled monoclonal antibodies against epidermal growth factor receptor and placental alkaline phosphatase, *J. Nucl. Med.*, 30, 1636, 1989.
115. Bloom, H. J. G., Peckman, M. J., Richardson, A. E., Alexander, P. A., and Payne, P. M., Glioblastoma multiforme: a controlled trial to assess the value of specific active immunotherapy in patients treated by radical surgery and radiotherapy, *Br. J. Cancer*, 27, 253, 1973.
116. Mahaley, M. S., Jr., Bigner, D. D., Dudka, L. F., Wilds, P. R., Williams, D. H., Bouldin, T. W., Whitaker, J. N., and Bynum, J. M., Immunobiology of primary tumors. Part 7: Active immunization of patients with anaplastic human glioma cells: a pilot study, *J. Neurosurg.*, 59, 201, 1983.
117. Merchant, R. E., Grant, A. J., Merchant, L. H., and Young, H. F., Adoptive immunotherapy for recurrent glioblastoma multiforme using lymphokine activated killer cells and recombinant interleukin-2, *Cancer*, 62, 665, 1988.

15 Gene Therapy for Adult Cancers: Advance in Immunologic Approaches Using Cytokines

Hideaki Tahara and Michael T. Lotze

I. INTRODUCTION

Cancer is fundamentally a genetic disorder. Although theoretically possible, gene therapy approaches that can "fix" the genetic disorders of cancer cells do not appear readily applicable. Cancer gene therapy is different from therapies being developed for the treatment of hereditary genetic defects due to the varied origins and characteristics of cancer cells. First, gene(s) responsible for causing cancer are still under investigation. No single gene has been shown to be fully responsible for the transformation and progression of cancer cells. To the contrary, there is much evidence that multiple steps of genetic alteration are responsible for cell transformation. It is unlikely that correction of multiple defects could be readily reversible. Second, all of the existing cancer cells need to be affected by the gene therapy. Efforts have been made in transferring transcriptional regulatory factor genes or antisense oncogenes to correct abnormal proliferative properties of tumor cells. However if the therapeutic gene was not delivered to every cell, the cancer would recur rapidly, similar to what is observed with the outcome of chemotherapy with partial responses. It is very difficult or impossible with currently available vector systems to introduce genes into all of the cancer cells, even at the primary tumor site. Furthermore, cancer is quite likely to be a systemic disease with distant metastasis, requiring therapy of multiple sites. These facts indicate that therapy for cancer, except for brain cancer, which rarely has distant metastasis, should consist of modalities capable of treating metastases that have spread from the primary site.

A variety of other gene therapy applications have been proposed for cancer treatment.

1. *Combination with chemotherapy.* (a) Transfer gene(s) protective against the adverse effects of the existing chemotherapeutic agents into the noncancerous cells; transfer multiple drug

0-8493-2440-8/95/$0.00+$.50

resistance (MDR) gene into bone marrow stem cells with subsequent administration of cytotoxic reagents. (b) Transfer genes encoding enzymes that can convert nontoxic prodrug to cytotoxic drugs, such as transfer of the herpes simplex thymidine kinase gene into cells to confer sensitivity to ganciclovir[1] or transfer of the bacterial gene for cytosine deaminase into the cells to confer lethal sensitivity to 5-fluorocytosine.[2]

2. *Combination with radiotherapy.* Use of radioprotective genes to reduce the adverse effect of radiation to the normal cells.
3. *Immunologic approaches.* Transfer genes for immunoregulatory molecules (i.e., cytokines, MHC antigen, viral antigen, etc.) into cells at a vaccine site (often into tumor cells or adjacent to tumor cells) or lymphocytes to stimulate immune reaction specific to tumor cells.

In this chapter, we will focus on the current status of immunologic approaches, especially those relevant to development of cancer vaccines.

II. OVERVIEW OF CYTOKINE GENE THERAPY

Cytokines are pleiotropic mediators that can modulate and shape the quality and intensity of the immune response, either activating and augmenting or alternatively suppressing some immunologic events. Cytokines are occasionally autocrine or endocrine but largely paracrine hormones, and they are produced mainly by lymphocytes (lymphokines) and monocytes (monokines). Anticancer therapy using these agents as biologic modifiers presumably takes advantage of these immunostimulatory effects. Antitumor activities of cytokines can be classified in several categories: (1) direct inhibition of tumor growth (α-interferon), (2) reversal of the anergy-inducing effects of tumor cells and expansion of new T-cell effectors (IL-2), (3) augmentation of the effector function of T-cells to recognize MHC presented peptide epitopes on tumor cells (GM-CSF), (4) enhanced recruitment of cells to inflammatory sites (IL-4). As a therapy, cytokines can be delivered systemically (subcutaneous, i.v.), regionally (intraperitoneal, intralymphatic) or locally (intra-arterial delivery, cytokine gene therapy). Within these strategies, some of the rationale for cytokine gene therapy is dictated by the notions that local delivery of cytokines most closely mimics the natural immune response and that many cytokines cannot be tolerated when administered at high systemic levels required for an effective response.

Inadvertently, TNF transfection of tumors was initially shown to delay tumor growth when tested for its systemic effect, including induction of cachexia, anemia, and inflammation.[3] Following this initial investigation, the mature phase of cytokine gene therapy began in earnest only a few years ago with the reports of IL-4 transfection[4] and IL-2 transfection.[4-6] The establishment of the technique of local cytokine delivery through genetic manipulation opened the way for the study of antitumor effects of other immunostimulatory cytokines; it also delayed tumor establishment *in vivo* when delivered locally. Cytokines potentially useful in gene therapy are listed in Table 1.

In this chapter, we will review sequentially each of the cytokines tested and analyze their efficacy from the standpoint of three separate and increasingly difficult objectives of cytokine gene delivery described as follows: (1) to abrogate establishment of tumor (establishment model), (2) to immunize naive animals against wild-type tumor (immunization model), and (3) to treat animals with established tumors (treatment model). This

Table 1

Cytokines Tested in Gene Therapy of Cancer

Cytokine	MW (kDa)	Normal Cell Source	Target Cells	Effects
IL-1	17	Macrophages	Most mononuclear cells, endothelial cells	Immunoaugmentation (inflammatory and hematopoietic)
IL-2	15	T-cells, LGL	T-cells, B-cells	Activates T-cells, NK-cells, and macrophages
IL-3	14–28	T-cells	Myeloid cells	Hematopoietic growth factor, promotes growth of early myeloid progenitor cell
IL-4	20	T-cells, mast cells	T-cells, B-cells, eosinophils, endothelial cells, macrophages	Promotes IgE reactions, activates target-cells, immunostimulation
IL-6	22–30	Fibroblasts, T-cells	B-cells, endothelial cells	Augments inflammation; B-cell growth factor; augments polyclonal immunoglobulin production
IL-7	25	Stromal cells (thymus and bone marrow)	T-cells, B-cells	Generates pre-B and pre-T-cells, lymphocyte growth factor; lymphopoetin
IL-10	20	T-cells, macrophages, some tumor cells	B-cells	Stimulates B-cells, antiinflammatory, DTM inhibition immunosuppression, monocytes
IL-12	70	Macrophages, B-cells, PMNs	NK-cells, T-cells	Stimulates IFN-γ production and proliferation and cytotoxicity of target cells, promotes TH1 cellular immune response
TGF-β	14	Platelets, bone, some human cells	Most mammalian cells	Immunosuppression, wound healing, bone remodeling
TNF-α	17	T-cells, macrophages	Most mononuclear cells, endothelial cells	Inflammation, immuno-enhancing, tumoricidal, augments vascular thromboses and tumor necrosis
IFN-g	20–25	Th1 cells, NK-cells	NK-cells, macrophages	Stimulates target-cells; induces MHC antigens and other proteins, antiviral antiproliferative and immunomodulation
G-CSF	18–22	Monocytes	Neutrophils	Differentiation of mature neutrophils, myeloid growth factor
GM-CSF	23	Various cell types	Granulocyte-macrophage populations	Regulation of survival, differentiation, proliferative and functional activities

type of understanding is helpful for examining the current status of these cytokines. It should be noted that exact comparisons among individual cytokines have usually not been carried out, since antitumor effects were often examined over a limited dose range of cytokine expression using different tumor systems.

III. SUMMARY OF INDIVIDUAL CYTOKINE GENE THERAPY

A. Interleukin 2 (IL-2)

The first biologic agent reported to be applied in cytokine gene therapy was IL-2.[7] IL-2 is a growth factor that stimulates the proliferation of cytotoxic T-cells, helper T-cells, and NK-cells and stimulates the cytolytic activity of both T- and NK-cells known as lymphokine-activated killer (LAK) activity.[9-11] Bubenik showed that local production of IL-2, achieved by retroviral transduction of fibroblasts with the IL-2 gene, substantially inhibited the establishment and growth of human HeLa tumor cells in athymic BALB/c mice.[7] Additionally, they showed that splenocytes from animals with such treatment were cytotoxic for HeLa and other tumor targets. These results suggested that LAK cells were important for the antitumor effects observed in that study. Subsequently, Fearon demonstrated that locally secreted IL-2 could delay the establishment and growth of two poorly immunogenic murine tumors, CT26 colon cancer and B16 melanoma.[5] These effects were correlated with increased cytolytic T-cell function in treated animals. Furthermore, long-term protective immunity to parental B16 cells could be generated by immunizing with the IL-2-transfected B16 tumor. These results suggested a possible T-helper deficiency in the antitumor response of untreated mice. Similarly, Gansbacher confirmed, using the murine fibrosarcoma CMS-5, that local IL-2 administration via gene transfection of tumor cells could abrogate tumorigenicity as well as induce long-lasting protective immunity against subsequent challenge with the non-transfected parental tumor cells.[6] Recently, efficacy of IL-2 gene therapy was shown in treatment models. Connor has shown that IL-2-transfected MBT-2 (a FANFT-induced murine bladder tumor) cells could be effectively used to treat C3H mice with pre-existing, palpable, well-established bladder tumors.[12] Seven days following intravesical instillation of parental nontransfected 2×10^4 MBT-2 cells, mice were treated with 5×10^6-irradiated IL-2-secreting MBT-2 cells three times at weekly intervals. Treatment resulted in complete tumor regression in three of five mice and delayed tumor progression in two of five mice by 20 days. These three animals with complete tumor regression remained tumor free for more than eight weeks and were resistant to subsequent challenge with nontransfected parental MBT-2 cells. The other treatment of mice with IFN-γ-secreting cells, cisplatin, or combination with IL-2-secreting tumors had a less pronounced effect in decreasing tumor growth or enhancing survival. When tumor-free mice were rechallenged with intravesicle instillation of nontransfected parental MBT-2 cells, no tumor growth was observed in nine of nine mice tested. These results demonstrated that effective immunologic memory had been established using this strategy.

B. Interleukin 3 (IL-3)

IL-3 is a pluripotent hematopoietic growth factor that is produced by T-cells and increases the maturation and development of cells of the myeloid lineage. IL-3 has been used clinically to treat myelosuppression in patients receiving chemotherapy.[13] Pulaski et al. recently have shown that IL-3 gene transferred into the BALB/c murine line 1 lung carcinoma could lead to tumor rejection by enhancing the development of $CD4^+$-dependent cytotoxic T-lymphocytes (CTL).[14] Tumor-infiltrating lymphocytes (TIL) obtained from tumors transfected with IL-3 were shown to induce CTL similar to that from IL-2-producing tumors, and mRNA for IL-2 was observed within the IL-3-producing tumors but not within nontransfected parental tumors. These results suggested that IL-3 might indirectly induce CTL by stimulating release of CTL growth factors from T-helper cells. Despite the similarity of the TIL from IL-3-secreting tumors with that from IL-2 and the presence of IL-2 in TIL from IL-3-transfected tumors, TIL isolated from tumors producing IL-2 and those producing IL-3 differed in their ability to induce TIL. These isolated from the IL-2 transfected tumors lysed YAC-1 very effectively (considered to be nonspecific effector activity), while TIL from the IL-3-transfected tumors did not. A more recent study from the same group has concluded that IL-3 inhibits the development of nonspecific killer cells both *in vitro* and *in vivo* only when IL-3 is produced early in the development of such nonspecific killers.[15] They showed that addition of IL-3 had a strong inhibitory effect on the generation of nonspecific killer cells when IL-3 was added at the initiation of IL-2-stimulated culture or 1 day later but not when added on 2 or 3 days after the initiation of IL-2 stimulation. These results for inducing specific CTL are favorable for inducing effective immunity against the tumor, but results for immunity have not yet been reported.

C. Interleukin 4 (IL-4)

IL-4, a T_H2 cytokine, was initially identified as a B-cell growth factor that induced isotype switching in B-cells.[16] IL-4 also serves as a T-cell growth factor and enhances the cytolytic activity of T-cells and NK-cells, especially in mice.[17] It also up-regulates the expression of adhesion molecules such as VCAM that leads to the binding and retention of macrophages, eosinophils, and memory T-cells at sites of inflammation.[18] In contrast to IL-2, IL-4 can activate and augment both antibody and T-cell responses, as well as nonspecific inflammation, making it a unique molecule for treating cancer. Tepper and Leder transfected the IL-4 gene into two different murine tumor cell lines, the BALB/c plasmacytoma J558L and the mammary adenocarcinoma K4851, and observed that high IL-4-producing cells could elicit an antitumor effect in BALB/c mice.[4] In this study, they found that early infiltrates at tumor inoculation sites lacked T-cells and consisted mainly of eosinophils and macrophages. They also found that tumor formation was blocked even when these IL-4 gene-transfected tumors were injected into athymic nude mice. These results and another study from the same group suggested that the observed initial antitumor effect was T-cell-independent but eosinophil-dependent.[19] In this study, eosinophils of C57BL/6 mice were depleted using MAb RB6-8C5, a monoclonal antibody that binds to surface antigens present on mature granulocytes (thus depleting neutrophils as well as eosinophils), and both low (10,000 U per 10^6 cells per 48 hr) and high (20,000 U per 10^6 cells per 48 hr) IL-4-producing B16 tumor cells were inoculated into C57BL/c mice. Failure of tumor growth was noted in animals bearing high IL-4-producing tumor plus Mab RB6-8C5, whereas tumor formation was restored in animals bearing low IL-4 producers and MAb RB6-8C5. In both cases, addition of anti-IL-4 antibody restored tumorigenicity, and the growth of the implanted tumor was *inversely* correlated with eosinophilic inflammation at the tumor inoculation site.

A more recent report by the same group supports this observation.[20] They reported that there was a significant inhibition of the growth of two glioma lines in nude mice when LT-1 (IL-4 secreting J558L) tumor cells were admixed with them and coadministered. Subcutaneous growth of the rat glioma cell line C6-BAG was markedly inhibited compared to controls when mixed with LT-1. Subcutaneously injected U87 human glioma cells were completely rejected when mixed with LT-1 cells, while contra lateral controls (U87 + J558L) developed very large tumors. Histologic study of C6-BAG or U87 glioma with LT-1 tumor revealed tumor necrosis and an eosinophilic infiltrate containing occasional macrophages at 3 days postinjection, while glioma with nontransfected J558L controls showed aggressive tumor growth without necrosis or infiltrate. *In situ* intracerebral administration of U87 glioma with LT1 cells also inhibited tumor growth compared to U87 with nontransfected J558L controls. Six of twelve (50%) animals that received LT-1 cells survived at 9.5 weeks postimplantation, while all the control animals were dead at that point. Up to 13 weeks after the injection, 4 of 12 (33%) from the LT-1 group were healthy and neurologically normal. Hematoxylin and eosin staining of control brains at 4 days postinoculation revealed glial tumor infiltrating into normal tissue with a small necrotic area and no evidence of an inflammatory response. However, LT-1-treated brain showed that the majority of the tumor was necrotic with granulocytic infiltrate throughout, and this infiltrate proved to consist almost entirely of eosinophils using Giemsa staining.

Blankenstein reported an interesting study using the IL-4-dependent CTLL derivative, CT.4S, to assess its function as an autocrine growth factor.[21] In contrast to studies with other "autocrine" growth factors such as IL-6, IL-4-transfected CT.4S cells proliferated in the absence of exogenous IL-4 *in vitro*, but these cells were unable to grow in mice. These results suggested that IL-4 induced local immune processes that inhibit the growth of CT.4S cells. In unpublished studies (personal communication), he has shown that the antitumor effect of IL-4 transfection was IFN-γ-dependent using injection of neutralizing antibodies. Golumbek and Pardoll have reported similar findings using Renca (renal carcinoma) and CT26 (colon carcinoma) cell lines.[22] Using IL-4-transfected tumor cells, they were able to show complete inhibition of tumor establishment even at high tumor doses (2×10^7). Subsequent challenges with parental tumors were completely rejected. Furthermore, Renca tumor cells engineered to secrete large amounts of IL-4 (15,000 U per 10^6 per 24 hr) caused regression of 6- to 9-day established tumors, but tumor cells secreting lower doses (3300 U/10^6/24 hr) did not. Even though the early tumor infiltrate consisted primarily of macrophages and granulocytes, the sustained antitumor response was due to $CD8^+$ T-cells that appeared to be associated with long-lasting, tumor-specific immunity. Recently, it was found that simple irradiation of this tumor also induced similar antitumor effects.[23] The debate still exists whether the radiated tumor cells are better as a tumor antigen compared to nonradiated live cells.

Modesti and Forni have shown that activated TS/A tumor rejection caused by systemic IL-4 administration consisted of a multicell-mediated reaction involving granulocytes, macrophages, and lymphocytes, which causes direct membrane and cytoplasmic damage to the tumor cell.[24] In contrast to the previous study by Tepper and Leder,[4] IL-4-activated tumor rejection in their studies does not occur in T-cell-depleted mice. This study suggested the possibility that the high amount of IL-4 used in the Tepper experiments resulted in a strong, immediate, nonspecific response that did not require T-cell effects. We have recently initiated an IL-4 gene therapy protocol based on these studies as well as our own.[25] In this protocol, fibroblasts capable of producing IL-4 are used at the site of a vaccine. We have treated patients with fibroblasts producing up to 50,000 U/24 hr admixed with tumor and observed eosinophil, macrophage, and $CD4^+$ cells infiltrating the vaccine site on immunohistochemistry (unpublished observation).

D. Interleukin 6 (IL-6)

IL-6 is produced by a wide variety of cells, including T_H2 cells, macrophages, and monocytes. IL-6 has been reported as a critical component in the acute phase inflammatory response. Like IL-1,[26] it is an inducer of B-cell maturation and proliferation, and T-cell maturation.[27,28] Blankenstein showed that IL-6 gene transfection into J558L tumors (produced 500 U/ml) had no effect on tumor growth when injected into mice,[29] but Sun et al. reported that IL-6 transfection caused slow growth of B16 tumors compared to control cells, and that this delayed growth was found to be related to nonspecific inflammatory mechanisms.[30] Interestingly, tumor cell adhesion to matrix protein-coated plastic surfaces was enhanced with IL-6 gene transfer and was associated with a reduction in tumor angiogenesis. In a comprehensive study of the effect of local IL-6 on the Lewis lung carcinoma cell line D122, inhibition of tumor growth was inversely related to the amount of IL-6 produced by tumor cells.[31] Tumor suppression was also observed in nude mice, although this was not as vigorous as in normal mice, suggesting that both T-cell-independent and dependent antitumor mechanisms exist. Moreover, IL-6 production markedly activated the anti-D122 cytolytic activity of CTLs, enhanced macrophage accumulation at tumor sites, and inhibited the lung metastasis of B16. Long-term immunity against local tumor growth and tumor metastasis was achieved against nontransfected parental D122 tumors when irradiated IL-6-producing D122 cells were used for immunization in immunocompetent animals but not in nude mice. This result indicates that antitumor memory presumably requires immunocompetent T-cells.

E. Interleukin 7 (IL-7)

IL-7 was originally described as a B-cell growth factor for early B-cells[32] and subsequently reported to support the proliferation of mature $CD4^+$ and $CD8^+$ T-cells,[33-35] induce cytotoxic T-cell differentiation,[36-38] and enhance generation of LAK activity. Hock showed that IL-7 could induce substantial T-cell-dependent antitumor effects in a plasmacytoma (J558L).[39] The suppression of tumor growth was not apparent when nude mice were used as recipients, but depletion of $CD8^+$ T-cells in normal mice had no effect on tumor rejection, even though $CD8^+$ cells were detected in tumor infiltrates. These results suggested that IL-7-induced anti-J558L immunity is dependent on the induction of CD4+ T-cells. CR3+ cells, presumably macrophages, were also required for tumor suppression in addition to CD4+ T-cells. These results were confirmed in similar experiments with a murine mammary adenocarcinoma cell line TS/A.

F. Interferon Gamma (IFN-γ)

IFN-γ is a pleiotropic cytokine produced primarily by TH1-cells and NK-cells that induces the expression of MHC class I and class II antigens to enhance antigen presentation. IFN-γ also activates macrophages, enhancing MHC molecule and Fc receptor expression, increasing antibody-dependent cellular cytotoxicity, hydrogen peroxide and nitric oxide production, and monocyte tumoricidal activity, presumably secondary to TNF production. IFN-γ markedly enhances the expression of TNF receptors on a variety of cell types, plays an important role in the inflammatory process, and is required for the full development of cytotoxic T-cells. Miyatake showed that IFN-γ transfection suppressed the tumor establishment of the murine neuroblastoma C1300 by augmenting antitumor immunity.[40] Gansbacher also showed that local IFN-γ abrogates CMS-5 fibrosarcoma tumorigenicity and can induce strong long-lasting antitumor immunity.[6] This inhibition of CMS-5 correlated with marked increase (10- to 30-fold over control) in the expression of MHC class I antigen. Cytotoxicity assays using splenocytes from immune mice show that antitumor immunity was associated with enhanced specific CTL activity against the CMS-5 tumor. In addition, IFN-γ-secreting CMS-5 cells but not unmodified tumor cells were efficiently and specifically lysed by splenocytes from mice inoculated with the wild-type CMS-5 tumor. Restifo also suggested that the mechanism of local IFN-γ induced antitumor immunity is due to enhanced antigen presentation associated with an increase in MHC class I antigen using MCA 101 and MCA 102 murine methylcholantrene-induced sarcomas.[41] In contrast, Porgador et al., using 3LL lung carcinoma, reported that stronger antitumor immunity was found in those tumors that were high expressors of IFN-γ, even though the MHC class I antigen was at an equally high level.[42] Based on this result, they suggested that IFN-γ-induced antitumor is derived primarily from the induction of cytotoxic T-cells.

G. Tumor Necrosis Factor α (TNF-α)

TNF-α is produced primarily by macrophages and certain T-cells and possesses potent direct and indirect antitumor effects via its pleiotropic functions described below. It enhances the function of monocytes and macrophages,[43,44] increases neutrophil cytotoxic activation,[45] and augments T-cell proliferation and activation.[46-50] When MCA-205, a murine sarcoma, was transfected to secrete TNF-α, it showed significant reduction of tumor size after an initial (7 to 8 day) period of growth without apparent cachexia of the animals.[51] TNF-α-producing MCA-205 cells admixed with parental cells also inhibited tumor growth in a similar manner, suggesting that local production of TNF-α is sufficient for tumor regression. T-cell depletion studies showed that tumor suppression required both CD4+ and CD8+ T-cells since *in vivo* depletion of either subset inhibits tumor regression. Blankenstein showed that with TNF-α gene transfer to the murine J558L plasmacytoma abrogated tumor establishment in 60% of animals and significantly delayed tumor growth in the remaining 40% of animals compared to parental controls.[29] Histologic examination of the tumor showed an abundant infiltrate of macrophages, and the administration of antitype 3 complement receptor (CR3), which inhibits inflammatory cell migration, abolished tumor suppression. Teng also showed that the TNF-α gene transfer into 1591-RE, a UV-induced murine skin tumor, leads to significant inhibition of tumor growth in mice.[52] Injection of these TNF-α-producing tumor cells into nude mice also results in suppression of tumor growth. These results suggested that the initial antitumor effects of TNF-α is T-cell-independent and that the ability of TNF-α to inhibit tumor growth in T-cell-deficient mice indicates that TNF-α may be useful for treating both "antigenic" and "nonantigenic" tumors.

H. Granulocyte-Colony Stimulating Factor (G-CSF)

Granulocyte colony-stimulating factor (G-CSF) was first identified by its ability to induce differentiation of the murine myelomonocytic leukemia cell line WEHI-3B.[53] G-CSF is less pleiotropic compared to other cytokines and primarily activates polymorphonuclear cells by stimulating release of neutrophilic granules and enhancement of neutrophil migration to the site of cytokine production.[54] Colombo et al. genetically engineered the murine colon adenocarcinoma C-26 to produce G-CSF in order to examine the role of neutrophils in antitumor immunity.[55] Local G-CSF-inhibited tumor growth in immunocompetent mice. This effect was also observed in nude and NK-depleted mice. Hence, G-CSF's antitumor effect was considered to be T-cell independent. Histologic studies showed that the tumor infiltrate is composed mainly of neutrophils. In contrast to other cytokines tested in this manner, the antitumor effect of G-CSF has not been shown to result in antitumor effect of the nontransfected parental tumor cells admixture with G-CSF-producing tumor, which resulted in no rejection.

I. Granulocyte-Macrophage Colony Stimulating Factor (GM-CSF)

Dranoff et al. screened a variety of murine tumor systems to examine a variety of promising cytokines for cancer gene therapy using their retroviral vectors that carry available cDNAs for cytokines.[23] They have observed that inoculation of live B16 melanoma transfected with both GM-CSF and IL-2 could generate potent systemic protection against nontransfected B16 tumor challenge. A variety of other cytokines, including IL-2 alone and IL-4 , IL-6, and IFN-γ in combination with IL-2, resulted in rejection of initial tumor but could not induce significant long-term immunity in most animals. They also showed that inoculation of irradiated B16, CT-26, CMS-5, and RENCA tumor cells transfected with GM-CSF alone induced complete suppression of growth of a subsequent challenge using nontransfected tumor cells. In their system, GM-CSF appeared to be the most potent of the molecules they tested (IL-2, 4, 5, 6, IL-1RA, IFN-γ, TNF-α, ICAM-1, CD2).

J. Interleukin 12 (IL-12)

IL-12, formerly termed natural killer-cell stimulatory factor (NKSF)[56] or cytotoxic lymphocyte maturation factor (CLMF),[57] is a disulfide-linked heterodimeric cytokine composed of a 35 kDa light chain (p35) and a 40 kDa heavy chain (p40).[56,57] The two cDNAs encoding the p35 and p40 chains of IL-12 from both the mouse and human have been cloned. Unlike most other cytokines, simultaneous transfection of mammalian cells with two different genes is necessary for production of biologically active IL-12.[58,59] This cytokine exerts a variety of biologic effects on human T- and NK-cells *in vitro*. These include the ability to synergize with IL-2 in augmenting allogeneic CTL response, LAK activity,[57] and INF-γ production from peripheral blood lymphocytes.[56,60] IL-12 also directly stimulates the production of IFN-γ and other cytokines from peripheral blood T- and NK-cells,[60,61] enhances the lytic activity of NK-cells,[56,61] and promotes the expansion of activated NK-cells and activated T-cells ($CD4^+$ and $CD8^+$ subsets).[62] IL-12 has also been shown to induce primarily a T_H1 (cellular immune) response *in vitro*.[63] In addition to the *in vitro* results, recent studies with the administration of recombinant murine IL-12 to normal mice revealed that IL-12 enhances NK and CTL activity and induces IFN-γ production *in vivo*.[64]

Based on these results, which suggested profound immunologic effects of IL-12 alone or in combination with IL-2, we have initiated murine therapeutic studies to investigate the possible antitumor effect of IL-12. These experiments demonstrated that systemic application of IL-12 can suppress and prolong the survival of tumor-bearing mice, even when IL-12 treatment is started as late as 14 days after tumor inoculation. Such therapeutic effects were also reported in similar models by Brunda et al.[65] In their studies, therapeutic intervention with systemic administration of IL-12 can be initiated as late as day 28 after injection of tumor cells (M5076 reticulum cell sarcoma), resulting in inhibition of tumor growth, reduction in the number of metastases, and increase in survival time. Although IL-12 has demonstrated potent antitumor effects when injected systemically, induction of long-term immunity is less frequent and variable from experiment to experiment.[65,66] Interestingly, Brunda observed that the best results with systemic IL-12 administration (complete regression of the tumor and induction of protective immunity against tumor rechallenge) was observed following peritumoral injections of IL-12 in a subcutaneous tumor model using Renca cells.[65] Since IL-12 is secreted only by "professional" antigen-presenting cells (i.e., macrophages and B-cells), it is possible that local administration of IL-12 at the site of tumor might approximate the role of IL-12 in eliciting an endogenous immune response. More recent evidence suggests that IL-12 plays a critical costimulatory role with B7 on APC in inducing proliferation and IFN-γ production from T_H1 clones (G. Trinchiari, personal communication). Thus, IL-12 may be the ideal cytokine for inducing tumor-specific immunity when administered at the tumor site.

To test the application of IL-12 paracrine therapy described above, we have established animal models using fibroblasts genetically engineered to secrete murine IL-12 and examined its effects on tumor growth.[67] NIH3T3 cells were transfected with expression plasmids carrying both the murine p35 and p40 genes of murine IL-12, using the calcium phosphate method to express 100 to 240 U per 10^6 cells per 48 hr of IL-12.[68] The effects of paracrine secretion of IL-12 on tumor establishment and vaccination models were examined using the poorly immunogenic murine melanoma cell line BL-6 in C57BL/6 (B6) mice. To determine the effects of IL-12 on tumor formation, nonirradiated BL-6 cells were subcutaneously inoculated into B6 mice admixed with NIH3T3 cells transfected with both subunits of mIL-12(3T3-IL-12) or with cells transfected with only the neomycin phosphotransferase gene (3T3-Neo). Compared to mice injected with BL-6 alone, the day of emergence of detectable tumors was significantly delayed in mice injected with BL6 admixed with 3T3-IL-12, but not in mice with BL6 admixed with 3T3-Neo. Effectiveness in this system was related to the amount of IL-12 expressed by the 3T3-IL-12. To determine whether locally secreted IL-12 at the tumor site induces effective antitumor immunity, 10^6 irradiated tumor cells mixed with 3T3-IL-12 or 3T3-Neo were injected as a vaccine, and the response to a tumor challenge was subsequently examined. With a tumor challenge of less than 1×10^5 nonirradiated BL-6 cells, significant delay of establishment of tumor was noted with a relatively small amount of IL-12 secretion (1.2 U per 5×10^5 cells per 48 hr). Larger amounts of secreted IL-12 provided no additional therapeutic benefit. These results suggest that local delivery of IL-12 inhibits tumor growth in a dose-dependent manner and leads to the development of an antitumor immune response.

Our previous system used allogeneic 3T3 cells as a means to deliver IL-12 continuously at the site of tumor. However, these cells probably induce an allogeneic immune response, thus effectively eliminating long-term IL-12 production. This makes direct interpretation of our results somewhat complicated. Development of efficient vectors for transfection of appropriate autologous cells will enable us to ask questions raised by our studies and also test more clinically relevant systems. Thus, we have constructed a retroviral vector that allows us to express genes of both subunits of mIL-12 with a selectable marker. The requirement for simultaneous expression of p35 and p40 for production of biologically active IL-12[58] presented unique problems in the creation of these vectors. The existing strategies using regulated splicing or heterogeneous promoters raised substantial concern about discordant production of the two chains from the transfected cells. To overcome this problem, we developed vectors utilizing an internal ribosome entry site (IRES) sequence. The IRES sequence used was obtained from the 5′ nontranslated region of equine encephalomyocarditis virus (EMCV). Other picornaviruses have similar sequences that allow cap-independent translation initiation from the internal ribosome entry site. A single polycistronic transcript from two separate genes can thus be transcribed following a single promoter, and a second gene can be translated in a cap-independent manner. A modified MFG-based retroviral vector was constructed using an IRES fragment of EMCV, obtained from the pCITE plasmid (Novagen, Madison, WI). The resultant construct is shown in Figure 1-A (DFG-mIL-12). To obtain a retrovirus that carries both the mIL-12 genes and the neomycin phosphotranspherase gene as a selectable marker, *neo* gene (pMC1*neo* Poly A, Strategene, La Jolla, CA) was subcloned into 3′ of the p35 cDNA in DFG-mIL-12 with another IRES sequence (termed TFG-mIL-12-*neo*, Figure 1-B). TFG- and DFG-human IL-12 has also been constructed in a similar manner to prepare for this clinical application.[69] Retroviral supernatant was generated using these proviral constructs and the yCRIP packaging cell line.[70]

Expression of these retroviral vectors was measured using NIH3T3 cells. These cells were infected with the supernatant of the producer cell line, and their supernatant was measured for expression of the transfected genes (Table 2). These results demonstrate the

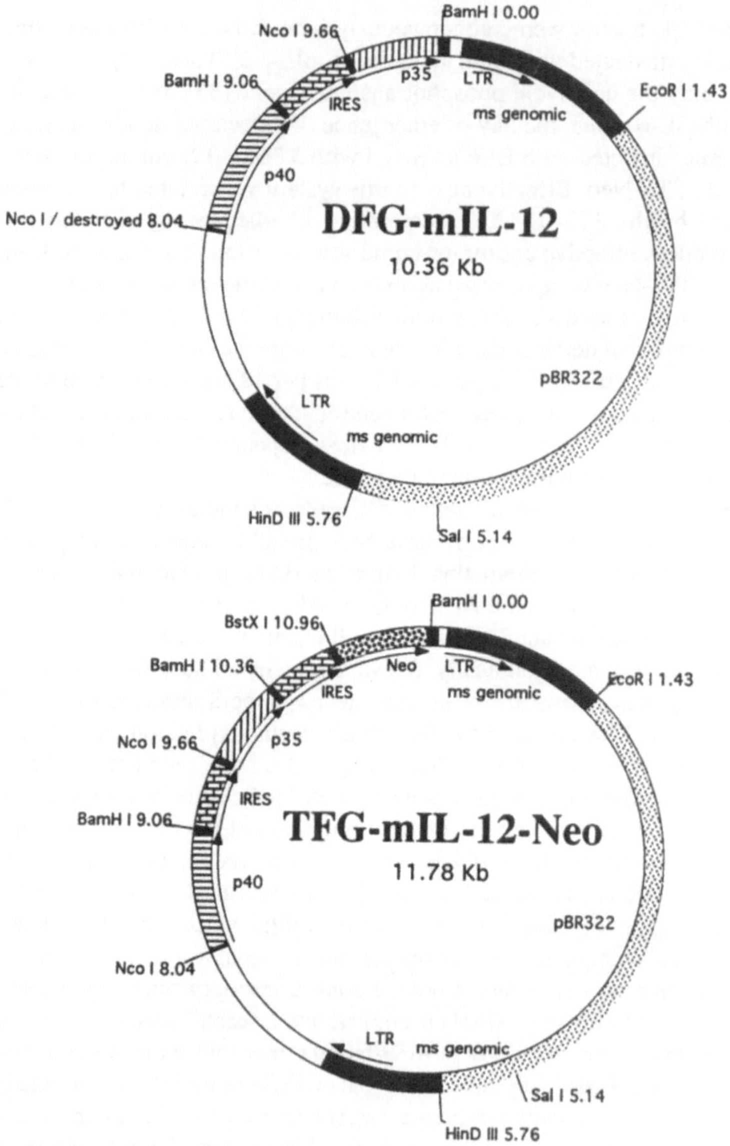

FIGURE 1. Retroviral vectors carrying IL-12 cDNAs. (Top) DFG-mIL-12. Both p35 and p40 of murine IL-12 can be expressed. (Bottom) TFG-mIL-12. In addition to both p35 and p40 genes, the neomycin phosphotransferase gene can be expressed as a selectable marker.

ability of our vectors to transfect and express large quantities of biologically active IL-12 from fibroblasts efficiently. Using the TFG-mIL-12-*neo* vector, we are currently testing the therapeutic benefit of paracrine IL-12 secretion (Tahara et al., submitted). Colombo et al. also constructed a similar retroviral vector carrying IL-12 genes using IRES sequence and are currently testing its effect (personal communication).

Table 2

Expression of the Transfected NIH-3T3 Cells

Vector(s)	Transfection Method	Selection	IL-12 Secretion[a]
BL-pSV-p35 and p40	Ca-Phosphate	+ (Colony #8)	2.1
DFG-mIL-12	Infection	—	5
TFG-mIL-12	Infection	+ (bulk)	133

[a] Measured as (ng per 10^6 cells per 48 hr).

IV. COMPARISON AMONG THE EFFECTS OF CYTOKINES TESTED IN GENE THERAPY SETTING

The cytokines described in Section III have the property of suppressing tumor establishment of various tumor cell lines, but not all of them resulted in long-term immunity, which is considered to be necessary for clinical application of this strategy. To date, protective effects have been reported with IL2, 4, 6, 12, IFN-γ, TNF-α, and GM-CSF, as shown in Table 3. In the development of long-term immunity against tumor cells, T-cell-mediated responses against tumors and the generation of memory T-cells appear to be crucial.

Using an IL-2 gene therapy approach, Gansbacher demonstrated that memory CTLs were generated, and these animals could reject a second dose of tumor cells.[71] This ability of IL-2 to generate memory CTL was also observed by Fearon using a murine colon cancer,[5] Connor using a murine bladder tumor,[12] and Tsai using a mouse mammary tumor.[72] Using IL-4, long-term immunity was reported to be also T-cell-dependent, although the initial antitumor response associated with local IL-4 production appeared to be mediated primarily through eosinophils and macrophages.[22] We have reported that long-term protective effects were also observed when live or irradiated MC 38 (murine colon carcinoma) was intradermally inoculated admixed with syngeneic fibroblasts transfected with IL-4 (immune animals remain tumor-free 100 days post 10^7 tumor challenge.[25]

With TNF-α, only Asher et al. have been able to show long-term immunity, which was related to the presence of CD4 and CD8 antitumor effector cells. Blankenstein et al. demonstrated that TNF-α-caused antitumor effects were primarily mediated by a macrophage response using the J558L tumor,[29] and Teng et al. showed that inhibition of tumor growth was T-cell-independent and preserved in nude mice.[52] No long-term immunity studies were reported. With IFN-γ, Restifo has reported that increased MHC Class I by IFN-γ could elicit $CD8^+$ TIL that can then be used therapeutically to treat animals with established metastases.[41] Associated with this mechanism, IFN-γ has been considered to be highly effective in generating long-term immunity to tumors, presumably due to its enhancement of effective antigen presentation and increased CTL activity. Porgador et al. observed that IL-6-mediated long-term immunity against Lewis

Table 3

Comparison of Cytokine Gene Therapy in Murine Models

Cytokine	Reference	Tumor	Strain	I	II	III
IL-2	Bubenick 1988	HeLa	nu/nu	x		
	Gansbacher 1990	CMS-5	BALB/c	x	x	
	Fearon 1990	CT26	BALB/c	x	x	
		B16	C57BL/6	x	x	
	Connor 1993	MBT-2	C3H/HeJ	x	x	x
	Tsai 1993	4TO7	BALB/c	x	x	
	Forni 1992	TS/A	BALB/c	x	x	
IL-3	Lord 1993	Murine Line 1	BALB/c	x		
IL-4	Tepper 1989	CT26	BALB/c	x	x	
	Golumbek 1991	Renca	BALB/c	x	x	x
	Modesti 1993	TS/A	BALB/c	x	x	
	Tepper 1993	U87	nu/nu	x		
IL-6	Blankenstein 1991	J558L	BALB/c	x		
	Sun 1992	B16	C57BL/6	x		
	Porgador 1992	D122	C57BL/6	x	x	
IL-7	Hock 1991	J558L	BALB/c	x		
		TS/A	BALB/c	x		
IL-12	Tahara 1994	BL6	C57BL/6	x	x	
	Tahara (submitted)	MCA207	C57BL/6	x	x	x
		MCA102	C57BL/b	x	x	
		B16	C57BL/b	x	x	
		TS/A	BALB/c	x	x	
G-CSF	Colombo 1991	CMS-5	BALB/c	x		
IFN-γ	Watanabe 1989	C1300	C57BL/6	x		
	Gansbacher 1990	CMS-5	BALB/c	x	x	
	Restifo 1992	MCA 101	C57BL/6	x		
	Porgador 1992	MCA 102				
GM-CSF	Dranoff 1993	B16	C57BL/6	x	x	x
		Lewis lung carcinoma				
		WP-4				
		RENCA				
		CMS-5				
		CT-26	BALB/c	x	x	
TNF-α	Asher 1991	MCA205	C57BL/6	x	x	
	Teng 1991	1591-RE	nu/nu	x		
	Blankenstein 1991	J558L	BALB/c	x		

Note: I: Delay/inhibition in tumor establishment models; II: Tumor growth suppression in vaccine/immunization models; III: Therapeutic effect for pre-existing tumor models.

Lung carcinoma was achieved only in immunocompetent animals but not in nude mice, again suggesting that such immunity is T-cell-dependent.[31] Dranoff showed that tumor cells transfected with GM-CSF with or without IL-2 can induce long-term T-cell-dependent immunity in a variety of tumor models and only GM-CSF-,[23] IL-4-, and IL-6-induced long-term immunity in their B16 tumor system. More recently, IL-12 was shown to have the ability to induce tumor immunity in BL6 melanoma system.[53]

To date, it is still an open question which cytokine is "the most potent" in gene therapy approach. The effect might be different in different tumor systems, protocols, target cells for transfection, and expression levels of cytokines. Detailed studies and consideration for these points should be performed before coming to any definitive conclusions. It might be possible that "the most potent" cytokine is different in each tumor type due to its biology. In most clinical situations, the therapeutic measure for cancer treatment should have an antitumor effect against established tumors. Hence, cytokine cancer therapy should have such effects before being considered for clinical application. Golumbek et al. demonstrated this feasibility using Renca cells transfected with the IL-4 gene.[22] Our group also supports these findings using the MCA-105 tumor. Growth of the tumor could be delayed up to 13 days with IL-4-transfected fibroblast immunization therapy.[25] Vaccination using 10^6 irradiated MCA-105 cells admixed with 10^6 fibroblasts producing more than 4000 U per ml per 24 hr of IL-4 were administered on days 7, 10, and 14 post tumor inoculation). Furthermore, tumor growth could be delayed up to 20 days by coadministration of 100,000 units of IL-2 b.i.d. on days 10 to 14 post tumor inoculation. Therapeutic benefits have been also reported with IL-2 and GM-CSF.[12,23]

These animal studies have facilitated the clinical application of cytokine gene therapy. Although the feasibility of cytokine cancer therapy has been supported with these results in animal models, the actual clinical application is still in its infancy, as described in following section.

V. CURRENT CLINICAL PROTOCOLS

As of November 4, 1993, the Recombinant DNA Advisory Committee (RAC) had approved 59 clinical protocols involving gene manipulation. Among them, 36 and 23 of them are gene therapy and gene marking protocols, respectively. Twenty-six of these protocols are for cancer treatment, and twelve of these use a cytokine gene therapy approach. These protocols include IL-2, TNF-α, IFN-γ, GM-CSF, and IL-4 gene therapies (Table 4). IL-2 and TNF-α studies for the treatment of melanoma are being conducted by Steve Rosenberg at the National Cancer Institute. Bernd Gansbacher is heading the IL-2 and IFN-γ clinical trials at the Memorial Sloan-Kettering Cancer Institute for patients with melanoma or renal cell carcinoma. At St. Jude's, Brenner is overseeing an IL-2 clinical protocol in patients with neuroblastoma. Our group at the Pittsburgh Cancer Institute has begun IL-4 gene therapy for patients with either metastatic melanoma, renal cell cancer, colorectal cancer, or breast cancer, and we are in the process of the modification of this protocol to include both local IL-4 production (by cytokine gene transfection) as well as systemic administration of IL-2 protein. We (Lotze and Tahara) also recently submitted a protocol for IL-12 gene therapy for cancer.

Table 4

RAC Approved Gene Therapy Protocols for Cancer Treatment (as of November 4, 1993)

Cytokine	Protocol #	Investigator	Institution	Date of RAC Approval (Submission)	Date of Approval	Description NIH
IL-2	9110-011	Rosenberg	NCI	10/07/91	10/15/91	Tumor vaccine to generate TILs
	9206-021,022	Gansbacher	MSKCC	06/02/92	08/14/92	Allogeneic HLA-A2 and melanoma vaccine
	9206-018	Brenner	St. Jude	06/01/92	08/14/92	Autologous neuroblastoma vaccine
	9309-053	Cassileth	U. of Miami	09/09/93	Pending	Autologous tumor vaccine in limited stage small-cell lung cancer
	9309-056	Das-Gupta	U. of IL	09/09/93	Pending	Allogeneic melanoma immunization for patients with unresectable melanoma
	9309-058	Economou	UCLA	09/10/93	Pending	Vaccine for melanoma
IL-4	9209-033	Lotze	U. of Pitt	09/15/92	02/05/93	Autologous skin fibroblasts admixed with tumor

IFN-γ	9306-043	Seigler	Duke	06/07/93	09/03/93	Autologous tumor cells for disseminated malignant melanoma
TNF	9007-003	Rosenberg	NCI	07/31/90	09/06/90	Autologous TILs
	9110-010	Rosenberg	NCI	10/07/91	10/15/91	Autologous cancer cells
GM-CSF	9303-040	Simons	Johns Hopkins	03/01/93	Pending	Autologous tumor cells for metastatic melanoma
IGF[a]	9306-052	Ilan	Case Western	06/08/93	Pending	Brain tumors, using episome-based antisense cDNA
MDR	9306-044	Deisseroth	U. of Texas	06/07/93	Pending	Chemoprotection during therapy for ovarian cancer
	9306-51	Hesdorffer	Columbia U	06/08/93	09/03/93	Chemoprotection during therapy for breast, ovarian, and brain cancer
	9309-054	O'Shaughnessy	NIH	09/09/93	10/07/93	Chemoprotection during therapy for breast cancer
HLA B-7	9306-045	Nabel	U. of MI	06/07/93	09/03/93	Immunotherapy by direct gene transfer
TK	9303-037	Culver	U. of IA	03/01/93	04/16/93	TK treatment for brain cancer
	9306-050	Raffel	Children's LA	06/08/93	09/03/93	TK treatment for astrocytoma
	9309-055	Kun	St. Jude	09/09/93	10/07/93	TK treatment for brain cancer

[a] IGF: Insulin-like growth factor I.

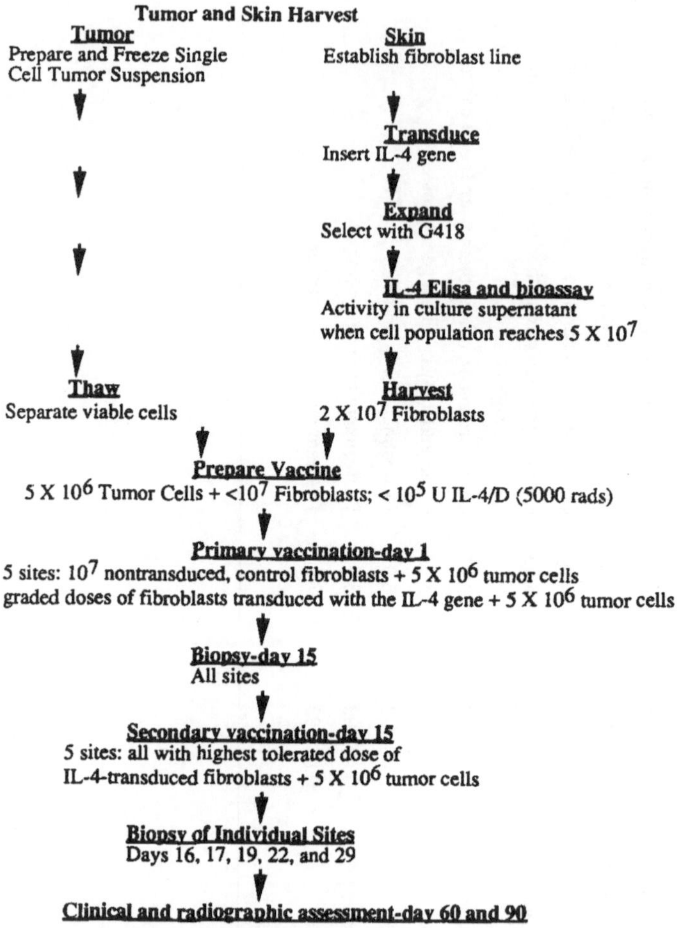

FIGURE 2. IL-4 gene therapy protocol schema.

As an example, IL-4 gene therapy protocol will be described in detail (Figure 2). In our current protocol, irradiated tumor cells are intradermally inoculated to the patients admixed with cultured autologous fibroblasts transfected with IL-4. The fibroblasts serve as the carrier of the IL-4 cDNA, since the paracrine effects of IL-4 using this strategy have been shown to be comparable to ones using tumor cells transfected with IL-4 paracrine to stimulate antitumor immunity. The fibroblast carrier has distinct advantages over the direct transfection of tumor cells. Human fibroblast cells are easier to obtain, culture, transfect, and select than many human tumor cells that are proposed to be used in some cytokine gene therapy protocols. We have successfully transfected both murine and human fibroblasts with the respective species-specific genes for IL-4 and have shown that there is substantial production of IL-4. We have treated six patients with this protocol and are currently evaluating the outcome of patients treated from both a clinical and immunologic standpoint. Early results showed infiltration mainly of eosinophils and macrophages at the vaccine site, as observed in animal models (Lotze et al., unpublished observation). Although the therapeutic benefit of cytokine gene therapy has not yet been demonstrated at this early stage, it appears that the accumulation of the results from clinical application will be valuable in terms of developing this innovative approach.

VI. FUTURE OF CANCER GENE THERAPY

Current efforts in cancer gene therapy are somewhat limited by the gene transfer methods and available molecules. If vector systems can be improved to have much greater efficiency, especially in *in vivo* application with the ability of targeting, clinical application of cytokine gene therapy will be much easier, and other strategies, including introducing tumor suppressor gene, will be available. Methods for regulating expression of transfected gene also widen the chance of appropriate application of therapeutic molecules. Along with these innovations, a greater understanding of the biology of the cancer cell or discovery of new cytokines might also provide the possibility of developing more sophisticated approaches.

REFERENCES

1. Culver, K. W., Ram, Z., Wallbridge, S., Ishii, H., Oldfield, E. H., and Blaese, R. M., In vivo gene transfer with retroviral vector-producer cells for treatment of experimental brain tumors, *Science*, 256, 1550, 1992.
2. Mullen, C. A., Kilstrup, M., and Blease, M., Transfer of the bacterial gene for cytosine deaminase to mammalian cells confers lethal sensitivity to 5-fluorocytosine: a negative selection system, *Proc. Natl. Acad. Sci. U.S.A.*, 89, 33, 1992.
3. Tracy, K. J., Wei, H., Manogue, K. R., et al., Cachectin/tumor necrosis factor induces cachexia, anemia and inflammation, *J. Exp. Med.*, 167, 1211, 1988.
4. Tepper, R. I., Pattengale, P. K., and Leder, P., Murine interleukin-4 displays potent anti-tumor activity in vivo, *Cell*, 57, 503, 1989.
5. Fearon, E. R., Padoll, D. M., Itaya, T., Golunbek, P., Uvilsky, H. I., Simons, J. W., Karasuyama, H., Vogelstein, B., and Frost, P., Interleukin-2 production by tumor cells bypasses T-helper function in the generation of an antitumor response, *Cell*, 60, 397, 1990.
6. Gansbacher, B., Bannerji, R., Daniels, B. K., Cronin, K., and Gilboa, E., Retroviral vector-mediated γ-IFN gene transfer into human cells generates potent and long lasting antitumor immunity, *Cancer Res.*, 50, 7820, 1990.
7. Bubinik, J., Voitenok, N. N., Kieler, J., Prassolov, V. S., Chumakov, P. M., Bubenikova, D., Simova, J., and Jandlova, T., Local administration of cells containing an inserted IL-2 gene and producing IL-2 inhibits growth of human tumors in nu/nu mice, *Immunol. Lett.*, 19, 279, 1988.
8. Zubler, R. H., Erard, F., Lee, R. K., Van Laer, M., Mingori, C., Moretta, L., and MacDonald, H. R., Mutant EL-4 thymoma cells polyclonally activate murine and human B cells via direct cell interaction, *J. Immunol.*, 134, 3662, 1985.
9. Fernadez-Botvan, R., Sandis, V. M., Mosmann, J. R., and Vitetta, E. S., Lymphokine-mediated regulation of the proliferative response of clones of T helper 1 and T helper 2 cells, *J. Exp. Med.*, 168, 543, 1988.

10. Cutun, M. C., Aregon, I., Sherman, F., Loudon, R., Clark, S. C., Penissia, B., and Trinchieri, G., Production of hematopoietic colony-stimulating factors by human natural killer cells, *J. Exp. Med.*, 169, 569, 1989.
11. Rosenberg, S. A. and Lotze, M. T., Cancer immunotherapy using interleukin-2 activated lymphocytes, *Annu. Rev. Immunol.*, 4, 681, 1986.
12. Connor, J., Bannerji, R., Saito, S., Heston, W., Fair, W., and Gilboa, E., Regression of bladder tumors in mice treated with interleukin 2 gene-modified tumor cells, *J. Exp. Med.*, 177, 1127, 1993.
13. Ganser, A., Undemann, A., Seipeh, G., Ottomann, O. G., Herrmann, F., Eder, M., Frisch, J., Schulz, G., Mertelsmann, R., Hoelzer, D., Clinical effects of recombinant human interleukin-3, *Am. J. Clin. Oncol.*, 1, 551, 1991.
14. Pulaski, B. A., McAdam, A. J., Hutter, E. K., Biggar, S., Lord, E. M., and Frelinger, J. G., Interleukin 3 enhances development of tumor-reactive cytotoxic cells by a CD4-dependent mechanism, *Cancer Res.*, 53, 2112, 1993.
15. McAdam, A. J., Yeh, K. Y., Pulaski, B. A., Freilinger, J. G., and Lord, E. M., IL-3 inhibits the generation of non-specific killers by IL-2, *J. Immunother.*, in press.
16. Mosmann, T. R., Bond, M. W., Coffman, R. L., Ohara, J., and Paul, W. E., T-cell and most cell lines respond to B-cell stimulatory factor 1, *Proc. Natl. Acad. Sci. U.S.A.*, 83, 5654, 1986.
17. Paul, W. E. and Ohara, J., *Annu. Rev. Immunol.*, 5, 429, 1987.
18. Schleimer, R. P., Strbinsky, S. A., Kaiser, J., Bickel, C. A., Kunk, D. A., Tomioka, K., Newman, W., Luscinskas, F. W., Gimbrone, M. A., Jr., McIntyre, B. W., et al., IL-4 induces adherence of human eosinophils and basophils but not neutrophils to endothelium. Association with expression of VCAM-1, *J. Immunol.*, 148, 1086, 1992.
19. Tepper, R. I., Coffman, R. L., and Leder, P., An eosinophil dependent mechanism for the anti-tumor effect of interleukin-4, *Science*, 257, 548, 1992.
20. Yu, J. S., Mei, M. X., Chiocca, A., Marhiza, R. L., and Tepper, R. I., Treatment of glioma by engineered interleukin-4-secreting cells, *Cancer Res.*, 53, 3125, 1993.
21. Blankenstein, T., Li, W. A., Muller, W., and Diamenstein, T., Retroviral interleukin-4 gene transfer into an interleukin-4-dependent cell line results in autocrine growth but not in tumorgenicity, *Eur. J. Immunol.*, 20, 935, 1990.
22. Golumbek, P. T., Lazenby, A. J., Levitsky, H. I., Jaffee, L. M., Karasyarna, H., Baker, M., and Pardoll, D., Treatment of established renal cancer by tumor cells engineered to secrete interleukin-4, *Science*, 254, 713, 1991.
23. Dranoff, G., Jaffee, E., Lazenby, A., Golumbek, P., Levitsky, H., Brosek, K., Jackson, V., Hamada, H., Pordell, D., and Mulligan, R., Vaccination with irradiated tumor cells engineered to secrete murine granulocyte macrophage colony-stimulating factor stimulates potent, specific and long lasting anti tumor immunity, *Proc. Natl. Acad. Sci. U.S.A.*, 90, 3539, 1993.
24. Modish, A., Masuelli, L., Modice, A., D'Orazi, G., Scarpa, S., Bosco, M. C., Forni, G., Ultrastructural evidence of the mechanisms responsible for interleukin-4-activated rejection of a spontaneous murine adenocarcinoma, *Intl. J. Cancer*, 53, 988, 1993.
25. Pippin, B. A., Rosenstein, M., Jacob, W. F., Chiang, Y., and Lotze, M. T., Local IL-4 delivery enhances immune reactivity to murine tumors: gene therapy in combination with IL-2, *Cancer Gene Ther.*, 1, 1, 1994.
26. Hirano, T., The biology of interleukin-6, in, *Interleukins: Molecular Biology and Immunology, Chemical Immunology*, Kishimoto, T., Ed., S. Karger, Basel, 1992.
27. Kishimoto, T., Factors affecting B cell growth and differentiation. *Annu. Rev. Immunol.*, 3, 133, 1985.
28. Lotze, M., Jirik, F., Kaboundis, R., Isoukas, C., Hirano, T., Kishimoto, T., Carson, D. A., BSF-2/IL-6 is a costimulant for human monocytes and T lymphocytes, *J. Exp. Med.*, 167, 1253, 1988.

29. Blankenstein, T., San, Z., Uberla, K., Miller, W., Rosen, H., Volk, H. D., Diamanstein, T., Tumor suppression after tumor cell-targeted tumor necrosis factor gene transfer, *J. Exp. Med.*, 173, 1047, 1991.
30. Sun, W. H., Kreisle, R. A., Phillips, A. W., and Erschler, W. B., In vivo and in vitro characteristics of interluekin-6-transfected B16 melanoma cells, *Cancer Res.*, 52, 5412, 1992.
31. Porgador, A., Tzehoval, E., Katz, A., Vaclai, E., Rivel, M., Feldman, M., and Eisenbach, L., Interleukin-6 gene transfection into Lewis Lung carcinoma tumor cells suppresses the malignant phenotype and confers immunotherapeutic competence against parental metastic cells, *Cancer Res.*, 52, 3679, 1992.
32. Namen, A. E., Schmierer, A. E., Mach, C. J., Overell, R. W., Park, L. S., Urdal, D. L., and Mochizuki, D. Y., B-cell precursor growth-promoting activity, *J. Exp. Med.*, 167, 988, 1988.
33. Morrissey, P. J., Goodwin, R. G., Nordan, R. P., Anderson, D., Goskin, K. H., Cosura, D., Sims, J., Lupton, S. A. B., Reed, S. G., Mochizuk, D., Eisenman, J., Conlon, A. J., and Namen, A. E., Recirant interleukin-7, pre-B cell growth factor has costimulatory activity on T-cells, *J. Exp. Med.*, 169, 707, 1989.
34. Chazen, G. D., Peniera, G. M. B., LeCros, G., Gillis, S., and Shevach, E. M., Interleukin-7 is a T cell growth factor, *Proc. Natl. Acad. Sci. U.S.A.*, 86, 5923, 1989.
35. Grabstein, K. H., Namen, A. E., Shanebeck, K.. Voice, R. F., Reed, S. G., and Widmer, M. B., Regulation of T cell proliferation by IL-7, *J. Immunol.*, 144, 3015, 1990.
36. Lynch, D. H. and Miller, R. E., Induction of murine lymphokine-activated killer cells by recombinant IL-7, *J. Immunol.*, 145, 1983, 1990.
37. Alderson, M. R., Sassenfeld, H. M., and Widmer, M. B., Interleukin 7 enhances cytolytic T lymphocyte generation and induces lymphokine-activated killer cells from human peripheral blood, *J. Exp. Med.*, 172, 577, 1990.
38. Hickman, C. J., Crim, J. A., Mostowski, H. S., and Siegel, J. P., Regulation of human cytotoxic T lymphocyte development by IL-7, *J. Immunol.*, 145, 2415, 1990.
39. Hock, H., Dorsch, M., Diamantstein, T., and Blankenstein, T., Interleukin 7 induces CD4+ T cell-dependent tumor rejection, *J. Exp. Med.*, 174, 1291, 1991.
40. Miyatake, S., Nishihara, K., Kiruchi, H., Yamashita, J., Namba, Y., Hanaoka, M., and Watanabe, Y., Efficient tumor suppression by glioma-specific murine cytotoxic T lymphocytes transfected with interferon g gene, *JNCI*, 82, 217, 1990.
41. Restifo, N. P., Spiess, P. J., Karp, S. E., Mule, J. J., and Rosenberg, S. A., A nonimmunogenic sarcoma transduced with the cDNA for interferon gamma elicits CD8+ T cells against the wild-type tumor: correlation with antigen presentation capability, *J. Exp. Med.*, 175, 1423, 1992.
42. Porgador, A., Bannerhji, R., Watanabe, Y., Feldman, M., Gilboa, E., and Eisenbach, L., Antimetastatic vaccination of tumor-bearing mice with two types of IFN-γ gene-inserted tumor cells, *J. Immunol.*, 150, 1458, 1993.
43. Philip, R. and Epstein, L., Tumor necrosis factor as immunomodulator and mediator of monocyte toxicity induced by itself, gamma-interferon and interleukin-1, *Nature*, 323, 86, 1986.
44. Talmadge, J. E., Phillips, E. H., Schneider, M., Rave, T., Pennington, R., Bauersox, O., and Lenz, B., Immunomodulary properties of recombinant murine and human tumor necrosis factor, *Cancer Res.*, 48, 544, 1988.
45. Shalaby, M. R., Aggerwal, B. B., Rinderknecht, E., Svedersky, P., Finkle, B. S., and Palladino, M. A., Activation of human polymorphonuclear neutrophil functions by interferon-gamma and tumor necrosis factors, *J. Immunol.*, 135, 2069, 1985.
46. Scheurich, P., Thoma, B., Ucer, V., and Pfizenmair, K., Immunoregulatory activity of recombinant human tumor necrosis factor (TNF)-alpha: induction of TNF receptors on human T cells and INF alpha mediated enhancement of T cell responses, *J. Immunol.*, 138, 1786, 1987.

47. Hackett, R. J., Davis, L. S., and Lipsky, P. E., Comparative effects of tumor necrosis factor and IL-1 beta on mitogen induced T cell activation, *J. Immunol.*, 140, 2639, 1988.
48. Ranges, G. E., Figari, I. S., Espevik, T., and Palladino, M. A., Inhibition of cytotoxic T cell development by transforming growth factor-beta and reversal by recombinant tumor necrosis factor-alpha, *J. Exp. Med.*, 166, 991, 1987.
49. Owen-Schaub, L. B., Gutterman, J. U., and Grimm, A. E., Synergy of tumor necrosis factor and interleukin-2 in the activation of human cytotix lymphocytes; effect of tumor necrosis factor alpha and interleukin-2 in the generation of human lymphokine-activated killer cell cytotoxicity, *Cancer Res.*, 48, 788, 1988.
50. Robinet, E., Branellec, D., Termijtelen, A. M., Blay, J. Y., Gay, F., and Chouaib, S., Evidence for tumor necrosis factor-alpha involvement in the optimal induction of class I allospecific cytotoxic T cells, *J. Immunol.*, 144, 4555, 1990.
51. Asher, A. L., Mule, J. J., Kasid, A., Reskfo, N. P., Salo, J. C., Reichert, C. M., Jaffee, G., Fendly, B., Kriegler, M., and Rosenberg, S. A., Murine tumor cells transduced with the gene for tumor necrosis factor-a: evidence for immune effects of tumor necrosis factor against tumors, *J. Immunol.*, 146, 3227, 1991.
52. Teng, M. N., Park, B. H., Koeppen, H. K. W., Tracey, K. J., Fendly, B. M., and Schreiber, H., Longterm inhibition of tumor growth by tumor necrosis factor in the absence of cachexia or T-cell immunity, *Proc. Natl. Acad. Sci. U.S.A.*, 88, 3535, 1991.
53. Nicola, N. A., Metcalf, D., Matsumoto, M., and Johnson, G. R., Purification of a factor inducing differentiation in murine myelmonobytic leukemia cells: identification as granulocyte colony-stimulating factor (G-CSF), *J. Biol. Chem.*, 258, 9017, 1983.
54. Metcalf, D., Nicola, N. A., and Begley, C. J., Identification of the human analogue of a regulator that induces differentiation in murine leukaemic cells, *Nature*, 314, 625, 1985.
55. Colombo, M. P., Ferrari, G., Stoppacciaro, A., Parenza, M., Rodolfo, M., Mavilio, F., and Parmiani, G., Granulocyte colony-stimulating factor gene transfer suppresses tumorigenicity of a murine adenocarcinoma in vivo, *J. Exp. Med.*, 173, 889, 1991.
56. Kobayashi, M., Fitz, L., Ryan, M., Hewick, R. M., Clark, S. C., Chang, S., Loudon, R., Sherman, F., Perussia, B., and Trinchieri, G., Identification and purification of natural killer cell stimulatory factor (NKSF), a cytokine with multiple biologic effects on human lymphocytes, *J. Exp. Med.*, 170, 827, 1989.
57. Stern, A. S., Podlaski, F. J., Hulmes, J. D., Pan, Y. C. E., Quinn, P. M., Wolitzky, A. G., Familletti, P. C., Stremlo, D. L., Truitt, T., Chizzonite, R., and Gately, M. K., Purification to homogeneity and partial characterization of cytotoxic lymphocyte maturation factor from human B-lymphoblastoid cells, *Proc. Natl. Acad. Sci. U.S.A.*, 87, 6808, 1990.
58. Wolf, S. F., Temple, P. A., Kobayashi, M., Young, E., Dicig, M., Lowe, L., Dzialo, R., Fitz, L., Ferenz, C., Hewick, R. M., Kelleher, K., Herrmann, S. H., Clark, S. C., Azzoni, L., Chan, S. H., Trinchieri, G., and Perussia, B., Cloning of cDNA for natural killer cell stimulatory factor, a heterodimeric cytokine with multiple biologic effect on T and natural killer cells, *J. Immunol.*, 146, 3074, 1991.
59. Gubler, U., Chua, A. O., Schoenhout, D. S., Dwyer, C. M., McComas, W., Motyka, R., Nabavi, N., Wolitzky, A. G., Quinn, P. M., Familletti, P. C., and Gately, M. K., Coexpression of two distinct genes is required to generate secreted bioactive cytotoxic lymphocyte maturation factor, *Proc. Natl. Acad. Sci. U.S.A.*, 88, 4143, 1991.
60. Chan, S. H., Perussia, B., Gupta, J. W., Kobayashi, M., Pospisil, M., Young, H. A., Wolf, S. F., Young, D., Clark, S. C., and Trinchieri, G., Induction of interferon-γ production by natural killer cell stimulatory factor: characterization of the responder cells and synergy with other inducers, *J. Exp. Med.*, 173, 869, 1991.

61. Naume, B., Gately, M., and Espevik, T., A comparative study of IL-12 (cytotoxic lymphocyte maturation factor)-, IL-12- and IL-7-induced effects on immunomagnetically purified CD56+ NK cells, *J. Immunol.*, 148, 2429, 1992.
62. Gately, M. K., Desai, B. B., Wolitzky, A. G., Quinn, P. M., Dwyer, C. M., Podlaski, F. J., Familletti, P. C., Sinigalia, F., Chizzonite, R., Gubler, U., and Stern, A. S., Regulation of human lymphocyte proliferation by a heterodimeric cytokine, IL-12 (cytotoxic lymphocyte maturation factor), *J. Immunol.*, 147, 874, 1991.
63. Hsieh, C.-S., Macatonia, S., Tripp, C., Wolf, S., O'Garra, A., and Murphy, K., Development of Th1 CD4+ T cells through IL-12 produced by listeria-induced macrophages, *Science*, 260, 547, 1993.
64. Gately, M. K., Warrier, R. R., Honasoge, S., Faherty, D. A., Connaughton, S. E., Anderson, T. D., Sarmiento, U., Hubbard, B. R., and Murphy, M., Administration of recombinant IL-12 to normal mice enhances cytolytic lymphocyte activity and induces production of IFN-g *in vivo*, *Int. Immunol.*, 6, 157, 1994.
65. Brunda, M. J., Luistro, L., Warrier, R. R., Wright, R. B., Hubbard, B. R., Murphy, M., Wolf, S. F., and Gately, M. K., Antitumor and antimetastatic activity of interleukin-12 against murine tumors, *J. Exp. Med.*, 178, 1223, 1993.
66. Nastala, C. L., Edington, H., Storkus, W., McKinney, T. G., Tahara, H., Brunda, M. J., Gately, M. K., Schreiber, R., and Lotze, M. T., Recombinant interleukin-12 administration induces tumor regression in association with interferon-gamma and nitric oxide production, *J. Immunol.*, 153, 1697, 1994.
67. Tahara, H., Zeh, H. J. Z., III, Stokus, W. J., Pappo, I., Watkins, S. C., Gubler, U., Wolf, S. F., Robbins, P. D., and Lotze, T. M., Fibroblasts genetically engineered to secrete interleukin-12 can suppress tumor growth and induce anti-tumor immunity to a murine melanoma *in vivo*, *Cancer Res.*, 54, 182, 1994.
68. Sambrook, J., Fritsch, E. F., and Maniatis, T., *Molecular Cloning, a Laboratory Manual*, 2nd ed., Cold Spring Harbor Laboratory Press, Cold Spring Harbor, NY, 1989.
69. Zitvogel, L., Tahara, H., Gubler, U., Wolf, S. F., Gately, M. K., Robbins, P. D., and Lotze, M. T., Construction and characterization of retroviral vectors expressing biologically active human IL-12, *Hum. Gene Ther.*, in press, 1994.
70. Danos, O. and Mulligan, R. C., Safe and efficient generation of recombinant retroviruses with amphotropic and ecotropic host ranges, *Proc. Natl. Acad. Sci. U.S.A.*, 85, 6460, 1988.
71. Gansbacher, B., Zier, K., Daniels, B., Cronin, K., Bannerji, R., and Gilboa, E., Interleukin-2 gene transfer into tumor cells abrogates tumorigenicity and induces protective immunity, *J. Exp. Med.*, 172, 1217, 1990.
72. Tsai, J., Gansbacher, B., Tait, L., Miller, F. R., and Heppner, G. H., Induction of anti-tumor immunity by interleukin-2 gene-transfected mouse mammary tumor cells versus transduced mammary stromal fibroblasts, *JNCI*, 85, 546, 1993.

Index

B

C

G

H

I

K

L

M

N

O

P

R

S

T